Cast of Characters (in order of appearance)
Chapters 1–6

Symbol	Meaning/Description
N	Number of cases
f	Frequency
$\%$	Percentage
cf	Cumulative frequency
$c\%$	Cumulative percentage
m	Midpoint
PR	Percentile rank
Mo	Mode
Mdn	Median
$\overline{X}$	Mean
R	Range
IQR	Inter-Quartile Range
s^2	Variance
s	Standard deviation
μ	Population mean
σ^2	Population variance
σ	Population standard deviation
P	Probability
z	z score
ME	Margin of error
CI	Confidence interval
$\sigma_{\overline{X}}$	Standard error of the sample mean (true)
$s_{\overline{X}}$	Standard error of the sample mean (estimated)
α	Level of significance
df	Degrees of freedom
t	t ratio
P	Sample proportion
π	Population proportion
s_P	Standard error of the sample proportion

Chapters 7–12

Symbol	Meaning/Description
$\bar{X}_1 - \bar{X}_2$	Difference between (sample) means
$\sigma_{\bar{X}_1 - \bar{X}_2}$	Standard deviation of the distribution of the difference between means
$s_{\bar{X}_1 - \bar{X}_2}$	Standard error of the difference between means
s_D	Standard deviation of the distribution of the (before/after) difference between means
$s_{\bar{D}}$	Standard error of the (before/after) difference between means
$P*$	Pooled sample proportion
$s_{P_1 - P_2}$	Standard error of the difference between proportions
SS	Sum of squares
MS	Mean square
F	F ratio
HSD	Honestly significant difference
q	Studentized range
χ^2	Chi-square
r	Pearson's correlation
ρ	Population correlation
$r_{XY.Z}$	Partial correlation
a	Y-intercept
b	Slope
e	Error term
r^2	Coefficient of determination
$1 - r^2$	Coefficient of nondetermination
R^2	Multiple coefficient of determination
L	Log-odds (logit)
r_s	Spearman's rank order correlation
G	Gamma
ϕ	Phi coefficient
C	Contingency coefficient
V	Cramér's V

Elementary Statistics in Social Research

ELEVENTH EDITION

Jack Levin
Northeastern University

James Alan Fox
Northeastern University

David R. Forde
University of Alabama

Allyn & Bacon

Boston   New York   San Francisco
Mexico City   Montreal   Toronto   London   Madrid   Munich   Paris
Hong Kong   Singapore   Tokyo   Cape Town   Sydney

Executive Editor: *Jeff Lasser*
Editorial Assistant: *Lauren Macey*
Senior Marketing Manager: *Kelly May*
Production Supervisor: *Elizabeth Gale Napolitano*
Production Management and Composition: *Progressive Publishing Alternatives*
Manufacturing Buyer: *Debbie Rossi*
Cover Designer: *Kristina Mose-Libon*

Library of Congress Cataloging-in-Publication Data

Levin, Jack, 1941–
    Elementary statistics in social research / Jack Levin, James Alan Fox, David R. Forde.
—11th ed.
        p.   cm.
    Includes index.
    ISBN-13: 978-0-205-57069-0
    ISBN-10: 0-205-57069-0
    1. Social sciences—Statistical methods. 2. Statistics. I. Fox, James Alan.   II. Forde,
David R. (David Robert), 1959–   III. Title.
    HA29.L388 2010
    519.5024'3—dc22

                                                                            2008054348

Printed in the United States of America
10 9 8 7 6 5 4 3 2 1   HAM   13 12 11 10 09

Allyn & Bacon
is an imprint of

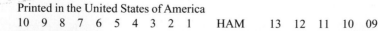

www.pearsonhighered.com

ISBN 13: 978-0-205-57069-0
ISBN 10:   0-205-57069-0

Contents

Preface

The Eleventh Edition of *Elementary Statistics in Social Research* provides an introduction to statistics for students in sociology and related fields, including political science, criminal justice, and social work. This book is not intended to be a comprehensive reference for statistical methods. On the contrary, our first and foremost objective has always been to provide an accessible introduction for a broad range of students, particularly those who may not have a strong background in mathematics.

Like its predecessors, the Eleventh Edition contains a number of pedagogical features. Most notably, step-by-step illustrations of statistical procedures continue to be located at important points throughout the text. We have again attempted to provide clear and logical explanations for the rationale and use of statistical methods in social research, and we have again included a large number of end-of-chapter questions and problems. Finally, we have again ended each part of the text with a section entitled "Looking at the Larger Picture," which carries the student through the entire research process.

For more than three decades, *Elementary Statistics in Social Research* has undergone extensive refinements and improvements in responding to instructor and student feedback. This edition also departs, in important ways, from earlier versions of the text. We have expanded our introduction to multiple regression, including a discussion of dummy variables, and to logistic regression. We have also added the inter-quartile range in Chapter 4, *t* tests with unequal variances in Chapter 7, and two-way analysis of variance in Chapter 8. Moreover, we have updated data and tightened or eliminated material, where needed. Finally, thanks to the addition of David R. Forde as a co-author, this edition is much more SPSS friendly. It contains a number of end-of-chapter problems to be worked out in SPSS using data sets available on-line and an appendix in which procedures for applying SPSS software are carefully explained. For those instructors who do not teach SPSS, however, these new aspects of the book can be easily excluded.

We have checked and verified all numerical exercises and illustrations to minimize the frustrations that students feel when they encounter computational errors. In addition, a solutions manual is available for instructors in which all problems are carried out in detail. Finally, because perfection is something to strive for, we encourage instructors and students to e-mail corrections or comments to j.levin@neu.edu.

The organization of the book remains unchanged. Following a detailed overview in Chapter 1, the text is divided into five parts: Part I (Chapters 2 through 4) introduces the student to the most common methods for describing and comparing data. Part II (Chapters 5 and 6) serves a transitional purpose. Beginning with a discussion of the basic concepts of probability, it leads the student from the topic of the normal curve as an important descriptive device to the use of the normal curve as a basis for generalizing from samples to populations.

Continuing with this decision-making focus, Part III (Chapters 7 through 9) contains several well-known tests of significance for differences between groups. Part IV (Chapters 10 through 12) includes procedures for obtaining correlation coefficients, and an introduction to regression analysis. Finally, Part V (Chapter 13) consists of an important chapter in which students learn, through example, the conditions for applying statistical procedures to research problems.

The text provides students with background material for the study of statistics. A review of basic mathematics, statistical tables, a list of formulas, and a glossary of terms are located in appendixes at the end of the book. Finally, the companion website provides data sets for completing the SPSS exercises as well as *ABCalc*, an easy-to-use statistics calculator developed for this textbook.

The ancillary package for this edition includes:

For Instructors:

- **Instructors Manual and Test Bank.**
- **Pearson MyTest Computerized Test Bank.**
- **NEW! Solutions Manual.** This supplement includes worked-out solutions to all end-of-chapter exercises.

For Students:

- **NEW! MySocKit (www.mysockit.com).** MySocKit is an online supplement for *Elementary Statistics in Social Research,* Eleventh Edition, that includes material previously published in the Student Workbook and Companion Website, plus a number of new features. It contains a series of exercises and computer applications; SPSS data sets used in those exercises; a tutorial on basic SPSS commands; an Excel statistics calculator; flashcards for learning key terms and formulas; and worked-out solutions to the odd-numbered end-of-chapter exercises. MySocKit is available with this text at no charge when a MySocKit Student Access Code Card is packaged on request.
- **Student SPSS Software.** The most current release of SPSS for Windows (on CD-ROM) can be packaged with the text at a significant discount from the retail version of Student SPSS. Ask your Pearson Arts and Sciences representative for details.

We thank the following reviewers of the Eleventh Edition for their insightful and helpful suggestions: Bruce P. Chadwick, Columbia University; Michael G. Elsamar, Boston University; Ann Hunter, University of Idaho; Juan Onesimo Sandoval, Northwestern University; and Paul T. von Hippel, Ohio State University. We are grateful to Jenna Savage and Sarah Damberger of Northeastern University, who spent countless hours verifying calculations and answers. Jenna Savage also was responsible for many excellent additional end-of-chapter problems and Sarah Damberger worked tirelessly on developing the new solutions manual. We are indebted to the Literary Executive of the late Sir Ron A. Fisher, F.R.S.; to Dr. Frank Yates, F.R.S.; and to Longman Group, Ltd. London, for permission to reprint Tables III, IV, V, and VI from their book *Statistical Tables for Biological, Agricultural, and Medical Research,* 6th ed., 1974. Finally, we acknowledge the important role of our personal computers and laptops, without "whose" assistance this revision would not have been possible.

Jack Levin
James Alan Fox
David R. Forde

1

Why the Social Researcher Uses Statistics

A little of the social scientist can be found in all of us. Almost daily, we take educated guesses concerning the future events in our lives in order to plan for new situations or experiences. As these situations occur, we are sometimes able to confirm or support our ideas; other times, however, we are not so lucky and must face the sometimes unpleasant consequences.

Consider some familiar examples: We might invest in the stock market, vote for a political candidate who promises to solve domestic problems, play the horses, take medicine to reduce the discomfort of a cold, throw dice in a gambling casino, try to psych out our instructors regarding a midterm, or accept a blind date on the word of a friend.

Sometimes we win; sometimes we lose. Thus, we might make a sound investment in the stock market, but be sorry about our voting decision; win money at the craps table, but discover we have taken the wrong medicine for our illness; do well on a midterm, but have a miserable blind date; and so on. It is unfortunately true that not all of our everyday predictions will be supported by experience.

The Nature of Social Research

Similar to our everyday approach to the world, social scientists attempt to explain and predict human behavior. They also take "educated guesses" about the nature of social reality, although in a far more precise and structured manner. In the process, social scientists examine characteristics of human behavior called *variables*—characteristics that differ or vary from one individual to another (for example, age, social class, and attitude) or from one point in time to another (for example, unemployment, crime rate, and population).

Not all human characteristics vary. It is a fact of life, for example, that the gender of the person who gave birth to you is female. Therefore, in any group of individuals, gender of mother is the *constant* "female." A biology text would spend considerable time discussing why only females give birth and the conditions under which birth is possible, but a social scientist would consider the mother's gender a given, one that is not worthy of study because it never varies. It could not be used to explain differences in the mental health of children because all of their mothers are females. In contrast, a mother's age, race, and mental health are variables: In any group of individuals, they will differ from person to person and can be the key to a greater understanding of the development of the child. A researcher therefore might study differences in the mental health of children depending on the age, race, and mental health of their mothers.

In addition to specifying variables, the social researcher must also determine the *unit of observation* for the research. Usually, social scientists collect data on individual persons. For example, a researcher might conduct interviews to determine if the elderly are victimized by crime more often than younger respondents. In this case, an individual respondent is the unit to be observed by the social scientist.

However, researchers sometimes focus their research on *aggregates*—that is, on the way in which measures vary across entire collections of people. For example, a researcher might study the relationship between the average age of the population and the crime rate in various metropolitan areas. In this study, the units of observation are metropolitan areas rather than individuals.

Whether focusing on individuals or aggregates, the ideas that social scientists have concerning the nature of social reality are called *hypotheses*. These hypotheses are frequently expressed in a statement of the relationship between two or more variables: at minimum, an *independent variable* (or presumed cause) and a *dependent variable* (or presumed effect). For example, a researcher might hypothesize that socially isolated children watch more television than children who are well integrated into their peer groups, and he or she might conduct a survey in which both socially isolated and well-integrated children are asked questions regarding the time they spend watching television (social isolation would be the independent variable; TV-viewing behavior would be the dependent variable). Or a researcher might hypothesize that the one-parent family structure generates greater delinquency than the two-parent family structure and might proceed to interview samples of delinquents and nondelinquents to determine whether one or both parents were present in their family backgrounds (family structure would be the independent variable; delinquency would be the dependent variable).

Thus, not unlike their counterparts in the physical sciences, social researchers often conduct research to increase their understanding of the problems and issues in their field. Social research takes many forms and can be used to investigate a wide range of problems. Among the most useful research methods employed by social researchers for testing their hypotheses are the experiment, the survey, content analysis, participant observation, and secondary analysis. For example, a researcher may conduct an experiment to determine if arresting a wife batterer will deter this behavior in the future, a sample survey to investigate political opinions, a content analysis of values in youth magazines, a participant observation of an extremist political group, or a secondary analysis of government statistics on unemployment. Each of these research strategies is described and illustrated in this chapter.

The Experiment

Unlike everyday observation (or, for that matter, any other research approach), the *experiment* is distinguished by the degree of *control* a researcher is able to apply to the research situation. In an experiment, researchers actually manipulate one or more of the independent variables to which their subjects are exposed. The manipulation occurs when an experimenter assigns the independent variable to one group of people (called an *experimental group*), but withholds it from another group of people (called a *control group*). Ideally, all other initial differences between the experimental and control groups are eliminated by assigning subjects on a random basis to the experimental and control conditions.

For example, a researcher who hypothesizes that frustration increases aggression might assign a number of subjects to the experimental and control groups at random by flipping a coin ("heads" you're in the experimental group; "tails" you're in the control group), so that the groups do not differ initially in any major way. The researcher might then manipulate frustration (the independent variable) by asking the members of the experimental group to solve a difficult (frustrating) puzzle, whereas the members of the control group are asked to solve a much easier (nonfrustrating) version of the same puzzle. After all subjects have been given a period of time to complete their puzzle, the researcher might obtain a measure of aggression by asking them to administer "a mild electrical shock" to another subject (actually, the other subject is a confederate of the researcher who never really gets shocked, but the subjects presumably do not know this). If the willingness of subjects to administer an electrical shock is greater in the experimental group than in the control group, this difference would be attributed to the effect of the independent variable, frustration. The conclusion would be that frustration does indeed tend to increase aggressive behavior.

In 1995, three University of Wisconsin researchers—Joanne Cantor, Kristen Harrison, and Marina Krcmar—conducted an experiment to study the effect of Motion Picture Association of America ratings (G, PG, PG-13, R) on children's decisions to watch a particular movie. The researchers manipulated the independent variable, motion picture ratings, by asking a sample of boys—aged 10 to 14—to select the one film they wished to watch from a list of three movies that had previously been judged to be equally appealing. In all cases, two of the movies were rated PG; only the rating of the third film, *The Moon-Spinners,* was varied. On a random basis, one-quarter of the boys were told it had been rated G, one-quarter PG, one-quarter PG-13, and one-quarter R. If the film's rating had no effect on their preferences, about 33% of the boys should have chosen to watch *The Moon-Spinners,* regardless of the rating it had been assigned.

The results showed something else: When the film *The Moon-Spinners* was rated G, none of the boys chose it. When the film got a PG rating, 38.9% of the boys selected it over the other two films. But when *The Moon-Spinners* was rated PG-13 or R, at least 50% wanted to see it rather than the other motion pictures on the list. Apparently, at least for boys aged 10 to 14, declaring a film "off limits" or "restricted" makes the movie more appealing. Conversely, rating a film G only serves to reduce its popularity.

One final result deserves to be mentioned. Using an identical experimental procedure, the researchers determined that a sample of girls in the same age group were apparently not effected by the media equivalent of "forbidden fruit." Only 11% of the girls selected the R-rated version of *The Moon-Spinners;* more than 29% chose the film when it had a G rating.

The Survey

As we have seen, experimenters actually have a direct hand in creating the effect that they seek to achieve. By contrast, *survey* research is *retrospective*—the effects of independent variables on dependent variables are *recorded* after—and sometimes long after—they have occurred. Survey researchers typically seek to reconstruct these influences and consequences by means of verbal reports from their respondents in self-administered questionnaires, face-to-face interviews, or telephone interviews.

Surveys lack the tight controls of experiments: Variables are not manipulated and subjects are not assigned to groups at random. As a consequence, it is much more difficult to establish cause and effect. Suppose, for instance, in a survey measuring fear of crime, that a researcher finds that respondents who had been victims of crime tend to be more fearful of walking alone in their neighborhoods than those who had not been victimized. Because the variable *victimization* was not manipulated, we cannot make the logical conclusion that victimization *causes* increased fear. An alternative explanation that the condition of their neighborhoods (poverty, for example) produces both fear among residents and crime in the streets is just as plausible.

Surveys also have advantages precisely because they do not involve an experimental manipulation. As compared with experiments, survey research can investigate a much larger number of important independent variables in relation to any dependent variable. Because they are not confined to a laboratory setting in which an independent variable can be manipulated, surveys can also be more *representative*—their results can be generalized to a broader range of people.

In 2000, for example, two Stanford University researchers interested in assessing the social consequences of Internet use conducted surveys with a national probability sample of the adult population, including both Internet users and nonusers. Norman Nie and Lutz Erbing contacted 4,113 respondents in 2,689 households around the country and asked them to report how many hours they spend on the Internet and in various social activities.

Results obtained by Nie and Erbing consistently indicated that regular Internet users (defined as those who spend at least five hours per week on the Web) are more isolated than nonusers. More specifically, of regular Internet users, some 25% said that they spend less time with family and friends, 8% less time attending social events outside the home, and 25% less time shopping in stores. In addition, more than 25% of workers who are also regular Internet users reported that the Internet has increased the amount of time they work at the office.

About the only undeniably beneficial change associated with Internet use was the finding that 14% of regular Internet users spend less time commuting in traffic!

At the August 2007 meeting of the American Sociological Association, Oregon State University's Scott Akins presented the results of a study in which he and his colleagues surveyed 6,713 adult residents of Washington State including 1,690 persons who identified themselves as Hispanic. Members of their sample were questioned about their use of illicit drugs and their ethnic identity. They were asked to indicate their marital status, educational level, socioeconomic status, and place of residence, urban versus rural.

Holding constant these other factors, Akins and his collaborators determined that illicit drug use increased among recent Hispanic immigrants as they remained longer in the United States and became more acculturated into American society. That is, to the extent that Hispanic immigrants became acculturated, they replaced their traditional cultural beliefs, language, and social patterns with those of their host society. Specifically, when asked whether they had used illicit drugs in the previous month, less than 1% of nonacculturated Hispanics indicated that they had. But 7.2% of acculturated Hispanics (not unlike 6.4% of white residents) responded in the affirmative when asked the same question about their drug use.

Content Analysis

As an alternative to experiments and surveys, *content analysis* is a research method whereby a researcher seeks objectively to describe the content of previously produced messages. Researchers who conduct a content analysis have no need directly to observe behavior or to question a sample of respondents. Instead, they typically study the content of books, magazines, newspapers, films, radio broadcasts, photographs, cartoons, letters, verbal dyadic interaction, political propaganda, or music.

In 2001, for example, James A. Fox, Jack Levin, and Jason Mazaik performed a content analysis of celebrities depicted in *People* magazine cover stories. The researchers sought to determine how the celebrities chosen to be featured by the most popular celebrity magazine in the United States (circulation: 3,552,287 per issue) had changed over almost three decades. Using appropriate coding sheets, each of the more than 1,300 covers of issues of *People* from its inception in 1974 through 1998 was scrutinized for various characteristics of the celebrity and the overall tone of the cover presentation.

Results obtained by Fox, Levin, and Mazaik indicated that the basis for *People* celebrities appearing in a cover story has, over the decades, become dramatically more negative. In 1974, during its first year of publication, less than 3% of all celebrities were featured for negative reasons, such as drug or alcohol dependence, child abuse, or the commission of a violent crime. Instead, most celebrities were on the cover because they had accomplished a positive goal—either by overcoming a personal problem or by accomplishing a career objective. By 1988 and continuing as a pattern through the 1990s, however, there was a major reversal in tone, so that almost half of all cover stories focused not on celebrities' positive accomplishments, but on their untoward characteristics. Along with musicians, athletes, and political figures, *People* for the first time also featured murderers and rapists.

Michael Welch, Melissa Fenwick, and Meredith Roberts, in 1998, conducted a content analysis of crime experts' quotes published in feature newspaper articles from 1992 to 1995 in the *New York Times,* the *Washington Post,* the *Los Angeles Times,* and the *Chicago*

BOX 1.1 • *Practical and Statistical: One Lump or Two?*

The important role of randomization in experimental research cannot be overstated, and whenever it is absent, one must examine closely whether the experimental and control groups are in fact comparable. Without randomization, there may be extraneous factors, rather than the independent variable itself, which could have produced the observed difference between the two groups.

Several years ago, for example, a study that uncovered a link between coffee consumption and cancer of the pancreas received a great deal of media coverage because of its frightening implications. Most people who saw the headlines were unprepared to question the validity of the findings. They had heard that the study was performed by researchers at a prestigious university and was published in a highly regarded journal. To them that meant the study must be correct.

These same laypersons may have erroneously assumed that the study had been an experiment on laboratory animals, such as rats. After all, rats are used all the time to test products before they are deemed safe for humans. One could imagine, for example, that 100 rats had been divided into two groups at random, that the experimental group was fed (or injected with) large doses of caffeine, and that signs of pancreatic cancer were observed in this group significantly more often than in the control group.

One would think this had been the research approach in its ideal form: (1) because of random assignment, the differential cancer rate could not be blamed on any other factor, since random assignment made it very probable that the groups were comparable at the outset; (2) unlike human beings, the rats had little choice in their participation—they couldn't say "no thanks" or "fill it to the rim" nor could they ask for Nutrasweet which might contaminate the results; and (3) the effects of lifelong consumption of coffee could be approximated in a matter of months by giving he rats very large doses of caffeine.

Given all these advantages, there is little wonder that rat studies are often used to test the safety of food additives. However, this coffee and cancer study was not an experiment with rats, but a survey of humans. The researchers had compared pancreatic cancer patients with other subjects in terms of how much coffee they reported to have used in the past. Aside from any problems of recall, one must seriously consider alternative explanations for why pancreatic cancer patients had a significantly higher rate of coffee consumption than the other subjects, besides concluding that the independent variable (coffee) had caused the dependent variable (cancer). Perhaps both variables (coffee consumption and cancer risk) are associated with nervousness and stress. That is, stress might lead one to drink more coffee as well as to be more prone to cancer.

One can easily see the great limitation to surveys—it is not always possible to isolate cause and effect as it is in an experiment. In an experiment all competing explanations are negated by the tight controls built into the research. On the other hand, experiments are not always preferable or even possible, depending on one's research interest. First, one must always question whether findings based on animal studies will hold true had human beings been used for subjects. There are major differences between rats and humans that prevent us from generalizing rat studies to people. Even experiments with human beings are sometimes limited in the extent to which we can generalize beyond the experiment. Experimental subjects asked to solve a puzzle in a laboratory in order to produce frustration might react quite differently to naturally produced stress.

Finally, certain hypotheses simply cannot be tested experimentally. For example, suppose one were interested in the effects of divorce on children's school performance. Rat studies would be useless in this case. Similarly, an experiment with human beings—such as telling a random selection of children that their parents were splitting up—would be outside the boundaries of ethical practice. A survey of children of divorced and married parents would be the only feasible way to study the problem.

Tribune. The researchers found that these four major newspapers presented a distorted media image of crime and what to do about it. White-collar, corporate, and political offenses were almost completely missing. Instead, most of the articles focused on street crimes—murders, assaults, robberies—committed by low-income individuals. In addition, most of the quotes by politicians and criminal justice practitioners supported "get tough" crime control policies. By contrast, professors and researchers quoted were more likely to address the social and economic causes of crime and to advocate rehabilitation, decriminalization, gun control, and criminal justice reform.

Participant Observation

Another widely used research method is *participant observation,* whereby a researcher "participates in the daily life of the people under study, either openly in the role of researcher or covertly in some disguised role, observing things that happen, listening to what is said, and questioning people, over some length of time."[1]

In 1991, for example, at a time when both the popularity as well as the criticism of Heavy Metal music had reached an all-time high, sociologist Deena Weinstein conducted an extensive study of Heavy Metal music, including a participant observation of metal concerts. Based on her research backstage, on concert lines, in concert audiences, and on tour buses, Weinstein argued that the detractors of Heavy Metal, lacking in knowledge of its roots or culture, often give the art form merely an unsophisticated and inaccurate reading. Out of ignorance, they denounce metal music as a symptom of a sick society and unfairly associate it with satanism, violence, and deviant behavior. By contrast, Weinstein's research, which begins with the viewpoint of fans rather than detractors of metal music, supports a more positive view: She defends Heavy Metal as a legitimate artistic expression of youth rebellion that deserves tolerance, if not respect, even from those who dislike it.

Secondary Analysis

On occasion, it is possible for the social researcher not to gather his or her own data but to take advantage of data sets previously collected or assembled by others. Often referred to as *archival data,* such information comes from government, private agencies, and even colleges and universities. The social researcher is therefore not the primary or first one to analyze the data; thus, whatever he or she does to examine the data is called *secondary analysis.* This approach has an obvious advantage over firsthand data collection: It is relatively quick and easy but still exploits data that may have been gathered in a scientifically sophisticated manner. On the other hand, the researcher is limited to what is available, and has no say as to how variables are defined and measured.

The April 2008 issue of the *American Sociological Review* contains an article in which the effect of age on happiness is researched. The author, Yang Yang, a University of Chicago sociologist, conducted a secondary analysis of data from the National Opinion Research Center's General Social Survey (GSS), which reports the results of face-to-face

---

[1]Howard S. Becker and Blanche Geer, "Participant Observation and Interviewing," in *Qualitative Methodology,* ed. William J. Filstead (Chicago: Markham, 1970), p. 133.

TABLE 1.1   *Government Websites Containing Social Science Data*

Website/Agency	Types of Data	URL
FEDSTATS	Links to data and reports from over 70 federal agencies	www.fedstats.gov
Bureau of the Census	Population, families, business, income, housing, voting	www.census.gov
Bureau of Justice Statistics (BJS)	Crime offenders, victims, justice system	www.ojp.usdoj.gov/bjs/
Bureau of Labor Statistics (BLS)	Employment, unemployment, prices, wages	www.bls.gov
Bureau of Transportation Statistics (BTS)	Travel, aviation, boating, trucking, roads, highways	www.bts.gov
National Center for Health Statistics (NCHS)	Births, illness, injury, deaths, health care, nursing homes	www.cdc.gov/nchs/
National Center for Education Statistics (NCES)	Elementary, secondary, higher education	www.nces.ed.gov

interviews with a representative sample of as many as 3,000 respondents collected every year between 1972 and 2004. In order to measure their degree of happiness, the GSS asked respondents: "Taken all together, how would you say things are these days—would you say that you are very happy, pretty happy, or not too happy?"

Yang found that his respondents became happier with advancing age, supporting the notion that people mature and develop in positive ways as they grow older. Moreover, happiness was greater during periods of economic prosperity.

Generations also differed in terms of degree of happiness, with baby boomers—those born between 1946 and 1964—being the least happy of any generational group. Yang's results indicated also that among 18-year-old respondents, white women are the happiest overall, followed by white men, black women, and black men. However, these race and gender differences almost disappear as respondents mature into old age. Apparently, happiness continues to increase even into the eighth decade of life.

In this text, we occasionally make use of archival data sources. In Chapter 2, for example, we present and analyze birth rates, homicide rates, unemployment figures, and income data drawn from various government agencies. Table 1.1 provides a list of useful government websites from which various kinds of data related to health, housing, population, crime, education, transportation and the economy can be found and downloaded.

Why Test Hypotheses?

Social science is often referred to, quite unfairly, as the study of the obvious. However, it is desirable, if not necessary, to test hypotheses about the nature of social reality, even those that seem logical and self-evident. Our everyday commonsense observations are generally based on narrow, often biased preconceptions and personal experiences. These can lead us to accept without criticism invalid assumptions about the characteristics of social phenomena and behavior.

To demonstrate how we can be so easily misled by our preconceptions and stereotypes, consider what we "know" about mass murderers—those individuals who simultaneously kill at least four victims. According to popular thinking (and media portrayals), mass murderers are typically insane individuals who go berserk or run amok, expressing their anger in a spontaneous and impulsive outpouring of aggression. Moreover, they are usually regarded as total strangers to their victims, who are unlucky enough to be in the wrong place at the wrong time—at a shopping mall, on a commuter train, or in a fast-food restaurant.

The foregoing conception of mass murderers may seem clear-cut and obvious. Yet, compiling detailed information from FBI reports about 697 mass killers over the period from 1976 to 1995, Fox and Levin found instead that mass murderers are rarely insane and spontaneous—they know exactly what they are doing and are not driven to kill by voices of demons. Random shootings in a public place are the exceptions; most mass murders occur within families or among acquaintances. Typically, mass murderers target spouses and all of their children, or bosses and their co-workers. Far from being impulsive, most mass killers are methodical and selective. They usually plan their attacks and are quite selective as to the victims they choose to kill. In an office massacre, for example, a mass killer might choose to murder only those co-workers and supervisors whom the murderer blames for losing an important promotion or getting fired.

Until recently, even criminologists all but ignored mass killings, perhaps believing that mass murder was merely a special case of homicide (albeit, by definition, yielding a larger body count), explainable by the same theories applied to single-victim incidents and therefore not deserving of special treatment. From this point of view, mass murder occurs in the same places, under the same circumstances, and for the same reasons as single-victim murder.

Comparing FBI reports of single-victim homicides with mass murders reveals quite a different pattern. The location of mass murder differs sharply from that of homicides in which a single victim is slain. First, mass murders do not tend to cluster in large cities as do single-victim crimes; rather, mass killings are more likely to occur in small-town or rural settings. Moreover, while the South (and the deep South in particular) is known for its high rates of murder, this does not hold for mass murder. In comparison to single-victim murder, which is highly concentrated in urban inner-city neighborhoods and in the deep South where arguments are often settled through gunfire, mass murder more or less reflects the general population distribution.

Not surprisingly, the firearm is the weapon of choice in mass-murder incidents, even more than in single-victim crimes. Clearly, a handgun or rifle is the most effective means of mass destruction. By contrast, it is difficult to kill large numbers of people simultaneously with physical force or even a knife or blunt object. Furthermore, although an explosive device can potentially cause the death of large numbers of people (as in the 1995 bombing of the Oklahoma City federal building), its unpredictability would be unacceptable for most mass killers who target their victims selectively. In addition, far fewer Americans are proficient in the use of explosives, as compared with guns.

The findings regarding victim–offender relationships are perhaps as counterintuitive as the weapon-use results may be obvious. Contrary to popular belief, mass murderers infrequently attack strangers who just happen to be in the wrong place at the wrong time. In fact, almost 40% of these crimes are committed against family members, and almost as many involve other victims acquainted with the perpetrator (for example, co-workers). It is well known that murder often involves family members, but this is especially pronounced among massacres.

The differences in circumstance underlying these crimes are quite dramatic. Although more than half of all single-victim homicides occur during an argument between the victim and the offender, it is relatively rare for a heated dispute to escalate into mass murder.

Some of the most notable differences between homicide types emerge in the offender data. Compared to those offenders who kill but one, mass murderers are especially likely to be male, are far more likely to be white, and are somewhat older (middle-aged). Typically, the single-victim offender is a young male and slightly more often black than white.

Victim characteristics are, of course, largely a function of the offender characteristics, indicating that mass killers generally do not select their victims on a random basis. For example, the victims of mass murder are usually white simply because the perpetrators to whom they are related or with whom they associate are white. Similarly, the youthfulness and greater representation of females among the victims of mass murder, as compared to single-victim homicide, stem from the fact that a typical mass killing involves the breadwinner of the household who annihilates the entire family— his wife and his children.

The Stages of Social Research

Systematically testing our ideas about the nature of social reality often demands carefully planned and executed research in which the following occur:

1. The problem to be studied is reduced to a testable hypothesis (for example, "one-parent families generate more delinquency than two-parent families").
2. An appropriate set of instruments is developed (for example, a questionnaire or an interview schedule).
3. The data are collected (that is, the researcher might go into the field and conduct a poll or a survey).
4. The data are analyzed for their bearing on the initial hypotheses.
5. Results of the analysis are interpreted and communicated to an audience (for example, by means of a lecture, journal article, or press release).

As we shall see in subsequent chapters, the material presented in this book is most closely tied to the data-analysis stage of research (see number 4 earlier), in which the data collected or gathered by the researcher are analyzed for their bearing on the initial hypotheses. It is in this stage of research that the raw data are tabulated, calculated, counted, summarized, rearranged, compared, or, in a word, *organized,* so that the accuracy or validity of the hypotheses can be tested.

Using Series of Numbers to Do Social Research

Anyone who has conducted social research knows that problems in data analysis must be confronted in the planning stages of a research project, because they have a bearing on the nature of decisions at all other stages. Such problems often affect aspects of the research design and even the types of instruments employed in collecting the data. For this reason, we constantly seek techniques or methods for enhancing the quality of data analysis.

Most researchers would agree on the importance of *measurement* in analyzing data. When some characteristic is measured, researchers are able to assign to it a series of numbers according to a set of rules. Social researchers have developed measures of a wide range of phenomena, including occupational prestige, political attitudes, authoritarianism, alienation, anomie, delinquency, social class, prejudice, dogmatism, conformity, achievement, ethnocentrism, neighborliness, religiosity, marital adjustment, occupational mobility, urbanization, sociometric status, and fertility.

Numbers have at least three important functions for social researchers, depending on the particular *level of measurement* that they employ. Specifically, series of numbers can be used to

1. *classify* or *categorize* at the nominal level of measurement,
2. *rank* or *order* at the ordinal level of measurement, and
3. assign a *score* at the interval level of measurement.

The Nominal Level

The *nominal level of measurement* involves naming or labeling—that is, placing cases into categories and counting their frequency of occurrence. To illustrate, we might use a nominal-level measure to indicate whether each respondent is prejudiced or tolerant toward Latinos. As shown in Table 1.2, we might question the 10 students in a given class and determine that 5 can be regarded as (1) prejudiced and 5 can be considered (2) tolerant.

TABLE 1.2 *Attitudes of 10 College Students toward Latinos: Nominal Data*

Attitude toward Latinos	Frequency
1 = prejudiced	5
2 = tolerant	5
Total	10

Other nominal-level measures in social research are sex (male versus female), welfare status (recipient versus nonrecipient), political party (Republican, Democrat, and Libertarian), social character (inner-directed, other-directed, and tradition-directed), mode of adaptation (conformity, innovation, ritualism, retreatism, and rebellion), and time orientation (present, past, and future), to mention only a few.

When dealing with nominal data, we must keep in mind that *every case must be placed in one, and only one, category.* This requirement indicates that the categories must be nonoverlapping, or *mutually exclusive.* Thus, a respondent's race classified as white cannot also be classified as black; any respondent labeled male cannot also be labeled female. The requirement also indicates that the categories must be *exhaustive*—there must be a place for every case that arises. For illustrative purposes, imagine a study in which all respondents are interviewed and categorized by race as either black or white. Where would we categorize a Chinese respondent if he or she were to appear? In this case, it might be necessary to expand the original category system to include Asians or, assuming that most respondents will be white or black, to include an "other" category in which such exceptions can be placed.

The reader should note that nominal data are not graded, ranked, or scaled for qualities, such as better or worse, higher or lower, more or less. Clearly, then, a nominal measure of sex does not signify whether males are superior or inferior to females. Nominal data are merely labeled, sometimes by name (male versus female or prejudiced versus tolerant), other times by number (1 versus 2), but always for the purpose of grouping the cases into separate categories to indicate sameness or differentness with respect to a given quality or characteristic. Thus, even when a number is used to label a category (for example, 1 = white, 2 = black, 3 = other), a quantity is not implied.

The Ordinal Level

When the researcher goes beyond the nominal level of measurement and seeks to order his or her cases in terms of the degree to which they have any given characteristic, he or she is working at the *ordinal level of measurement.* The nature of the relationship among ordinal categories depends on that characteristic the researcher seeks to measure. To take a familiar example, one might classify individuals with respect to socioeconomic status as lower class, middle class, or upper class. Or, rather than categorize the students in a given classroom as *either* prejudiced *or* tolerant, the researcher might rank them according to their degree of prejudice against Latinos, as indicated in Table 1.3.

The ordinal level of measurement yields information about the ordering of categories, but does not indicate the *magnitude of differences* between numbers. For instance, the social researcher who employs an ordinal-level measure to study prejudice toward Latinos *does not know how much more prejudiced one respondent is than another.* In the example given in Table 1.3, it is not possible to determine how much more prejudiced Joyce is than Paul or how much less prejudiced Ben is than Linda or Ernie. This is because the intervals between the points or ranks on an ordinal scale are not known or meaningful. Therefore, it is not possible to assign *scores* to cases located at points along the scale.

TABLE 1.3 *Attitudes of 10 College Students toward Latinos: Ordinal Data*

Student	Rank
Joyce	1 = most prejudiced
Paul	2 = second
Cathy	3 = third
Mike	4 = fourth
Judy	5 = fifth
Joe	6 = sixth
Kelly	7 = seventh
Ernie	8 = eighth
Linda	9 = ninth
Ben	10 = least prejudiced

The Interval/Ratio Level

By contrast to the ordinal level, the *interval and ratio levels of measurement* not only indicate the ordering of categories but also the exact distance between them. Interval and ratio measures employ constant units of measurement (for example, dollars or cents, Fahrenheit or Celsius, yards or feet, minutes or seconds), which yield equal intervals between points on the scale.

Some variables in their natural form are interval/ratio level—for example, how many pounds you weigh, how many siblings you have, or how long it takes a student to complete an exam. In the social sciences, naturally formed interval/ratio measures might include the length of a prison sentence, the number of children in a family, or the amount of time—in minutes and hours—an individual spends on the job.

Other variables are interval/ratio because of how we scale them. Typically, an interval/ratio measure that we construct generates a set of scores that can be compared with one another. As currently used by social scientists, for example, a well-known measure of job satisfaction, employed by Tom W. Smith who directs the General Social Survey at the National Opinion Research Center, is treated as an interval variable. In this process, respondents are asked to indicate how satisfied they are with the work they do on a four-point rating scale consisting of 1 for someone who is "very dissatisfied," 2 for someone who is "a little dissatisfied," 3 for someone who is "moderately satisfied," and 4 for someone who is "very satisfied." The occupations are then placed in a hierarchy from lowest to highest, depending on the overall evaluations—the mean satisfaction score—they receive from a group of respondents who hold the jobs they are asked to judge. In one recent study, for example, the job title *clergy* received a rating of 3.79 (almost at the "very satisfied" level), whereas *waiters* received a 2.85 (close to the "moderately satisfied" level); *physical therapists* got a score of 3.72, whereas *roofers* received a 2.84.

TABLE 1.4 *Satisfaction Scores of Eight Jobs: Interval Data*

Job	Satisfaction Score
Clergy	3.79
Teachers	3.61
Authors	3.61
Psychologists	3.59
Butchers	2.97
Cashiers	2.94
Bartenders	2.88
Roofers	2.84

As depicted in Table 1.4, we are able to order a group of eight occupations in terms of their degree of satisfaction and, in addition, determine the exact distances separating one from another. This requires making the assumption that our measure of job satisfaction uses a constant unit of measurement (one satisfaction point). Thus, we can say that the job of clergy is the most satisfying on the list because it received the highest score on the measure. We can also say that authors are only slightly more satisfied than psychologists, but much more satisfied than bartenders and roofers, both of which received extremely low scores. Depending on the purpose for which a study is designed, such information might be important to determine, but is not available at the ordinal level of measurement.

The ratio level is the same as the interval level, but in addition presumes the existence of an absolute or true zero point. In contrast, an interval level variable may have an artificial zero value or even none at all.

For example, age meets the condition for the ratio level, because a zero represents birth, or the complete absence of age. In contrast, the Fahrenheit scale of temperature possesses an artificial zero point, because "zero degrees" does not represent the total absence of heat, even though it does not feel particularly warm. Similarly, the IQ scale has no zero point at all—that is, there is no such thing as a zero IQ—and therefore qualifies only as an interval scale. Thus, we cannot say that a person with an IQ of 150 is 50% more intelligent than someone with an average 100 IQ.

Similarly, a score of zero on a scale of occupational satisfaction, if it existed, would indicate a total absence of any satisfaction at all ("complete dissatisfaction"), and therefore potentially represents a ratio scale. As constructed by the author, however, the scale of occupational prestige illustrated previously has not been given a score of zero (a score of "1" indicates "very" but not complete dissatisfaction) and is therefore at the interval, not the ratio, level.

When it comes right down to it, it makes little practical difference whether a variable is interval or ratio level. There are many important statistical techniques that assume a standard distance between scale points (that is, an interval scale), but there are very few that require valid ratios between scale points (that is, a ratio scale). Thus, throughout the remainder of the book, we shall indicate whether a technique requires the nominal level, the ordinal level, or the interval level.

Different Ways to Measure the Same Variable

As noted earlier, the level of measurement of certain naturally occurring variables like gender or hair color is clear-cut, while others are constructed by how the social researcher defines them. In fact, the same variable can be measured at different levels of precision depending on the research objectives.

Figure 1.1, for example, illustrates several ways in which the variable "pain" might be measured by a researcher interested in health issues. At the lowest level—nominal—respondents could be classified as being in pain or pain-free, or they could be asked to

Nominal Level

Question: Are you currently in pain? Yes or No

Question: How would you characterize the type of pain? Sharp, Dull, Throbbing

Ordinal Level

Question: How bad is the pain right now? None, Mild, Moderate, Severe

Question: Compared with yesterday, is the pain less severe, about the same, or more severe?

Interval/Ratio Level

0–10 Numerical Scale

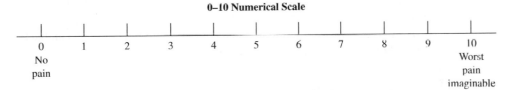

| 0 | 1 | 2 | 3 | 4 | 5 | 6 | 7 | 8 | 9 | 10 |
| No pain | | | | | | | | | | Worst pain imaginable |

Visual Analog Scale

No pain

Ask patient to indicate on the line where the pain is in relation to the two extremes. Quantification is only approximate; for example, a midpoint mark would indicate that the pain is approximately half of the worst possible pain.

Worst pain

Pain Faces Scale

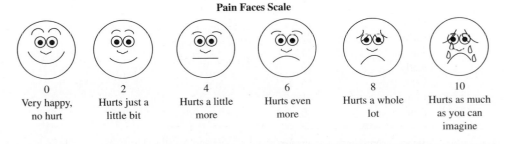

| 0 | 2 | 4 | 6 | 8 | 10 |
| Very happy, no hurt | Hurts just a little bit | Hurts a little more | Hurts even more | Hurts a whole lot | Hurts as much as you can imagine |

FIGURE 1.1 Different ways to measure pain

indicate what type of pain they are experiencing. Of course, the extent of pain could be measured in an ordinal scale ranging from none to severe or respondents could indicate whether their pain was better, worse, or the same. More precisely, the degree of pain could be reported numerically from 0 to 10, reflecting an interval/ratio level scale. Alternative ways of measuring pain at the interval/ratio level include having respondents indicate their degree of pain by marking on a continuum, from no pain to worst pain. Finally, the Pain Faces Scale might be used for children with limited verbal skills or foreign-speaking adults.

Treating Some Ordinal Variables as Interval

As we have seen, levels of measurement vary in terms of their degree of sophistication or refinement, from simple classification (nominal), to ranking (ordinal), to scoring (interval). Table 1.5 gives an example using interval data.

At this point, the distinction between the nominal and ordinal levels should be quite clear. It would be difficult to confuse the level of measurement attained by the variable "color of hair" (blond, redhead, brunette, and black), which is nominal, with that of the variable "condition of hair" (dry, normal, oily), which is ordinal.

The distinction between ordinal and interval, however, is not always clear-cut. Often, variables that in the strictest sense are ordinal may be treated as if they were interval when the ordered categories are fairly evenly spaced. Actually, an earlier example—the measure of job satisfaction—can be used to make this point. To treat this measure as interval rather than ordinal, it is necessary to assume that the distance between "very dissatisfied" and "a little dissatisfied" is roughly the same as the distance between "a little dissatisfied" and "moderately satisfied" and between "moderately satisfied" and "very satisfied." If we are unable to make the assumption of equal intervals between the points on the scale, then the satisfaction measure should be treated as an ordinal scale.

TABLE 1.5   *Attitudes of 10 College Students toward Latinos: Interval Data*

Student	Score[a]
Joyce	98
Paul	96
Cathy	95
Mike	94
Judy	22
Joe	21
Kelly	20
Ernie	15
Linda	11
Ben	6

[a]Higher scores indicate greater prejudice against Latinos.

To take another example, the following two variables (*rank of professor* and *attitude toward professor*) are both ordinal:

Scale Value	Rank of Professor	Attitude toward Professor
1	Distinguished professor	Very favorable
2	Full professor	Favorable
3	Associate professor	Somewhat favorable
4	Assistant professor	Neutral
5	Instructor	Somewhat unfavorable
6	Lecturer	Unfavorable
7	Teaching assistant	Very unfavorable

The *rank-of-professor* variable could hardly be mistaken for interval. The difference between *instructor* (5) and *lecturer* (6) is minimal in terms of prestige, salary, or qualifications, whereas the difference between *instructor* (5) and *assistant professor* (4) is substantial, with the latter generally requiring a doctorate and receiving a much higher salary. By contrast, the *attitude-toward-professor* variable has scale values that are roughly evenly spaced. The difference between *somewhat unfavorable* (5) and *unfavorable* (6) *appears* to be virtually the same as the difference between *somewhat unfavorable* (5) and *neutral* (4). In fact, this is true of most attitude scales ranging from *strongly agree* to *strongly disagree*.

Rather than split hairs, many researchers make a practical decision. Whenever possible, they choose to treat ordinal variables as interval, but only when it is reasonable to assume that the scale has roughly equal intervals. Thus, they would treat the *attitude-toward professor* variable as if it were interval, but they would never treat the *rank-of-professor* variable as anything other than ordinal. As you will see later in the text, treating ordinal variables that have nearly evenly spaced values as if they were interval allows researchers to use more powerful statistical procedures.

Further Measurement Issues

Whether a variable is measured at the nominal, ordinal, or interval level is sometimes a natural feature of the characteristic itself, and not at all influenced by the decisions that the social researcher makes in defining and collecting data. Hair color (black, brown, blonde, gray, and so on), race (black, white, Asian), and region of residence (Northeast, Mid-Atlantic, South, Midwest, Mountain, and West) are, for example, unquestionably nominal-level variables. A researcher, however, can still expand the meaning of basic characteristics like these in an attempt to increase the precision and power of his or her data. Hair color, for example, can be redefined in terms of shades (for example, from dark brown to platinum blonde) to elevate the level of measurement to ordinal status. Similarly, for the purpose of measuring geographic proximity to Southern culture, an ordinal-level "Southerness scale" might be developed to distinguish Mississippi and Alabama at one extreme, Kentucky and Tennessee next, followed by Maryland and Delaware, and then Connecticut and Vermont at the other extreme. Although it may be somewhat stretching the point, a researcher could also develop an interval-level Southerness scale, using the number of miles a state's center lies above or below the Mason–Dixon line.

More commonly, there are situations in which variables must be downgraded in their level of measurement, even though this might reduce their precision. To increase the response rate, for example, a telephone interviewer might redefine age, an interval-level variable, into ordinal categories such as toddler, child, teenager, young adult, middle-aged, and senior.

Another important measurement distinction that social researchers confront is between discrete and continuous variables. Discrete data take on only certain specific values. For example, family size can be expressed only in whole numbers from 1 on up (there is no such thing as 3.47 people in a family; it's either 1, 2, 3, 4, or more members). Family size therefore represents a discrete interval-level measure. Moreover, nominal variables (such as *New England states:* Massachusetts, Connecticut, Rhode Island, Vermont, Maine, and New Hampshire; *gender:* female and male; *religion:* Protestant, Catholic, Jewish, Muslim, Hindu), by virtue of their categorical nature, are always discrete.

Continuous variables, on the other hand, present an infinite range of possible values, although the manner in which we measure them may appear to be discrete. Body weight, for example, can take on any number of values, including 143.4154 pounds. Some bathroom scales may measure this weight to the nearest whole pound (143 pounds), and others may measure weight to the nearest half pound (143.5), and some even to the nearest tenth of a pound (143.4). Underlying whatever measuring device we use, however, is a natural continuum. Similarly, age is a continuous variable and theoretically could be measured in nanoseconds from birth on. Yet it is customary to use whole numbers (years for adults, weeks for infants) in recording this variable. As shown earlier, it is also a common practice arbitrarily to divide the continuum of age into categories such as toddler, child, teenager, young adult, middle-aged, and senior.

The Functions of Statistics

When researchers use numbers—they *quantify* their data at the nominal, ordinal, or interval level of measurement—they are likely to employ statistics as a tool of (1) *description* or (2) *decision making*. Let us now take a closer look at these important functions of statistics.

Description

To arrive at conclusions or obtain results, a social researcher often studies hundreds, thousands, or even larger numbers of persons or groups. As an extreme case, the U.S. Bureau of the Census conducts a complete enumeration of the U.S. population, in which millions of individuals are contacted. Despite the aid of numerous sophisticated procedures, it is always a formidable task to describe and summarize the mass of data generated from projects in social research.

To take a familiar example, the examination grades of 80 students have been listed in Table 1.6. Do you see any patterns in these grades? Can you describe these grades in a few words? In a few sentences? Can you tell if they are particularly high or low on the whole?

Your answer to these questions should be "no." However, using even the most basic principles of descriptive statistics, it is possible to characterize the distribution of the examination grades in Table 1.6 with a good deal of clarity and precision, so that overall

TABLE 1.6   *Examination Grades for 80 Students*

72	49	81	52	31
38	81	58	68	73
43	56	45	54	40
81	60	52	52	38
79	83	63	58	59
71	89	73	77	60
65	60	69	88	75
59	52	75	70	93
90	62	91	61	53
83	32	49	39	57
39	28	67	74	61
42	39	76	68	65
58	49	72	29	70
56	48	60	36	79
72	65	40	49	37
63	72	58	62	46

tendencies or group characteristics can be quickly discovered and easily communicated to almost anyone. First, the grades can be rearranged in consecutive order (from highest to lowest) and grouped into a much smaller number of categories. As shown in Table 1.7, this *grouped frequency distribution* (to be discussed in detail in Chapter 2) presents the grades within broader categories along with the number or *frequency (f)* of students whose grades fell into these categories. It can be readily seen, for example, that 17 students received grades between 60 and 69; only 2 students received grades between 20 and 29.

Another useful procedure (explained in Chapter 2) rearranges the grades graphically. As shown in Figure 1.2, the categories of grades are placed (from 20–29 to 90–99) along one line of a graph (that is, the *horizontal base line*) and their numbers or frequencies along another line (that is, the *vertical axis*). This arrangement results in a rather easily visualized

TABLE 1.7   *Examination Grades for 80 Students: A Grouped Frequency Distribution*

Grades	f
90–99	3
80–89	7
70–79	16
60–69	17
50–59	15
40–49	11
30–39	9
20–29	2

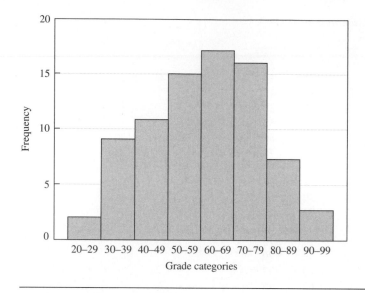

FIGURE 1.2   Graph of examination grades for 80 students

graphic representation in which we can see that most grades fall between 50 and 80, whereas relatively few grades are much higher or lower.

As elaborated on in Chapter 3, a particularly convenient and useful statistical method—one with which you are already more or less familiar—is to ask: What is the grade of the *average* person in this group of 80 students? The arithmetic average (or *mean*), which can be obtained by adding the entire list of grades and dividing this sum by the number of students, gives us a clearer picture of the overall group tendency or class performance. The arithmetic average in this example happens to be 60.5, a rather low grade compared against the class averages with which most students may be familiar. Apparently, this group of 80 students did relatively poorly as a whole.

Thus, with the help of statistical devices, such as grouped frequency distributions, graphs, and the arithmetic average, it is possible to detect and describe patterns or tendencies in distributions of scores (for example, the grades in Table 1.6) that might otherwise have gone unnoticed by the casual observer. In the present context, then, statistics may be defined as *a set of techniques for the reduction of quantitative data (that is, a series of numbers) to a small number of more convenient and easily communicated descriptive terms.*

Decision Making

For purposes of testing hypotheses, it is frequently necessary to go beyond mere description. It is often also necessary to make inferences—that is, to make decisions based on data collected on only a small portion or *sample* of the larger group we have in mind to study. Factors such as cost, time, and need for adequate supervision many times preclude taking a complete enumeration or poll of the entire group (social researchers call this larger group from which the sample was drawn a *population* or *universe*).

As we shall see in Chapter 6, every time social researchers test hypotheses on a sample, they must decide whether it is indeed accurate to generalize the findings to the entire population from which they were drawn. Error inevitably results from sampling, even sampling that has been properly conceived and executed. This is the problem of generalizing or *drawing inferences* from the sample to the population.[2]

Statistics can be useful for purposes of generalizing findings, with a high degree of confidence, from small samples to larger populations. To understand better this decision-making purpose of statistics and the concept of generalizing from samples to populations, let us examine the results of a hypothetical study that was conducted to test the following hypothesis:

Hypothesis: *Male college students are more likely than female college students to have tried marijuana.*

The researchers in this study decided to test their hypothesis at an urban university in which some 20,000 students (10,000 males and 10,000 females) were enrolled. Due to cost and time factors, they were not able to interview every student on campus, but did obtain from the registrar's office a complete listing of university students. From this list, every one-hundredth student (one-half of them male, one-half of them female) was selected for the sample and subsequently interviewed by members of the research staff. The interviewers asked each of the 200 members of the sample whether he or she had ever tried marijuana and then recorded the student's gender as either male or female. After all interviews had been completed and returned to the staff office, the responses on the marijuana question were tabulated by gender, as presented in Table 1.8.

TABLE 1.8   *Marijuana Use by Gender of Respondent: Case I*

Marijuana Use	Gender of Respondent	
	Male	*Female*
Number who have tried it	35	15
Number who have not tried it	65	85
Total	100	100

[2]The concept of *sampling error* is discussed in greater detail in Chapter 6. However, to understand the inevitability of error when sampling from a larger group, you may now wish to conduct the following demonstration. Refer to Table 1.6, which contains the grades for a population of 80 students. At random (for example, by closing your eyes and pointing), select a sample of five grades from the entire list. Find the average grade by adding the five scores and dividing by 5, the total number of grades. It has already been pointed out that the average grade for the entire class of 80 students was 60.5. To what extent does your sample average differ from the class average, 60.5? Try this demonstration on several more samples of a few grades randomly selected from the larger group. With great consistency, you should find that your sample mean will almost always differ at least slightly from that obtained from the entire class of 80 students. That is what we mean by *sampling error.*

TABLE 1.9 *Marijuana Use by Gender of Respondent: Case II*

Marijuana Use	Gender of Respondent	
	Male	*Female*
Number who have tried it	30	20
Number who have not tried it	70	80
Total	100	100

Notice that results obtained from this sample of 200 students, as presented in Table 1.8, are in the hypothesized direction: 35 out of 100 males reported having tried marijuana, whereas only 15 out of 100 females reported having tried marijuana. Clearly, in this small sample, males were more likely than females to have tried marijuana. For our purposes, however, the more important question is whether these gender differences in marijuana use are large enough to generalize them confidently to the much larger university population of 20,000 students. Do these results represent true population differences? Or have we obtained chance differences between males and females due strictly to sampling error—the error that occurs every time we take a small group from a larger group?

To illuminate the problem of generalizing results from samples to larger populations, imagine that the researchers had, instead, obtained the results shown in Table 1.9. Notice that these results are still in the predicted direction: 30 males as opposed to only 20 females have tried marijuana. But, are we still willing to generalize these results to the larger university population? Is it not likely that a difference of this magnitude (10 more males than females) would have happened simply by chance? Or can we confidently say that such relatively small differences reflect a real difference between males and females at that particular university?

Let us carry out the illustration a step further. Suppose that the social researchers had obtained the data shown in Table 1.10. Differences between males and females shown in the table could not be much smaller and still be in the hypothesized direction: 26 males in contrast to 24 females tried marijuana—only 2 more males than females. How many of

TABLE 1.10 *Marijuana Use by Gender of Respondent: Case III*

Marijuana Use	Gender of Respondent	
	Male	*Female*
Number who have tried it	26	24
Number who have not tried it	74	76
Total	100	100

us would be willing to call *this* finding a true population difference between males and females rather than a product of chance or sampling error?

Where do we draw the line? At what point does a sample difference become large enough so that we are willing to treat it as significant or real? With the aid of statistics, we can readily, and with a high degree of confidence, make such decisions about the relationship between samples and populations. To illustrate, had we used one of the statistical tests of significance discussed later in this text (for example, chi-square; see Chapter 9), we would already have known that *only those results* reported in Table 1.8 can be generalized to the population of 20,000 university students—that 35 out of 100 males but only 15 out of 100 females have tried marijuana is a finding substantial enough to be applied to the entire population with a high degree of confidence and is therefore referred to as a *statistically significant difference.* Our statistical test tells us there are only 5 chances out of 100 that we are wrong! By contrast, application of the same statistical criterion shows the results reported in Tables 1.9 and 1.10 are *statistically nonsignificant,* probably being the product of sampling error rather than real gender differences in the use of marijuana.

In the present context, then, statistics is *a set of decision-making techniques that aid researchers in drawing inferences from samples to populations and, hence, in testing hypotheses regarding the nature of social reality.*

An Important Note about Rounding

If you are like most students, the issue of rounding can be confusing. It is always a pleasure, of course, when an answer comes out to be a whole number because rounding is not needed. For those other times, however, when you confront a number such as 34.233333 or 7.126534, determining just how many digits to use in rounding becomes problematic.

For occasions when you need to round, the following rule can be applied: *Round a final answer to two more decimal digits than contained in the original scores.* If the original scores are all whole numbers (for example, 3, 6, 9, and 12), then round your final answer to two decimal places (for example, 4.45). If the original scores contain one decimal place (for example, 3.3, 6.0, 9.5, and 12.8), then round your answer to three decimal places (for example, 4.456). A discussion of *how* to round is given in Appendix B.

Many problems in this book require a number of intermediate steps before arriving at the final answer. When using a calculator, it is usually not necessary to round off calculations done along the way (that is, for intermediate steps). Your calculator will often carry many more digits than you will eventually need. As a general rule for intermediate steps, do not round until it comes time to determine your final answer.

Rules of thumb, of course, must be used with some degree of good judgment. As an extreme example, you would not want to round only to two decimal places in calculating the trajectory or thrust needed to send a missile to the moon; even a slight imprecision might lead to disaster. In doing problems for your statistics class, on the other hand, the precision of your answer is less important than learning the method itself. There may be times when your answer will differ slightly from that of your classmate or that contained in this book. For example, you may obtain the answer 5.55, whereas your classmate may get 5.56, yet you both may be correct. The difference is trivial and could easily have resulted from using two calculators with different memory capacities or from doing calculations in a different sequence.

In this text, we have generally followed this rule of thumb. In some illustrations, however, we rounded intermediate steps for the sake of clarity—but only to an extent that would not invalidate the final answer.

Summary

In the first chapter, we linked our everyday predictions about the course of future events with the experiences of social researchers who use statistics as an aid in testing their hypotheses about the nature of social reality. Almost daily, ordinary people take educated guesses about the future events in their lives. Unlike haphazard and biased everyday observations, however, researchers seek to collect *systematic* evidence in support of their ideas. For this purpose, and depending on their particular research objective, they might decide to conduct a survey, an experiment, participant observation, a content analysis or a secondary analysis. Depending on the particular level of measurement, series of numbers are often employed by social researchers to categorize (nominal level), rank (ordinal level), or score (interval/ratio level) their data. Finally, social researchers are able to take advantage of two major functions of statistics in the data-analysis stage of social research: description (that is, reducing quantitative data to a smaller number of more convenient descriptive terms) and decision making (that is, drawing inferences from samples to populations).

Terms to Remember

Hypothesis	Measurement
Variable	Level of measurement
Experiment	Nominal
Survey	Ordinal
Content analysis	Interval/Ratio
Participant observation	

Questions and Problems

1. A social researcher who joins a group of skinheads in order to study their recruiting tactics employs the method of research known as
   a. the experiment.
   b. the survey.
   c. content analysis.
   d. participant observation.
   e. secondary analysis.

2. A sociologist who studies gender differences in voting by comparing male and female reports originally collected by the U.S. Bureau of the Census employs the research method known as
   a. the experiment.
   b. the survey.
   c. content analysis.

    d. participant observation.

    e. secondary analysis.

3. Someone who ranks a list of cities from slowest to fastest pace of life is operating at the _____ level of measurement.

    a. nominal

    b. ordinal

    c. interval

4. A researcher who scores a set of respondents (from 0 to 10) in terms of their degree of empathy for accident victims is working at the _____ level of measurement.

    a. nominal

    b. ordinal

    c. interval

5. The statistical approach involved in generalizing from a sample of 25 patients to an entire population of the hundreds of patients in a particular hospital is known as

    a. description.

    b. decision making.

    c. content analysis.

    d. an experiment.

    e. secondary analysis.

6. A sociologist undertakes a series of studies to investigate various aspects of sports violence. For each of the following research situations, identify the research strategy (experiment, survey, content analysis, or participant observation) and the independent and dependent variables:

    a. Do male and female sports reporters describe combative sporting events (such as football) in the same way? To find out, the sociologist collects the game reports filed by a number of male and female newspaper writers on the day following the Super Bowl. He compares the aggressiveness contained in the adjectives used by the reporters to describe the game.

    b. Do children react differently after watching combative and noncombative sports? To find out, the sociologist randomly assigns school children to watch taped versions of either a hockey game (combative) or a swimming meet (noncombative). She then observes the aggressiveness of play demonstrated by the children immediately following their viewing the tapes.

    c. Are fans more aggressive when their team wins or loses? To find out, the sociologist spends his Saturdays in a sports bar that features the local college game on wide-screen television. He dresses in a team sweatshirt and becomes one of the crowd. At the same time, he observes the extent of arguing and fighting that goes on around him when the team is winning and losing.

    d. Do levels of personal aggressiveness influence the kinds of sporting events that people prefer to watch? To find out, the sociologist distributes a questionnaire to a random sample of adults. In addition to standard background information, the questionnaire includes a series of items measuring aggressiveness (for example, "How often do you get involved in heated arguments with neighbors or friends?") and a checklist of which sports the respondents like to watch.

7. Identify the level of measurement—nominal, ordinal, or interval—represented in each of the following questionnaire items:

    a. Your sex:

        1. _____ Female

        2. _____ Male

b. Your age:
   1. _____ Younger than 20
   2. _____ 20–29
   3. _____ 30–39
   4. _____ 40–49
   5. _____ 50–59
   6. _____ 60–69
   7. _____ 70 or older
c. How many people are in your immediate family? _____
d. Specify the highest level of education achieved by your mother:
   1. _____ None
   2. _____ Elementary school
   3. _____ Some high school
   4. _____ Graduated high school
   5. _____ Some college
   6. _____ Graduated college
   7. _____ Graduate school
e. Your annual income from all sources: _____ (specify)
f. Your religious preference:
   1. _____ Protestant
   2. _____ Catholic
   3. _____ Jewish
   4. _____ Other _____ (specify)
g. The social class to which your parents belong:
   1. _____ Upper
   2. _____ Upper-middle
   3. _____ Middle-middle
   4. _____ Lower-middle
   5. _____ Lower
h. In which of the following regions do your parents presently live?
   1. _____ Northeast
   2. _____ South
   3. _____ Midwest
   4. _____ West
   5. _____ Other _____ (specify)
i. Indicate your political orientation by placing an X in the appropriate space:
   LIBERAL ___ : ___ : ___ : ___ : ___ CONSERVATIVE
                1   2   3   4   5

8. For each of the following items, indicate the level of measurement—nominal, ordinal, or interval:
   a. A tailor uses a tape measure to determine exactly where to cut a piece of cloth.
   b. The speed of runners in a race is timed in seconds by a judge with a stopwatch.
   c. Based on attendance figures, a ranking of the Top 10 rock concerts for the year is compiled by the editors of a music magazine.
   d. A zoologist counts the number of tigers, lions, and elephants she sees in a designated wildlife conservation area.
   e. A convenience store clerk is asked to take an inventory of all items still on the shelves at the end of the month.
   f. The student life director at a small college counts the number of freshmen, sophomores, juniors, and seniors living in residence halls on campus.

g. Using a yardstick, a parent measures the growth of his child on a yearly basis.

h. In a track meet, runners in a half-mile race were ranked first, second, and third place.

9. A political scientist undertakes a series of studies to find out more about the voting population in her local town. For each of the following research situations, identify the research strategy (experiment, survey, content analysis, participant observation, or secondary analysis):

    a. Do males vote more than females? To find out, the researcher analyzes data collected by the U.S. Bureau of the Census.

    b. How many elderly people living in nursing homes vote? To find out, the researcher visits local nursing homes and questions elderly residents to find out how many voted in the last election.

    c. How organized is the election process? On Election Day, the researcher goes to a voting site, pretends to be just another voter, and observes how quickly and efficiently voters are moved through the voting process.

    d. Are people more likely to vote if they are well informed about the candidates? To find out, the researcher provides detailed information about both candidates to a random group of citizens over the age of 18 and compares their voter turnout on Election Day to that of a random group of citizens over 18 who did not receive the information.

10. A researcher who ranks a list of countries according to how much they have depleted their natural resources is working at the _____ level of measurement.

    a. nominal

    b. ordinal

    c. interval

11. Governments can be divided into three different types—unitary governments, federal governments, and confederations—depending on where the concentration of power is located. This would be considered which level of measurement?

    a. Nominal

    b. Ordinal

    c. Interval

12. A sociologist conducts a survey to determine the effects of family size on various aspects of life. For each of the following questionnaire items, identify the level of measurement (nominal, ordinal, or interval):

    a. Does family size affect school performance? Students are asked to circle their letter grade (A, B, C, D, or F) in various school subjects.

    b. Does family size differ by socioeconomic status? Parents are asked to provide their yearly income in dollars.

    c. Does parental health differ by family size? Parents are asked to rate their overall health on a scale from 1 to 5, with 1 being in very good health and 5 being in very poor health.

    d. Do the effects of family size differ with race and ethnicity? Respondents are asked to indicate if they are Black, White, Hispanic, Asian, or other.

13. To understand better the lives of homeless people, a researcher decides to live on the streets for one week disguised as a homeless person. Which of the following would describe this research strategy?

    a. Survey

    b. Content analysis

    c. Experiment

    d. Participant observation

    e. Secondary analysis

14. A high school health teacher distributes an anonymous survey among the students in her class to find out the average age at which these students are becoming sexually active. In this case, statistics are being used for which purpose?
    a. Decision making
    b. Experiment
    c. Secondary analysis
    d. Content analysis
    e. Description

15. Identify the level of measurement (nominal, ordinal, or interval) in each of the following items:
    a. American psychologist William Sheldon developed the idea that there are three major body types: ectomorph, endomorph, and mesomorph.
    b. In a study of short-term memory, a psychologist measures in seconds the time it takes for participants to remember words and numbers that were told to them an hour earlier.
    c. The same psychologist then groups the participants according to how good their short-term memory is, distributing them into five categories that range from "Very good short-term memory" to "Very poor short-term memory."
    d. Participants in a study about eating disorders are asked how many times they eat per day.
    e. Based on blood pressure readings, a psychologist ranks the stressfulness of various activities on a scale of 1 to 10, with 1 being the least stressful and 10 being the most stressful.
    f. In a study on color blindness, a psychologist counts the number of times that participants are able to identify the colors red, yellow, and blue in order to categorize them as either color blind or not color blind.
    g. A researcher interested in family relations focuses on the birth order of siblings.

16. For a very small group of his clients, a psychologist conducts a survey and determines that the most common phobia in the group is acrophobia (fear of heights). In this case, statistics are being used as a tool to perform which function?
    a. Experiment
    b. Secondary analysis
    c. Description
    d. Decision making
    e. Generalization

17. A psychologist is interested in studying how people experience grief. For each of the following situations, identify the research strategy (experiment, survey, content analysis, or participant observation) that she would be using:
    a. To find out how people cope with the loss of loved ones, the psychologist selects a random sample of people and distributes a questionnaire that asks them to provide information about their personal grieving experiences.
    b. The psychologist attends a grief-counseling meeting and pretends that she is one of the mourners (after having obtained permission from the grief counselor). In this way, she is able to observe firsthand how people express their grief.
    c. The grief counselor provides the psychologist with several anonymous journals in which people are urged to express their grief by writing down their thoughts and feelings in a stream-of-consciousness manner. The psychologist then reads through the various journal entries in an attempt to find patterns in the way that people experience grief.

SPSS Exercises

1. Identify the level of measurement—nominal, ordinal, interval/ratio—for each of the following variables from the General Social Survey:
   a. SEXEDUC
   b. TVHOURS
   c. RELIG
   d. NEWS

2. Identify the level of measurement—nominal, ordinal, interval/ratio—for each of the following variables from the Monitoring the Future Study:
   a. V13
   b. V194
   c. V1766
   d. V1779

3. Identify the level of measurement—nominal, ordinal, interval/ratio—for each of the following variables from the Best Places Study:
   a. CRIMEV
   b. SUICIDE
   c. MENTHLTH

4. The method of research used originally to conduct the General Social Survey is known as
   a. the experiment.
   b. the survey.
   c. content analysis.
   d. participant observation.
   e. secondary analysis.

5. The method of research you will use when you analyze the General Social Survey is known as
   a. the experiment.
   b. the survey.
   c. content analysis.
   d. participant observation.
   e. secondary analysis.

6. The method of research used to conduct the Best Places Study is known as
   a. the experiment.
   b. the survey.
   c. content analysis.
   d. participant observation.
   e. secondary analysis.

Looking at the Larger Picture: A Student Survey

The chapters of this textbook each examine particular topics at close range. At the same time, it is important, as they say, to "see the forest through the trees." Thus, at the close of each major part of the text, we shall apply the most useful statistical procedures to the same set of data drawn from a hypothetical survey. This continuing journey should demonstrate the process by which the social researcher travels from having abstract ideas to confirming or rejecting hypotheses about human behavior. Keep in mind both here and in later parts of the book that "Looking at the Larger Picture" is not an exercise for you to carry out, but a summary illustration of how social research is done in practice.

For many reasons, surveys have long been the most common data-collection strategy employed by social researchers. Through the careful design of a survey instrument—a questionnaire filled out by survey respondents or an interview schedule administered over the telephone or in person—a researcher can elicit responses tailored to his or her particular interests. The adage "straight from the horse's mouth" is as true for informing social researchers and pollsters as it is for handicapping the Kentucky Derby.

A rather simple yet realistic survey instrument designed to study smoking and drinking among high school students follows. The ultimate purpose is to understand not just the extent to which these students smoke cigarettes and drink alcohol, but the factors that explain why some students smoke or drink while others do not. Later in this book, we will apply statistical procedures to make sense of the survey results. But for now, it is useful to anticipate the kind of information that we can expect to analyze.

Suppose that this brief survey will be filled out by a group of 250 students, grades 9 through 12, in a hypothetical (but typical) urban high school. In this chapter, we introduced levels of measurement. Note that many of the variables in this survey are nominal—whether the respondent smokes or has consumed alcohol within the past month—as well as respondent characteristics, such as race and sex. Other variables are measured at the ordinal level—specifically, the extent of respondent's peer-group involvement, his or her participation in sports—exercise, as well as academic performance. Finally, still other variables are measured at the interval level—in particular, daily consumption of cigarettes as well as age and grade in school..

To experience firsthand the way that data are collected, you may decide to distribute this survey or something similar on your own. But just like on those television cooking shows, for our purposes here, we will provide at the end of each part of the text, "precooked" statistical results to illustrate the power of these techniques in understanding behavior. As always, it is important not to get caught up in details, but to see the larger picture.

Student Survey

Answer the following questions as honestly as possible. Do not place your name on the form so that your responses will remain completely private and anonymous.

1. What school grade are you currently in?
   _____

2. How would you classify your academic performance? Are you
   _____ an excellent, mostly A's student
   _____ a good, mostly B's student
   _____ an average, mostly C's student
   _____ a below average, mostly D's student

3. Within the past month, have you smoked any cigarettes?
   _____ Yes
   _____ No

4. If you are a smoker, how many cigarettes do you tend to smoke on an average day?
   _____ per day

5. Within the past month, have you had any beer, wine, or hard liquor?
   _____ Yes
   _____ No

6. If you have had beer, wine, or hard liquor in the past month, on how many separate occasions?
   _____ times

7. In terms of your circle of friends, which of the following would best describe you?

_____ I have lots of close friends.

_____ I have a few close friends.

_____ I have one close friend.

_____ I do not have any really close friends.

8. Does either of your parents smoke?

_____ Yes

_____ No

9. To what extent do you participate in athletics or exercise?

_____ Very frequently

_____ Often

_____ Seldom

_____ Never

10. What is your current age? _____ years old

11. Are you _____ Male _____ Female

12. How would you identify your race or ethnicity?

_____ White

_____ Black

_____ Latino

_____ Asian

_____ Other

Part I

Description

2

Organizing the Data

Collecting the data entails a serious effort on the part of social researchers who seek to increase their knowledge of human behavior. To interview or otherwise elicit information from welfare recipients, college students, drug addicts, gays, middle-class Americans, or other respondents requires a degree of foresight, careful planning, and control, if not actual time spent in the field.

Data collection, however, is only the beginning as far as statistical analysis is concerned. Data collection yields the raw materials that social researchers use to analyze data, obtain results, and test hypotheses about the nature of social reality.

Frequency Distributions of Nominal Data

The cabinetmaker transforms raw wood into furniture; the chef converts raw food into the more palatable versions served at the dinner table. By a similar process, the social

TABLE 2.1 *Responses of Young Boys to Removal of Toy*

Response of Child	*f*
Cry	25
Express anger	15
Withdraw	5
Play with another toy	5
	N = 50

researcher—aided by "recipes" called *formulas* and *statistical techniques*—attempts to transform raw data into a meaningful and organized set of measures that can be used to test hypotheses.

What can social scientists do to organize the jumble of raw numbers that they collect from their subjects? How do they go about transforming this mass of raw data into an easy to understand summary form? The first step is to construct a *frequency distribution* in the form of a table.

Suppose a researcher who studies early socialization is interested in the responses of young boys to frustration. In response to the removal of their toys, do they act with anger, or do they cry? How often do they find alternative toys? Do some children react by withdrawing? The researcher performs an experiment on 50 boys aged 2 by presenting and then removing a colorful toy.

Let us examine the frequency distribution of nominal data in Table 2.1. Notice first that the table is headed by a number and a title that gives the reader an idea as to the nature of the data presented—responses of young boys to removal of toy. This is the standard arrangement; every table must be clearly titled and, when presented in a series, labeled by number as well.

Frequency distributions of nominal data consist of two columns. As in Table 2.1, the left column indicates what characteristic is being presented (response of child) and contains the categories of analysis (cry, express anger, withdraw, and play with another toy). An adjacent column (headed *frequency,* or *f*) indicates the number of boys in each category (25, 15, 5, and 5, respectively) as well as the total number of boys (50), which can be indicated either by *N* = 50 or by including the word *Total* below the categories. A quick glance at the frequency distribution in Table 2.1 clearly reveals that more young boys respond by crying or with anger than by withdrawing or by finding an alternative object for play.

Comparing Distributions

Suppose next that the same researcher wishes to compare the responses of male and female children to the removal of a toy. Making comparisons between frequency distributions is a procedure often used to clarify results and add information. The particular comparison a researcher makes is determined by the question he or she seeks to answer.

In this example, the researcher decides to investigate gender differences. Are girls more likely than boys to find an alternative toy? To provide an answer, the researcher might

TABLE 2.2 *Response to Removal of Toy by Gender of Child*

	Gender of Child	
Response of Child	*Male*	*Female*
Cry	25	28
Express anger	15	3
Withdraw	5	4
Play with another toy	5	15
Total	50	50

repeat the experiment on a group of 50 girls and then compare the results. Let us imagine that the data shown in Table 2.2 are obtained. As shown in the table, 15 out of 50 girls but only 5 of the 50 boys responded by playing with another toy in the room.

Proportions and Percentages

When a researcher studies distributions of equal size, the frequency data can be used to make comparisons between the groups. Thus, the numbers of boys and girls who found alternative toys can be directly compared, because there were exactly 50 children of each gender in the experiment. It is generally not possible, however, to study distributions having exactly the same number of cases.

For more general use, we need a method of standardizing frequency distributions for size—a way to compare groups despite differences in total frequencies. Two of the most popular and useful methods of standardizing for size and comparing distributions are the proportion and the percentage.

The *proportion* compares the number of cases in a given category with the total size of the distribution. We can convert any frequency into a proportion P by dividing the number of cases in any given category f by the total number of cases in the distribution N:

$$P = \frac{f}{N}$$

Therefore, 15 out of 50 girls who found an alternative toy can be expressed as the following proportion:

$$P = \frac{15}{50} = .30$$

Despite the usefulness of the proportion, many people prefer to indicate the relative size of a series of numbers in terms of the *percentage,* the frequency of occurrence of a category per 100 cases. To calculate a percentage, we simply multiply any given proportion by 100. By formula,

TABLE 2.3 *Gender of Students Majoring in Engineering at Colleges A and B*

| Gender of Student | Engineering Majors | | | |
| | College A | | College B | |
	f	%	f	%
Male	1,082	80	146	80
Female	270	20	37	20
Total	1,352	100	183	100

$$\% = (100)\frac{f}{N}$$

Therefore, 15 out of 50 girls who responded by finding an alternative can be expressed as the proportion $P = 15/50 = .30$ or as a percentage $\% = (100)(15/50) = 30\%$. Thus, 30% of the girls located another toy with which to amuse themselves.

To illustrate the utility of percentages in making comparisons between large and unequal-sized distributions, let us examine the gender of engineering majors at two colleges where the engineering programs are of very different size. Suppose, for example, that College A has 1,352 engineering majors and College B has only 183 engineering majors.

Table 2.3 indicates both the frequencies and the percentages for engineering majors at Colleges A and B. Notice how difficult it is to determine quickly the gender differences among engineering majors from the frequency data alone. By contrast, the percentages clearly reveal that females were equally represented among engineering majors at Colleges A and B. Specifically, 20% of the engineering majors at College A are females; 20% of the engineering majors at College B are females.

Ratios and Rates

A less commonly used method of standardizing for size, the *ratio,* directly compares the number of cases falling into one category (for example, males) with the number of cases falling into another category (for example, females). Thus, a ratio can be obtained in the following manner, where f_1 = frequency in any category, and f_2 = frequency in any other category:

$$\text{Ratio} = \frac{f_1}{f_2}$$

If we were interested in determining the ratio of blacks to whites, we would compare the number of black respondents ($f = 150$) to the number of white respondents ($f = 100$) as 150/100. By canceling common factors in the numerator and denominator, it is possible to reduce a ratio to its simplest form, for example, $150/100 = 3/2$. (There are 3 black respondents for every 2 white respondents.)

The researcher might increase the clarity of this ratio by giving the base (the denominator) in a more understandable form. For instance, the *sex ratio* often employed by demographers who seek to compare the number of males and females in any given population is generally given as the number of males per 100 females.

To illustrate, if the ratio of males to females is 150/50, there are 150 males for 50 females (or reducing, 3 males for every 1 female). To obtain the conventional version of the sex ratio, we multiply the previous ratio by 100:

$$\text{Sex ratio} = (100)\frac{f\,\text{males}}{f\,\text{females}} = (100)\left(\frac{150}{50}\right) = 300$$

It then turns out that there are 300 males in the population for every 100 females.

Another kind of ratio—one that tends to be more widely used by social researchers   is known as a *rate*. Sociologists often analyze populations regarding rates of reproduction, death, crime, unemployment, divorce, marriage, and the like. However, whereas most other ratios compare the number of cases in any category or subgroup with the number of cases in any other subgroup, rates indicate comparisons between the number of *actual* cases and the number of *potential* cases. For instance, to determine the birth rate for a given population, we might show the number of actual live births among females of childbearing age (those members of the population who are exposed to the risk of childbearing and therefore represent potential cases). Similarly, to determine the divorce rate, we might compare the number of actual divorces against the number of marriages that occur during some period of time (for example, 1 year). Rates are often given in terms of a base having 1,000 potential cases. Thus, birth rates are given as the number of births per 1,000 females; divorce rates might be expressed in terms of the number of divorces per 1,000 marriages. If 500 births occur among 4,000 women of childbearing age,

$$\text{Birth rate} = (1,000)\frac{f\,\text{actual cases}}{f\,\text{potential cases}} = (1,000)\left(\frac{500}{4,000}\right) = 125$$

It turns out there are 125 live births per every 1,000 women of childbearing age.

There is nothing particularly special about calculating rates per potential case or per 1,000 potential cases. In fact, expressing rates per capita (that is, per person), per 1,000, or even per million simply comes down to the decision of what would be the most convenient basis. For example, expenditures for public education are usually expressed per pupil (as determined by average daily attendance, because attendance changes during the school year due to a number of factors including transfers and dropouts). To calculate this rate, we divide the total expenditure in dollars by the enrollment figure:

$$\text{Per capita (pupil) expenditure} = \frac{\text{expenditure for public schools}}{\text{number of pupils}}$$

Therefore, if a town spends $14 million for its public schools with a total enrollment of 2,280 students, the per pupil expenditure is

$$\frac{\$14{,}000{,}000}{2{,}280} = \$6{,}140$$

In contrast to the previous per capita (or per pupil) rate, some rates are calibrated on a per 100,000 basis. Suicide rates are derived from

$$\text{Suicide rate} = \frac{\text{number of suicides}}{\text{population}} \times 100{,}000$$

If a state has a population of 4.6 million residents and tallies 562 suicides in a year, the suicide rate per 100,000 is

$$\frac{562}{4{,}600{,}000} \times 100{,}000 = .000122 \times 100{,}000 = 12.2$$

Thus, in this state there were 12.2 suicides for every 100,000 residents.

It is important to note that we could have defined the rate as suicides per capita without multiplying the fraction (suicides over population) by the 100,000 scale factor. However, the rate of .000122 that would result, although correct, is very awkward because of its small size. So we magnify the rate to a more readable and digestible form by multiplying by 100,000 (moving the decimal point five places to the right), which then converts the per capita rate of .000122 to a per 100,000 rate of 12.2.

Rates are often used to make comparisons between different populations. For instance, we might seek to compare birth rates between blacks and whites, between middle-class and lower-class females, among religious groups or entire nations, and so on. Another kind of rate, *rate of change,* can be used to compare the same population at two points in time. In computing rate of change, we compare the actual change between time period 1 and time period 2 with the level at time period 1 serving as a base. Thus, a town population that increases from 20,000 to 30,000 between 1990 and 2005 experiences the following rate of change:

$$(100)\left(\frac{\text{time } 2f - \text{time } 1f}{\text{time } 1f}\right) = (100)\left(\frac{30{,}000 - 20{,}000}{20{,}000}\right) = 50\%$$

In other words, there was a population increase of 50% over the period 1990 to 2005.

Notice that a rate of change can be *negative* to indicate a decrease in size over any given period. For instance, if a town's population changes from 15,000 to 12,000 over a period of time, the rate of change is

$$(100)\left(\frac{12{,}000 - 15{,}000}{15{,}000}\right) = -20\%$$

Simple Frequency Distributions of Ordinal and Interval Data

Because nominal data are labeled rather than graded or scaled, the categories of nominal-level distributions do not have to be listed in any particular order. Thus, the data on marital status shown in Table 2.4 are presented in three different, yet equally acceptable, arrangements.

In contrast, the categories or score values in ordinal or interval distributions represent the degree to which a particular characteristic is present. The listing of such categories or score values in simple frequency distributions must be made to reflect that order.

For this reason, ordinal and interval categories are always arranged in order, usually from their highest to lowest values, but sometimes from their lowest to highest values. For instance, we might list the categories of social class from upper to lower or post the results of a biology midterm examination in consecutive order from the highest grade to the lowest grade.

Disturbing the order of ordinal and interval categories reduces the readability of the researcher's findings. This effect can be seen in Table 2.5, where both the "incorrect" and "correct" versions of a distribution of attitudes toward a proposed tuition hike on a college campus have been presented. Which version do you find easier to read?

TABLE 2.4 *The Distribution of Marital Status Shown Three Ways*

Marital Status	f	Marital Status	f	Marital Status	f
Married	30	Single	20	Previously married	10
Single	20	Previously married	10	Married	30
Previously married	10	Married	30	Single	20
Total	60	Total	60	Total	60

TABLE 2.5 *A Frequency Distribution of Attitudes toward a Proposed Tuition Hike on a College Campus: Incorrect and Correct Presentations*

Attitude toward a Tuition Hike	f	Attitude toward a Tuition Hike	f
Slightly favorable	2	Strongly favorable	0
Somewhat unfavorable	21	Somewhat favorable	1
Strongly favorable	0	Slightly favorable	2
Slightly unfavorable	4	Slightly unfavorable	4
Strongly unfavorable	10	Somewhat unfavorable	21
Somewhat favorable	1	Strongly unfavorable	10
Total	38	Total	38
INCORRECT		CORRECT	

Grouped Frequency Distributions of Interval Data

Interval-level scores are sometimes spread over a wide range (highest minus lowest score), making the resultant simple frequency distribution long and difficult to read. When such instances occur, few cases may fall at each score value, and the group pattern becomes blurred. To illustrate, the distribution set up in Table 2.6 contains values varying from 50 to 99 and runs almost four columns in length.

To clarify our presentation, we might construct a *grouped frequency distribution* by condensing the separate scores into a number of smaller categories or groups, each containing more than one score value. Each category or group in a grouped distribution is known as a *class interval,* whose *size* is determined by the number of score values it contains.

The examination grades for 71 students originally presented in Table 2.6 are rearranged in a grouped frequency distribution, shown in Table 2.7. Here we find 10 class intervals, each having size 5. Thus, the highest class interval (95–99) contains the five score values 95, 96, 97, 98, and 99. Similarly, the interval 70–74 is of size 5 and contains the score values 70, 71, 72, 73, and 74.

The frequencies are next to the class intervals in Table 2.7. This column tells us the number of cases or scores in each of the categories. Thus, whereas the class interval 95–99 spans five score values (95, 96, 97, 98, and 99), it includes three scores (95, 96, and 98).

The more meaningful column, particularly if comparisons to other distributions are considered (such as the final examination scores during a different term with a different number of students), is the percent column. This column is also called the *percentage distribution.* For example, we can see that 4.23% of the students scored in the 95–99 class interval.

Class Limits

Suppose you were to step on a digital bathroom scale and the number 123 appears on the display. Do you weigh exactly 123? Or isn't it more realistic to say that you weigh approx-

TABLE 2.6 *Frequency Distribution of Final-Examination Grades for 71 Students*

Grade	*f*	Grade	*f*	Grade	*f*	Grade	*f*
99	0	85	2	71	4	57	0
98	1	84	1	70	9	56	1
97	0	83	0	69	3	55	0
96	1	82	3	68	5	54	1
95	1	81	1	67	1	53	0
94	0	80	2	66	3	52	1
93	0	79	8	65	0	51	1
92	1	78	1	64	1	50	1
91	1	77	0	63	2		*N* = 71
90	0	76	2	62	0		
89	1	75	1	61	0		
88	0	74	1	60	2		
87	1	73	1	59	3		
86	0	72	2	58	1		

TABLE 2.7 *Grouped Frequency Distribution of Final-Examination Grades for 71 Students*

Class Interval	f	%
95–99	3	4.23
90–94	2	2.82
85–89	4	5.63
80–84	7	9.86
75–79	12	16.90
70–74	17	23.94
65–69	12	16.90
60–64	5	7.04
55–59	5	7.04
50–54	4	5.63
Total	71	100[a]

[a]The percentages as they appear add to only 99.99%. We write the sum as 100% instead, because we know that .01% was lost in rounding.

imately 123? Specifically, your weight is more than 122.5 but less than 123.5, and the scale rounds to the nearest whole number. When we construct class intervals of the weight range 120–129 pounds, we must include a "fudge factor" surrounding the whole numbers. Thus, this class interval for weight is actually from 119.5 (the low end of 120) to 129.5 (the high end of 129). The actual limits for this interval are 119.5 to 129.5. Thus, in reality, anyone whose exact weight is between 119.5 and 129.5 will be included in this interval. In practical terms, anyone whose exact weight is between 119.5 and 129.5 will "tip the scale" in whole numbers from 120–129.

Each class interval has an *upper limit* and a *lower limit*. At first glance, the highest and lowest score values in any given category seem to be these limits. Thus, we might reasonably expect the upper and lower limits of the interval 60–64 to be 64 and 60, respectively. In this case, however, we would be wrong, because 64 and 60 are actually not the limits of interval 60–64.

Unlike the highest and lowest score values in an interval, *class limits* are located at the point halfway between adjacent class intervals and also serve to close the gap between them (see Figure 2.1). Thus, the upper limit of the interval 90–94 is 94.5, and the lower limit of the interval 95–99 is also 94.5. Likewise, 59.5 serves as the upper limit of the interval 55–59 and as the lower limit of the interval 60–64.

Finally, as you can see from the figure, the distance between the upper and lower limits of a class interval determines its size. That is,

$$i = U - L$$

where i = size of a class interval
U = upper limit of a class interval
L = lower limit of a class interval

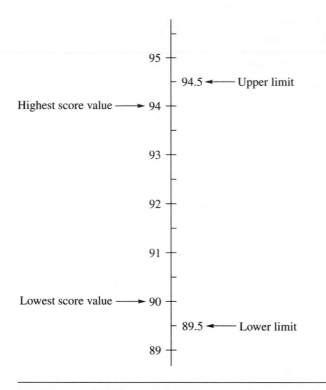

FIGURE 2.1   The highest and lowest score values versus the upper and lower limits of the class interval 90–94

For example, for the interval 90–94, the size *(i)* is 94.5 − 89.5 = 5. This corresponds to the value we obtain by simply counting the range of values within the interval (90, 91, 92, 93, and 94). To avoid any confusion, we recommend that you always calculate class interval size by subtracting the lower limit from the upper.

The Midpoint

Another characteristic of any class interval is its *midpoint (m)*, which we define as the middlemost score value in the class interval. A quick and simple method of finding a midpoint is to look for the point at which any given interval can be divided into two equal parts. Let's use some illustrations: 50 is the midpoint of the interval 48–52; 3.5 is the midpoint of the interval 2–5. The midpoint can also be computed from the lowest and highest score values in any interval. To illustrate, the midpoint of the interval 48–52 is

$$m = \frac{\text{lowest score value} + \text{highest score value}}{2} = \frac{48 + 52}{2} = 50$$

In a sense, the midpoint can be regarded as the spokesperson for all score values in a class interval. It is a single number that can be used to represent the entire class interval.

Guidelines for Constructing Class Intervals

Constructing class intervals is just a special way of categorizing data. As discussed earlier, categories, and thus class intervals, must be mutually exclusive (nonoverlapping) and exhaustive (a place for every case).

Beginning students generally find it difficult to construct class intervals on their own. Indeed, it is a skill that develops only with practice. However, there are some general guidelines that make the task easier. Note that these are only guidelines, which, under certain circumstances, can be violated.

To present interval data in a grouped frequency distribution, the social researcher must consider the number of categories he or she wishes to employ. Texts generally advise using as few as 3 or 4 intervals to as many as 20 intervals. In this regard, it would be wise to remember that grouped frequency distributions are employed to reveal or emphasize a group pattern. Either too many or too few class intervals may blur that pattern and thereby work against the researcher who seeks to add clarity to the analysis. In addition, reducing the individual score values to an unnecessarily small number of intervals may sacrifice too much precision—precision that was originally attained by knowing the identity of individual scores in the distribution. In sum, then, the researcher generally makes a decision as to the number of intervals based on the set of data and personal objectives, factors that may vary considerably from one research situation to another.

After deciding on the number of class intervals, a researcher must then begin constructing the intervals themselves. Two basic guidelines help make this task easier and should be followed whenever possible. First, it is preferable to make the size of class intervals a whole number rather than a decimal. This tends to simplify calculations in which size is involved. Second, it is preferable to make the lowest score in a class interval some multiple of its size. Customarily, for example, exam scores are categorized as 90–99, 80–89, and so on, so that the lowest scores (for example, 80 and 90) are multiples of 10.

Cumulative Distributions

It is sometimes desirable to present frequencies in a cumulative fashion, especially when locating the position of one case relative to overall group performance. *Cumulative frequencies (cf)* are defined as the total number of cases having any given score *or a score that is lower.* Thus, the cumulative frequency cf for any category (or class interval) is obtained by adding the frequency in that category to the total frequency for all categories below it. In the case of college board PSAT scores in Table 2.8, we see that the frequency f associated with the class interval 20–24 is 3. This is also the cumulative frequency for this interval, because no member of the group scored below 20. The frequency in the next class interval, 25–29, is 6, and the cumulative frequency for this interval is 9(6 + 3) Therefore, we learn that 6 students earned PSAT scores between 25 and 29, but that 9 students received scores of 29 *or lower.* We might continue this procedure, obtaining cumulative frequencies for all class intervals, until we arrive at the topmost entry, 75–79, whose cumulative frequency (336) is equal to the total number of cases, because no member of the group scored above 79.

In addition to cumulative frequency, we can also construct a distribution that indicates *cumulative percentage (c%)*, the percentage of cases having any score or a score that

TABLE 2.8 *Cumulative Frequency (cf) Distribution of PSAT Scores for 336 Students*

Class Interval	f	%	cf
75–79	4	1.19	336
70–74	24	7.14	332
65–69	28	8.33	308
60–64	30	8.93	280
55–59	35	10.42	250
50–54	55	16.37	215
45–49	61	18.15	160
40–44	48	14.29	99
35–39	30	8.93	51
30–34	12	3.57	21
25–29	6	1.79	9
20–24	3	.89	3
Total	336	100	

is lower. To calculate cumulative percentage, we modify the formula for percentage (%) introduced earlier in this chapter as follows:

$$c\% = (100)\frac{cf}{N}$$

where cf = cumulative frequency in any category
       N = total number of cases in the distribution

Applying the foregoing formula to the data in Table 2.8, we find that the percent of students who scored 24 or lower was

$$c\% = (100)\left(\frac{3}{336}\right)$$
$$= (100)(.0089)$$
$$= .89$$

The percent who scored 29 or lower was

$$c\% = (100)\left(\frac{9}{336}\right)$$
$$= (100)(.0268)$$
$$= 2.68$$

The percent who scored 34 or lower was

TABLE 2.9 *Cumulative Percentage (c%)*
Distribution of PSAT Scores for 336 Students
(Based on Table 2.8)

Class Interval	f	%	cf	c%
75–79	4	1.19	336	100.00
70–74	24	7.14	332	98.81
65–69	28	8.33	308	91.67
60–64	30	8.93	280	83.33
55–59	35	10.42	250	74.40
50–54	55	16.37	215	63.99
45–49	61	18.15	160	47.62
40–44	48	14.29	99	29.46
35–39	30	8.93	51	15.18
30–34	12	3.57	21	6.25
25–29	6	1.79	9	2.68
20–24	3	.89	3	.89
Total	336	100		

$$c\% = (100)\left(\frac{21}{336}\right)$$
$$= (100)(.0625)$$
$$= 6.25$$

A cumulative percentage distribution based on the data in Table 2.8 is shown in Table 2.9. Note that the $c\%$ distribution can also be obtained by summing the percent (%) distribution.

Percentile Ranks

Your statistics professor hands back the midterm exams. Knowing that a major part of your course grade depends on this test, you slowly turn over the booklet to reveal a red 77 with a circle around it. Should you quietly cheer and think about the celebration ahead? Or should you start contemplating extra-credit work you might propose to bring up your grade?

By the conventional standards that you learned in elementary and secondary school, you might immediately have translated the 77 into a C+, slightly above average. But in college, or at least in some classes, conventional standards can be tossed out the window. The score of 77 means *nothing,* without some sense of how the rest of the class performed. If most of the class scored in the 50s and 60s, then celebration can be scheduled for that evening. But if most of the class scored in the 80s and 90s, you may want to postpone the party.

To put it another way, the quality of the raw score 77 depends on how easy the test is. On a very tough exam, a 77 might be a commendable score, whereas on a simple test, you probably should have done better. Of course, the level of difficulty of an exam can only be gauged by how the class performed as a whole—that is, on the entire distribution of scores. Thus, the only realistic way to tell if your 77 was an excellent, good, average, or poor score is to compare it against the entire distribution of scores in the class.

"How does a 77 rank in terms of the entire class?" you ask your professor. She responds that you scored the same as or better than 60% of the class, indicating that your *percentile rank* was 60%.

The percentile rank of any given score, say, 77, is defined as the percentage of the cases in a distribution that falls at or below that score (for example, the percent of the class scoring 77 or lower). Percentile ranks are simple to compute if your professor provides the entire collection of *raw* scores. For example, in the following collection of 20 scores, your 77 would be *ranked* twelfth from the bottom. Thus, your *percentile rank* would be twelfth out of 20, or 60%:

$$\text{Twelfth out of } 20 = 60\%$$

94 92 91 88 85 84 80 79 77 76 74 74 71 69 65 62 56 53 48 40

↑ Twelfth score from the bottom

There are points in a distribution of scores whose percentile ranks are so important and commonly used that they are given specific names. *Deciles* are points that divide the distribution into 10 equally sized portions. Thus, if a score is located at the first decile (percentile rank = 10), we know that 10% of the cases fall at or below it; if a score is at the second decile (percentile rank = 20), then 20% of the cases fall at or below it, and so on. *Quartiles* are points that divide the distribution into quarters. If a score is located at the first quartile (percentile rank = 25), we know that 25% of the cases fall at or below it; if a score is at the second quartile (percentile rank = 50), 50% of the cases fall at or below it; and if a score is at the third quartile (percentile rank = 75), 75% of the cases fall at or below it (see Figure 2.2). Finally, as we will encounter again in the next chapter, the *median* is the point that divides the distribution of scores in two, half above it and half below it. Thus, the median corresponds to a percentile rank of 50, but it is also the fifth decile and second quartile.

Dealing with Decimal Data

Not all data come in the form of whole numbers. This should not disturb us in the least, however, because the procedures we have learned and will learn in later chapters apply to decimals as well as whole numbers. So that we get used to decimal data from the start, let's consider constructing a frequency distribution of the state unemployment data for June 2008, shown in Table 2.10. From the raw scores, we do not get a very clear picture of the nationwide patterns of unemployment. We are drawn to the extremes: The numbers range from a high of 8.5 (Michigan) to a low of 2.8 (South Dakota). Very little other information emerges until we construct a grouped frequency distribution.

Because there are only 50 cases, we would not want too many categories. An excessive number of class intervals would spread the cases too thinly. Determining the actual limits of the class intervals is the most difficult part of all. Satisfactory results come with a

Percentile Rank	Decile	Quartile
95		
90 =	9th	
85		
80 =	8th	
75 =		3rd
70 =	7th	
65		
60 =	6th	
55		
50 =	5th	2nd
45		
40 =	4th	
35		
30 =	3rd	
25 =		1st
20 =	2nd	
15		
10 =	1st	
5		

FIGURE 2.2　Scale of percentile ranks divided by deciles and quartiles

TABLE 2.10　*State Unemployment Rates, June 2008*

State	Unemployment Rate	State	Unemployment Rate
Alabama	4.7	Maine	5.3
Alaska	6.8	Maryland	4.0
Arizona	4.8	Massachusetts	5.2
Arkansas	5.0	Michigan	8.5
California	6.9	Minnesota	5.3
Colorado	5.1	Mississippi	6.9
Connecticut	5.4	Missouri	5.7
Delaware	4.2	Montana	4.1
Florida	5.5	Nebraska	3.3
Georgia	5.7	Nevada	6.4
Hawaii	3.8	New Hampshire	4.0
Idaho	3.8	New Jersey	5.3
Illinois	6.8	New Mexico	3.9
Indiana	5.8	New York	5.3
Iowa	4.0	North Carolina	6.0
Kansas	4.3	North Dakota	3.2
Kentucky	6.3	Ohio	6.6
Louisiana	3.8	Oklahoma	3.9

(continued)

TABLE 2.10   *Continued*

State	Unemployment Rate	State	Unemployment Rate
Oregon	5.5	Utah	3.2
Pennsylvania	5.2	Vermont	4.7
Rhode Island	7.5	Virginia	4.0
South Carolina	6.2	Washington	5.5
South Dakota	2.8	West Virginia	5.3
Tennessee	6.5	Wisconsin	4.6
Texas	4.4	Wyoming	3.2

Source: Bureau of Labor Statistics

great deal of trial and error as well as practice. There is no "right" setup of class intervals, but those in Table 2.11 might be a good place to start.

Once we have the skeleton for the frequency distribution (its class intervals and the frequencies), the rest is fairly straightforward. Percentages, cumulative frequencies, and cumulative percentages are obtained in the usual way. For other calculations such as midpoints, however, keep in mind that these data are expressed with one decimal digit. As a consequence, this digit is important in determining the interval size or the range of score values spanned by a class interval. For example, the size of the 4.0–4.4 interval is .5, because it contains the score values 4.0 through 4.4 inclusive. There are 5 score values between 4.0 and 4.4, each one-tenth apart, so the size is $(5)(1/10) = 5$.

TABLE 2.11   *Frequency Distribution of State Unemployment Rates, June 2008*

Class Interval	f
8.5–8.9	1
8.0–8.4	0
7.5–7.9	1
7.0–7.4	0
6.5–6.9	6
6.0–6.4	4
5.5–5.9	6
5.0–5.4	10
4.5–4.9	4
4.0–4.4	8
3.5–3.9	5
3.0–3.4	4
2.5–2.9	1
	$N = 50$

More on Class Limits

The class limits associated with a grouped frequency distribution serve as dividers between categories that are designed to avoid any ambiguity about where a particular score is to be placed. As discussed earlier, we can use as class limits a value halfway between the highest possible value in an interval and the lowest possible value in the next interval, as in the PSAT illustration presented earlier.

Suppose we are constructing a grouped frequency distribution of body weights of New York City firefighters in categories having 20-pound widths. The class limit separating the intervals 180–199 and 200–219 would be 199.5, assuming that weights are measured in whole pounds rather than fractional weights. If, however, a digital scale is used that provides weights in half-pound measurements (for example, 184.5, 203.0, and 218.5), the intervals would be 180.0–199.5 and 200.0–219.5 to accommodate half pounds, with a limit of 199.75 dividing the two adjacent weight class groups. Finally, if a more precise scale measuring weights in tenths of pounds is used, the groups would be 180.0–199.9 and 200.0–219.9, with the limit 199.95 separating the two groups.

There is more than one acceptable approach for establishing these limits, yet the best approach always comes down to matters of accuracy and feasibility. The halfway point dividing categories in the approach described above works well for discrete measurements, such as whole numbers, halves, or tenths, but characteristics having natural or meaningful steps from one group to the next can use an alternative strategy based on substantively meaningful class limits.

Suppose we have a grouped distribution of age of onset of breast cancer (see Table 2.12), which included a category ranging from 50 to 59 years old. We might use as limits 49.5 and 59.5 by halving the distance between this group and the ones below and above. This would create several problems, however. First, 59-year-olds (just like people of other ages) will report their age as 59 all the way up until their sixtieth birthday. Thus, a 59-year-old patient who is just short of her next birthday would technically not fall within the 49.5–59.5 range even though her age is 59. Second, in common usage, we refer to the 50–59 age values as the fifties, so it might seem odd to have the category extend down to 49.5. Finally, the midpoint for this interval would be 54.5, even though in a practical sense 55 should be at the middle.

TABLE 2.12 *Age of Onset of Breast Cancer in a Study of 150 Patients*

Age of Onset	f	%
70–79	9	6
60–69	18	12
50–59	42	28
40–49	51	34
30–39	30	10
Total	150	100

TABLE 2.13   *Two Approaches for Establishing Class Limits*

	Acceptable Method		Better Method	
Score Values	*Lower Limit*	*m*	*Lower Limit*	*m*
90 up to 100	89.5	94.5	90	95
80 up to 90	79.5	84.5	80	85
70 up to 80	69.5	74.5	70	75
60 up to 70	59.5	64.5	60	65
50 up to 60	49.5	54.5	50	55

A solution to this awkward approach is to use as an alternative, the ages 40, 50, 60, and 70 as the class limits, and treat them as inclusive lower limits. That is, the 50–59 category would be from 50 up to (but not including) 60. The 50 up to 60 interval would have a lower limit of 50, an upper limit 60 (or rather just under 60), a width of 10 years, and a midpoint of 55.

Let's consider another example, returning to the exam score situation encountered earlier. As shown in Table 2.13, there are two approaches for setting class limits for categorizing exam scores, but one is clearly better than the other. The better approach treats the lowest score (80) value as the lower limit and the lowest score value of the next higher category (90) as a noninclusive upper boundary. That is, the 80's range extends from 80 up to but not including 90. Of course, if exam scores are given as whole numbers, this will make little difference in the end. But if we were measuring a composite average for the entire course, a score of 89.63 is possible and technically should fall in the B range, not the A range (even though a kindly professor might bump the score up a bit).

The choice between using the first approach, distinguishing class intervals at the halfway point in setting class limits, or the second approach, using the lowest value as the lower limit and just below the next category as the upper limit, often comes down to personal preference, feasibility, and logical sense, not what is strictly right or wrong. With data that are always whole numbers, such as the count of people in a household or number of dates per month, the halfway point method will generally work best. But when dealing with data that are continuous and have natural breakpoints along the continuum, such as hours since last meal, one may be better off using the following breakpoints as class limits (just as a public garage does in charging for parking):

Hours since Last Meal	Class Limits	*m*
0–1	0 up to 2	1
2–3	2 up to 4	3
4–5	4 up to 6	5
6–7	6 up to 8	7
8–9	8 up to 10	9

TABLE 2.14   *Frequency Distribution of Family Income Data*

Income Category	f (families in 1,000s)	%
$100,000 and over	8,391	11.8
$75,000–$99,999	7,826	11.0
$50,000 $74,999	15,112	21.3
$35,000–$49,999	12,357	17.4
$25,000–$34,999	9,079	12.8
$15,000–$24,999	9,250	13.0
$10,000–$14,999	4,054	5.7
$5,000–$9,999	2,887	4.1
Less than $5,000	1,929	2.7
	$N = 70,885$	100.0

Flexible Class Intervals

Although we did not make a point of it earlier, you may have noticed that all the frequency distributions used so far have had class intervals all of equal size. There are occasions, however, in which this practice is not at all desirable. For example, if a student gets a perfect score of 100 on an exam, the topmost interval can be expanded to include it—that is, 90–100 instead of 90–99.

Table 2.14 presents a distribution of census data on family income, which is typical of distributions constructed with income data. As shown, grouped frequency distributions can have open-ended top or bottom class intervals, such as $100,000 and over. The other major departure from the simple distributions provided earlier is the use of class intervals of varying size. Note that, whereas the class intervals containing the lower incomes have a size of $5,000, the size of the class intervals is stretched for higher-income levels. What would have been the result had a fixed class interval size of $5,000 been maintained throughout the distribution? The $25,000–$34,999 class interval would have two categories, the $35,000–$49,999 class interval would have turned into three categories, and both the $50,000–$74,999 and $75,000–$99,999 class intervals would have become five categories each. The effect would be to make unnecessarily fine distinctions among the persons of higher income and to produce a needlessly lengthy frequency distribution. That is, in terms of standard of living, there is a big difference between the $5,000–$9,999 class interval and the $10,000–$14,999 class interval. In contrast, the difference between a $60,000–$64,999 category and a $65,000–$69,999 category would be relatively unimportant.

These new twists in frequency distributions should not cause you much difficulty in adapting what you have already learned in this chapter. Fortunately, the computations of cumulative distributions, percentile ranks, and the like do not change for frequency distributions with class intervals of unequal size or with unbounded top or bottom class intervals. The only modification involves calculating midpoints with unbounded top or bottom class intervals. Let's consider an example.

TABLE 2.15 *Frequency Distribution of Family Income Data (with Midpoints)*

Income Category	m	f	%
$100,000 and over	$125,000	8,391	11.8
$75,000–$99,999	$87,500	7,826	11.0
$50,000–$74,999	$62,500	15,112	21.3
$35,000–$49,999	$42,500	12,357	17.4
$25,000–$34,999	$30,000	9,079	12.8
$15,000–$24,999	$20,000	9,250	13.0
$10,000–$14,999	$12,500	4,054	5.7
$5,000–$9,999	$7,500	2,887	4.1
Less than $5,000	$2,500	1,929	2.7
		$N = 70,885$	100.0

Table 2.15 displays midpoints for a family income distribution using the lowest score values in each category as the class limit. Thus, for example, the $25,000–$34,999 income category, using $25,000 and $35,000 as class limits, yields a midpoint of $30,000.

But what do we do about the highest class interval ($100,000 and over), which has no upper limit? What should we plug into the formula? There is no hard and fast rule to apply, just common sense. The class intervals have become progressively wider with increasing income. Continuing with the same progression, we might pretend the highest interval, for most of the remaining families, to be $100,000–$149,999, which yields a midpoint of $125,000.

Cross-Tabulations

Frequency distributions like those discussed so far are seen everywhere. Publications from the Bureau of the Census consistently employ frequency distributions to describe characteristics of the U.S. population; presentation of the raw data—all the millions of observations—would, of course, be impossible.

We even see frequency distributions in daily newspapers; journalists, like social researchers, find tables a very convenient form of presentation. Most newspaper readers are capable of understanding basic percentages (even though they may forget how to compute them). A basic table of frequencies and percentages for some variable is usually sufficient for the level of depth and detail typically found in a newspaper. Social researchers, however, want to do more than just describe the distribution of some variable; they seek to explain why some individuals fall at one end of the distribution while others are at the opposite extreme.

To accomplish this objective, we need to explore tables more deeply by expanding them into two and even more dimensions. In particular, a *cross-tabulation* (or *cross-tab* for short) is a table that presents the distribution—frequencies and percents—of one variable (usually the dependent variable) across the categories of one or more additional variables (usually the independent variable or variables).

When the state of Massachusetts instituted a mandatory seat belt law, it called for a $15 fine for failure to comply. To measure compliance with the law, Fox and Tracy

TABLE 2.16 *Frequency Distribution of Seat Belt Use*

Use of Seat Belts	f	%
All the time	499	50.1
Most of the time	176	17.7
Some of the time	124	12.4
Seldom	83	8.3
Never	115	11.5
Total	997	100

completed a telephone survey of 997 residents of the Boston area concerning their use of seat belts and their opinions regarding the controversial law. For the primary question—the extent to which the respondent used his or her seat belt—the simple frequency distribution shown in Table 2.16 was obtained.

About half of the respondents in the survey (50.1%) stated that they wore their seat belts all the time. Two-thirds of the respondents (50.1% + 17.7% = 67.8%) stated that they wore their seat belts at least most of the time.

We are not content with just knowing the extent of seat belt compliance, however. To analyze the survey data more fully, we start by examining what types of people wear their seat belts—that is, what respondent characteristics are related to seat belt usage.

One of the more dramatic differences is between the males and females in the survey. A cross-tabulation can be employed to look at the differences between the sexes in terms of seat belt use. A cross-tabulation is essentially a frequency distribution of two or more variables taken simultaneously. The cross-tabulation given in Table 2.17 shows, for example, that there were 144 males who said they wore their seat belts all the time and 110 females who reported that they wore their seat belts most of the time.

The foundation for cross-tabulations was presented earlier in the chapter when the gender distributions of engineering majors across two colleges were compared. Cross-

TABLE 2.17 *Cross-Tabulation of Seat Belt Use by Gender*

Use of Seat Belts	Gender of Respondent		Total
	Male	*Female*	
All the time	144	355	499
Most of the time	66	110	176
Some of the time	58	66	124
Seldom	39	44	83
Never	60	55	115
Total	367	630	997

TABLE 2.18 *Cross-Tabulation of Seat Belt Use by Gender with Total Percents*

| Use of Seat Belts | Gender of Respondent | | Total |
	Male	*Female*	
All the time	144	355	499
	14.4%	35.6%	50.1%
Most of the time	66	110	176
	6.6%	11.0%	17.7%
Some of the time	58	66	124
	5.8%	6.6%	12.4%
Seldom	39	44	83
	3.9%	4.4%	8.3%
Never	60	55	115
	6.0%	5.5%	11.5%
Total	367	630	997
	36.8%	63.2%	100.0%

Row marginal (row totals)

Total sample size

Column marginal (column totals)

tabulations can be thought of as a series of frequency distributions (in this case, two of them) attached together to make one table. In this example, we have essentially a frequency distribution of seat belt use among males juxtaposed with a comparable frequency distribution of seat belt use for females.

As with one-variable frequency distributions, percentages give the results fuller meaning than frequencies alone. If we retain the same procedure as before, that is, dividing each frequency *(f)* by the sample size *(N)*,

$$\% = (100)\frac{f}{N}$$

we obtain the percentage results for the two variables jointly, as shown in Table 2.18. For example, the percentage of the sample that is female and wears seat belts all the time is obtained from dividing the number of female "all-the-time" wearers by the number of respondents in the sample overall:

$$(100)\left(\frac{355}{997}\right) = (100)(.356) = 35.6\%$$

Thus, 35.6% of the sample consists of females who wear seat belts all the time (see Table 2.18).

Frequency distributions of each variable separately can be found along the margins of a two-way cross-tabulation. These are called *marginal distributions*. That is, the right margin provides a frequency and percent distribution for seat belt use identical to what we had in Table 2.16. Because the seat belt variable is placed along the rows of the cross-tabulation, the frequencies and percents for seat belt use form the row totals. Likewise, the marginal distribution of gender is found in the bottom margin of the cross-tabulation. These frequencies and percents for males and females are the column totals, because gender is the variable heading the columns.

The percentages in Table 2.18 are called *total percents* (total%) because they are obtained by dividing each frequency by the total sample size:

$$total\% = (100)\frac{f}{N_{total}}$$

For instance, 14.4% of the sample consists of males who wear their seat belt all the time. Similarly, 11.0% of the sample consists of females who wear their seat belt most of the time.

There is, however, something very unsettling about these percentages. For example, the small value of the percentage of "never-wearer" males (6.0%) is ambiguous. It could reflect a low prevalence or representation of males, a low prevalence of seat belt usage in the sample overall, a low rate of seat belt use among males specifically, or even a low prevalence of males among never-wearers.

There are other approaches to calculating percentages that might resolve this ambiguity. One alternative would divide the number of male never-wearers by the number of never-wearers in the sample, the number of male seldom-wearers by the number of seldom-wearers, and so on, and do the comparable calculations for the females. In other words, we divide the frequencies in each row by the number of cases in that row (see Table 2.19). These percents are called *row percents:*

$$row\% = (100)\frac{f}{N_{row}}$$

TABLE 2.19   *Cross-Tabulation of Seat Belt Use by Gender with Row Percents*

| Use of Seat Belts | Gender of Respondent | | |
	Male	Female	Total
All the time	144	355	499
	28.9%	71.1%	100.0%
Most of the time	66	110	176
	37.5%	62.5%	100.0%
Some of the time	58	66	124
	46.8%	53.2%	100.0%
Seldom	39	44	83
	47.0%	53.0%	100.0%
Never	60	55	115
	52.2%	47.8%	100.0%
Total	367	630	997
	36.8%	63.2%	100.0%

For example, the percentage of always-users who are female is derived from dividing the number of female all-the-time-wearers by the number of all-the-time-wearers overall:

$$(100)\left(\frac{355}{499}\right) = (100)(.711) = 71.1\%$$

Thus, we find that 71.1% of the all-the-time users in the sample are females.

Row percentages give the distribution of the column variable for each value of the row variable. Thus, these percentages represent the gender distribution within each level of seat belt use. Also, row percentages sum to 100% across each row, including the column marginal in the bottom of the cross-tabulation.

Conversely, one could calculate percentages in the other direction. *Column percents* (column%) are obtained by dividing each frequency by the number of cases in that column:

$$col\% = (100)\frac{f}{N_{column}}$$

The percentage of females who always wear a seat belt is obtained, for example, by dividing the number of female all-the-time wearers by the number of females overall:

$$(100)\left(\frac{355}{630}\right) = (100)(.563) = 56.3\%$$

Thus, 56.3% of the females in the study said that they always wear a seat belt.

Column percents for our cross-tabulation are presented in Table 2.20. Note that the percentages here sum to 100% along each column. Thus, the percentages reflect the seat belt use distribution for each gender separately as well as overall.

TABLE 2.20   *Cross-Tabulation of Seat Belt Use by Gender with Column Percents*

| Use of Seat Belts | Gender of Respondent | | Total |
	Male	Female	
All the time	144	355	499
	39.2%	56.3%	50.1%
Most of the time	66	110	176
	18.0%	17.5%	17.7%
Some of the time	58	66	124
	15.8%	10.5%	12.4%
Seldom	39	44	83
	10.6%	7.0%	8.3%
Never	60	55	115
	16.3%	8.7%	11.5%
Total	367	630	997
	100.0%	100.0%	100.0%

Choosing among Total, Row, and Column Percents

We now have three sets of percentages—total, row, and column percents. You might wonder which are correct? In a mathematical sense, all are correct; that is, these were calculated in the correct way. But in terms of substantive meaning, certain percentages may be misleading or even useless.

First, as we noted previously, the total percents are sometimes ambiguous in their meaning, as in our cross-tabulation of seat belt use by gender. Next, according to the row percents, females predominate in every row, except for the "Never" subgroup in which the genders are nearly equal. What does this imply? Can we draw any conclusions, such as suggesting that males don't drive as much and consequently don't show up in any great proportions in any seat belt usage level? Obviously, this inference would be farfetched. The low representation of males in almost every category of use is simply a consequence of the low percentage of males in the sample in general (36.8%). Thus, that 71.1% of the all-the-time seat belt group are female seems a lot less overwhelming when we take into account that 63.2% of the total sample are female.

For our purposes, the most informative percentages are the column percents. We are interested in comparing males and females in terms of seat belt usage. That is, we want to know what percentage of the females wear their seat belts frequently compared to males. For example, 39.2% of the males say they always wear their belts compared to 56.3% of the females. Conversely, 16.3% of the males reported that they never wear a seat belt compared to only 8.7% of the females.

Fortunately, there is a rule of thumb to guide our choice between row and column percentages: *If the independent variable is on the rows, use row percents; if the independent variable is on the columns, use column percents.* In our example, we are concerned with the influence a respondent's gender has on seat belt behavior; gender is the independent variable. Because it is given on the columns, we should use column percents.

Another way of stating this rule may be more meaningful: If we wish to compare rows in a cross-tabulation, we need to use row percents; column percents are required for comparing columns. Again, in our example, we want to compare the males to the females in terms of seat belt use. Gender is the column variable, and the column percents provide the seat belt distributions for the males and the females separately. Thus, these column percents should be used for making the gender comparison.

In certain cases, it may not be easy to tell which is the independent variable. For example, in the cross-tabulation of husband's political party affiliation by wife's political party affiliation in Table 2.21, neither variable can clearly be said to be a result of the other. (*Note:* The figures in each cell of the table represent frequency, row percent, column percent, and total percent, respectively.) To some extent, the political affiliations of husband and wife may affect each other reciprocally, and in many cases party affiliation may have been set long before the couple even met. Similarity (or even dissimilarity) in political outlook may have been part of the attraction.

In terms of the data in Table 2.21, we could compute the percentage of democratic husbands who have democratic wives (70 out of 100, row percent 70.0%), or we could compute the percentage of democratic wives who are married to democratic husbands (70 out of 110, column percent 63.6%). Both would be meaningful depending on the researcher's particular interest. However, for cases like this in which there is no one variable that can be singled out as the cause of the other, total percents (which implicate neither one

TABLE 2.21   *Cross-Tabulation of Husband's Political Party Affiliation by Wife's Political Party Affiliation: Frequency and Row, Column, and Total Percents*

Frequency Row % Column % Total %	Wife's Political Party		Total
	Democrat	*Republican*	
Husband's political party			
Democrat	70	30	100
	70.0%	30.0%	52.6%
	63.6%	37.5%	
	36.8%	15.8%	
Republican	40	50	90
	44.4%	55.6%	47.4%
	36.4%	62.5%	
	21.1%	26.3%	
Total	110	80	190
	57.9%	42.1%	100.0%

as being the independent variable) are frequently used. For Table 2.21, in 36.8% of the marriages, both partners are Democrats (70 out of 190), and in 26.3%, both partners are Republicans (50 out of 190). Overall, in 63.1% of the marriages the husband and wife have the same political party affiliation.

The choices among total, row, and column percentages are as follows:

1. If the independent variable is on the rows, use row percents.
2. If the independent variable is on the columns, use column percents.
3. If there is no clear-cut independent variable, use total, row, or column percents, whichever is most meaningful for the particular research focus.

Graphic Presentations

Columns of numbers have been known to evoke fear, anxiety, boredom, apathy, and misunderstanding. Whereas some people seem to tune out statistical information presented in tabular form, they may pay close attention to the same data presented in graphic or picture form. As a result, many commercial researchers and popular authors prefer to use graphs as opposed to tables. For similar reasons, social researchers often use visual aids—such as pie charts, bar graphs, frequency polygons, line charts, and maps—in an effort to increase the readability of their findings.

Pie Charts

The *pie chart,* a circular graph whose slices add up to 100%, is one of the simplest methods of graphical presentation. Pie charts are particularly useful for showing the differences

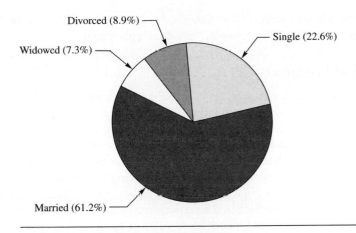

FIGURE 2.3 Pie chart of marital status

Source: Bureau of the Census

in frequencies or percentages among categories of a nominal-level variable. To illustrate, Figures 2.3 and 2.4 present the distribution of marital status for adults ages 18 and over. Notice that 22.6% of adults are single (never married), 61.2% are married, 7.3% are widowed, and 8.9% are divorced.

In many instances, a researcher may want to direct attention to one particular category in a pie chart. In this case, a researcher may wish to highlight the single adult group. To highlight this aspect of the pie chart, we can "explode" (move slightly outward) the section of the pie that is most noteworthy, as in Figure 2.4.

It is generally not advisable to use a pie chart for data that are classified in ordered categories, such as level of education ranging from "less than high school diploma" to "graduate degree." Even though educational levels increase as you progress around the

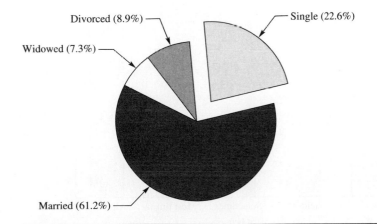

FIGURE 2.4 Pie chart of marital status (with exploded piece)

Source: Bureau of the Census

pie, eventually the highest category (for example, "graduate degree") would be followed, illogically, by the lowest level (for example, "less than high school").

Bar Graphs and Histograms

The pie chart provides a quick and easy illustration of data that can be divided into a few categories only. (In fact, some computer graphics software packages limit the number of possible pie sections.) By comparison, the *bar graph* can accommodate any number of categories at any level of measurement and, therefore, is far more widely used in social research.

Figure 2.5 illustrates a bar graph of the frequency distribution of seat belt use presented in Table 2.16. The bar graph is constructed following the standard arrangement: a horizontal base line (or *x* axis) along which the score values or categories (in this case, the levels of seat belt use) are marked off; and a vertical line (*y* axis) along the left side of the figure that displays the frequencies for each score value or category. (For grouped data, either the midpoints of the class intervals or the interval ranges themselves are arranged along the base line.) As we see in Figure 2.5, the taller the bar, the greater the frequency of the category.

Although some researchers prefer bar graphs of frequencies, bar graphs of percentages and horizontal bar graphs are often used. Figure 2.6, for example, shows a bar graph of the percentage distribution of seat belt use. Note that the graph is identical to the frequency bar graph except for the scale used along the *y* axis (percentages rather than frequencies). Also, bar graphs can be constructed either vertically or horizontally; the choice often comes down to a practical decision about which will fit better on the page. Generally, bar graphs with numerous categories are displayed better horizontally, with the categories identified along the left-hand axis and the bars extending across to the right.

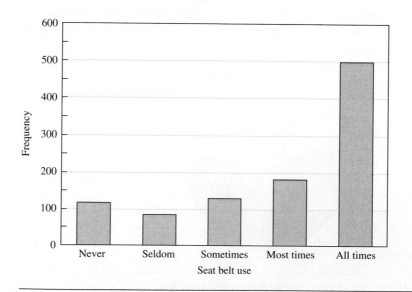

FIGURE 2.5   Bar graph of seat belt use (with frequencies)

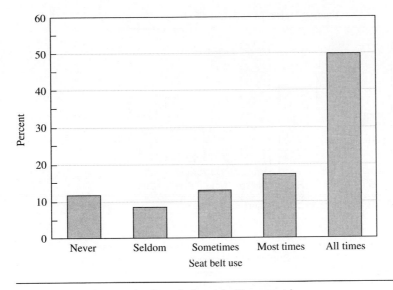

FIGURE 2.6   Bar graph of seat belt use (with percents)

The terms bar graph and *histogram* are often used interchangeably, although there is a small but important difference between the two graphical techniques. Bar graphs are typically used to display the frequency or percentage distribution of a discrete variable, especially at the nominal level. Because of the lack of continuity from category to category, a bar graph includes space between the bars to emphasize differentness, rather than continuity along a scale. Histograms, by contrast, are used to display continuous measures, especially at the interval level; the bars of the histogram are joined to emphasize continuity of the points along a scale. Ordinal-level data may be displayed in either fashion, depending on whether the researcher wishes to emphasize continuity (histogram) or discontinuity (bar graph).

The bar graph of distribution (a) in Figure 2.7 of student majors in a large sociology course includes separation between the bars because the appearance of continuity would be completely misleading. By contrast, distribution (b) in Figure 2.7 of student grade point averages represents a real continuum, and therefore a histogram with no separation between categories is appropriate.

Bar graphs and histograms can display the effect of one variable on another. For example, Figure 2.8 shows the seat belt use distribution by gender from the data in Table 2.17. It now makes a big difference whether we graph frequencies or percentages. The graph in Figure 2.8 is distorted because there are more females than males in the sample. As a result, most of the female bars are taller than the comparable male bars, blurring the effect of gender on seat belt use. We get a better depiction instead by graphing the column percents from Table 2.20. Thus, the bar graph in Figure 2.9 allows us to see not only the distribution of seat belt use but also how it is influenced by gender.

Bar graphs and histograms are also used to graph volumes and rates across population subgroups or across time, rather than just frequency and percent distributions. For example,

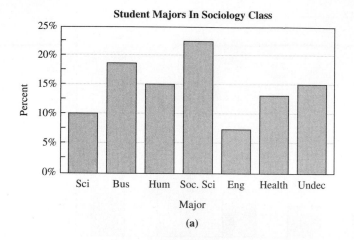

(a)

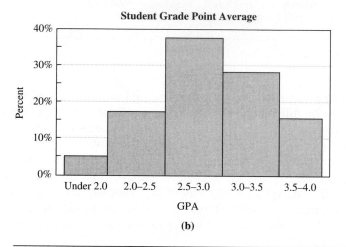

(b)

FIGURE 2.7   Comparing a bar graph and a histogram

birth rates (number of births per 1,000 women, see Table 2.22) are shown by age of mother in Figure 2.10. These rates are obtained by dividing the number of births to women of a particular age group by the number of women in that age group and then multiplying by 1,000. Because the two extreme categories have such small rates, the bars are barely visible. Therefore, to enhance the overall readability of the graph, we have labeled each bar with its value (which is always a good idea anyway).

Frequency Polygons

Another commonly employed graphic method is the *frequency polygon.* Although the frequency polygon can accommodate a wide variety of categories, it, like the histogram, tends to stress *continuity* along a scale rather than *differentness;* therefore, it is particularly

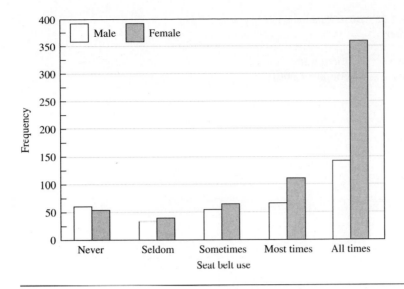

FIGURE 2.8   Bar graph of scat belt use by gender (with frequencies)

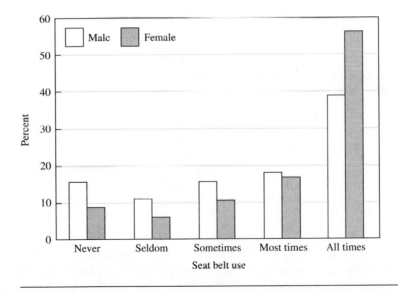

FIGURE 2.9   Bar graph of seat belt use by gender (with percents)

useful for depicting ordinal and interval data. This is because frequencies are indicated by a series of points placed over the score values or midpoints of each class interval. Adjacent points are connected with a straight line, which is dropped to the base line at either end. The height of each point or dot indicates frequency of occurrence.

TABLE 2.22 *Birth Rate by Age of Mother*

Age of Mother	Birth Rate (Births per 1,000)
10–14	1.2
15–19	54.4
20–24	110.4
25–29	113.1
30–34	83.9
35–39	35.3
40–44	6.8
45–49	.3

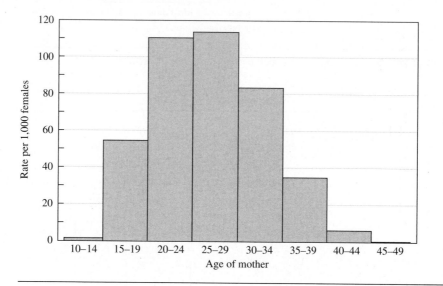

FIGURE 2.10 Histogram of birth rate per 1,000 females by age of mother

Table 2.23 shows a frequency distribution of examination scores for a class of 71 students. A frequency polygon for this distribution is then presented in Figure 2.11. Note that the frequencies of the class intervals are plotted above their midpoints; the points are connected by straight lines, which are dropped to the horizontal base line at both ends, forming a polygon.

To graph cumulative frequencies (or cumulative percentages), it is possible to construct a *cumulative frequency polygon.* As shown in Figure 2.12, cumulative frequencies are arranged along the vertical line of the graph and are indicated by the height of points above the horizontal base line. Unlike a regular frequency polygon, however, the straight line connecting all points in the cumulative frequency polygon cannot be dropped back to the base line, because the cumulative frequencies being represented are a product of successive additions. Any given cumulative frequency is never less (and is usually more) than

TABLE 2.23   *Grouped Frequency Distribution of Examination Grades*

Class Interval	f	cf
95–99	3	71
90–94	2	68
85 89	4	66
80–84	7	62
75–79	12	55
70–74	17	43
65–69	12	26
60–64	5	14
55–59	5	9
50–54	4	4
	N = 71	

the preceding cumulative frequency. Also unlike a regular frequency polygon, the points in a cumulative graph are plotted above the upper limits of class intervals rather than at their midpoints. This is because cumulative frequency represents the total number of cases *both within and below* a particular class interval.

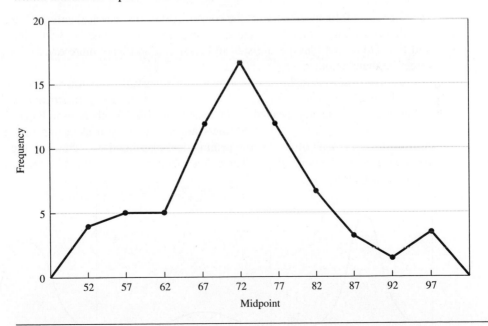

FIGURE 2.11   Frequency polygon for distribution of student examination grades

The Shape of a Frequency Distribution.   Frequency polygons can help us to visualize the variety of shapes and forms taken by frequency distributions. Some distributions are

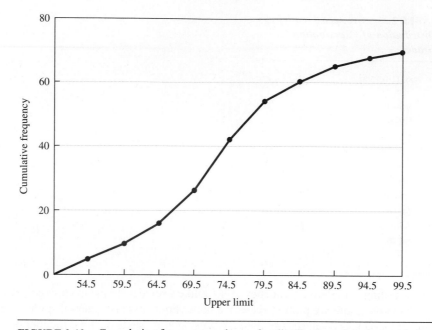

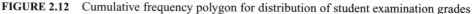

FIGURE 2.12   Cumulative frequency polygon for distribution of student examination grades

symmetrical—folding the curve at the center creates two identical halves. Therefore, such distributions contain the same number of extreme score values in both directions, high and low. Other distributions are said to be *skewed* and have more extreme cases in one direction than the other.

There is considerable variation among symmetrical distributions. For instance, they can differ markedly in terms of *peakedness* (or *kurtosis*). Some symmetrical distributions, as in Figure 2.13(a), are quite peaked or tall (called *leptokurtic*); others, as in Figure 2.13(b), are rather flat (called *platykurtic*); still others are neither very peaked nor very flat (called *mesokurtic*). One kind of mesokurtic symmetrical distribution, as shown in Figure 2.13(c), the *normal curve*, has special significance for social research and will be discussed in some detail in Chapter 5.

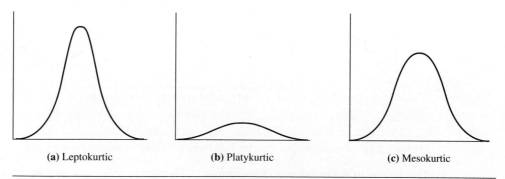

FIGURE 2.13   Some variation in kurtosis among symmetrical distributions

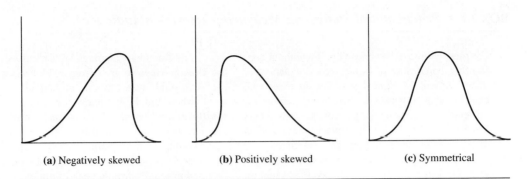

(a) Negatively skewed   **(b)** Positively skewed   **(c)** Symmetrical

FIGURE 2.14 Three distributions representing direction of skewness

There is a variety of skewed or asymmetrical distributions. When skewness exists so that scores pile up in one direction, the distribution will have a pronounced "tail." The position of this tail indicates where the relatively few extreme scores are located and determines the *direction* of skewness.

Distribution (a) in Figure 2.14 is *negatively skewed* (skewed to the left), because it has a much longer tail on the left than the right. This distribution shows that most respondents received high scores, but only a few obtained low scores. If this were the distribution of grades on a final examination, we could say that most students did quite well and a few did poorly.

Next, look at distribution (b), whose tail is situated to the right. Because skewness is indicated by the direction of the elongated tail, we can say that the distribution is *positively skewed* (skewed to the right). The final examination grades for the students in this hypothetical classroom will be quite low, except for a few who did well.

Finally, let us examine distribution (c), which contains two identical tails. In such a case, there is the same number of extreme scores in both directions. The distribution is not at all skewed, but is perfectly symmetrical. If this were the distribution of grades on the final examination, we would have a large number of more or less average students and few students receiving very high and low grades.

Line Charts

We saw previously that bar graphs and histograms could be used to display frequencies and percents from a distribution of scores as well as volumes and rates across groups, areas, or time. Frequency polygons can similarly be modified to display volumes and rates between groups or across time, although this method uses a *line chart*. In other words, frequency polygons show the *frequency* distribution of a set of scores on a single variable, whereas line charts display changes in a variable or variables between groups or over time.

In a line chart, the amount or rate of some variable is plotted and then these points are connected by line segments. Figure 2.15, for example, shows in line chart form the birth rates by age of mother that were previously displayed in a bar graph. As you can see by comparing Figures 2.15 and 2.10, it makes little difference which method is employed.

Whereas subgroup comparisons (such as the age groups subdividing the childbearing years) are plotted with bars or lines, time trend data are far more customarily depicted with

BOX 2.1 • *Practical and Statistical: Measuring Television Audience*

For most people, a day doesn't go by without confronting some sort of percentage calculation or interpretation. It is critical, therefore, to understand exactly what a particular percentage represents. Generally, the basis for the percentage—percentage of what?—holds the key. Percentages calculated on different bases can mean very different things.

Speaking of bases, different percentages were used to measure the audience tuned into an extra-inning playoff game between long-standing rivals, the Boston Red Sox and the New York Yankees. The rating for a broadcast is the percentage of all households watching the program, so the base for the calculation is the total number of households with televisions, whether or not they are turned on. Alternatively, the share for a broadcast is the percentage of households with televisions turned on that are tuned into the program, so only households watching television are included in the calculation.

The line chart displays both measures for the Boston viewing area from 8:15 P.M., when the first pitch was thrown, to just after 1:00 A.M., when the home team prevailed with a twelfth-inning home run. As the game progressed well past the bedtimes of most young fans and many working adults, the percentage of all households watching the game—the rating—dropped gradually from a peak of almost 40% at 10:00 P.M., when the Yankees took the lead, down to about 20%, when the Red Sox won the game. In contrast to the ratings nose dive, the share increased steadily from about 50% of all households watching any station at 10:00 P.M. up to nearly 70% by game's end. Thus, while many viewers of the game appeared to have turned off the TV and turned into bed, the game had more staying power than whatever else was competing for late-night viewers.

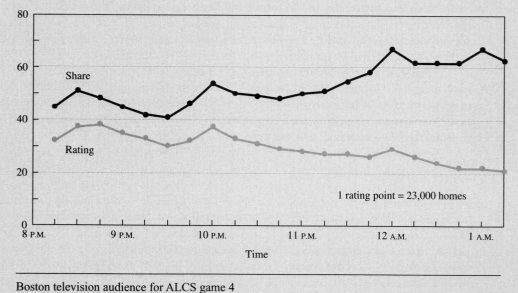

Boston television audience for ALCS game 4
Source: Nielsen Media Research

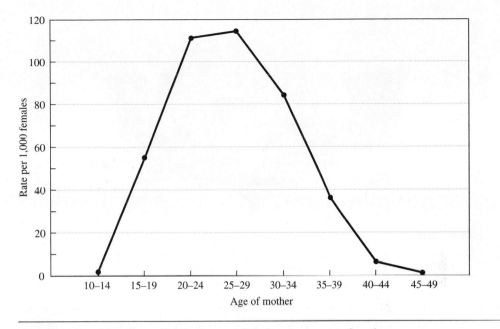

FIGURE 2.15   Line chart of birth rate per 1,000 females by age of mother

line charts. Figure 2.16, for example, shows the U.S. homicide rate (the number of homicides reported to the police per 100,000 residents) from 1950 to 2005. In the graph, we can clearly see a sharp and sudden upturn in the rate of homicides in the mid-1960s, an upward trend that continued until 1980, a downturn until the mid-1980s, a resurgence in the late 1980s, and another downturn in the 1990s before a leveling off since 2000. It is incumbent on the social researcher, of course, to attempt an explanation for these trends. Among those

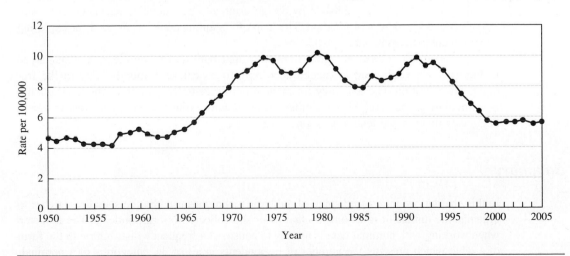

FIGURE 2.16   U.S. homicide rate, 1950–2005

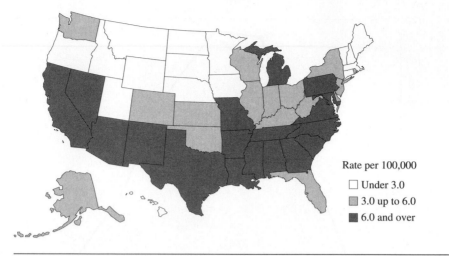

FIGURE 2.17   Map of state homicide rates, 2005

reasons advanced in the literature were an increase in racial violence, a rise in drug use and the emergence of crack cocaine, changes in sentencing, changes in police practices, changes in the size of the adolescent population, and increased access to firearms.

Maps

At one time, social researchers relied almost exclusively on pie charts, bar graphs, frequency polygons, and line charts. In recent years, however, as graphics software for the computer has matured, researchers have begun to employ other forms of graphic presentation. One type in particular—the map—has become quite popular in conjunction with the greater use of data collected and published by the government (for example, census data) as well as data coded by address or location. The map offers an unparalleled method for exploring geographical patterns in data.

For example, a three-category frequency distribution of homicide rates is displayed in Figure 2.17. Each state is shaded according to its category membership in the frequency distribution. The tendency for rates of homicide to be greater as one moves south is immediately apparent. This becomes the launching point for the social researcher to begin establishing and testing why this is so.

Summary

In this chapter, we introduced some of the basic techniques used by the social researcher to organize the jumble of raw numbers that he or she collects from respondents. The first step when working with nominal data is usually to construct a frequency distribution in the form of a table which presents the number of respondents in all of the categories of a nominal-level variable or compares different groups on the categories of the same variable. Comparisons between groups or time periods can also be made by means of proportions, percentages, and rates. For the purpose of presenting ordinal or interval data, there are simple, grouped, and cumulative frequency (and percentage) distributions. Frequency and percentage distributions can be extended to include two and even more dimensions. In a cross-tabulation, the table presents the distribution of frequencies or percentages of one variable (usually, the dependent variable) over the categories of one or more additional variables (usually, the independent variable). There are three possible ways to determine percentages for cross-tabulations: row percents, column percents, and total percents. The choice between row and column percents depends on the placement of the independent variable within the cross-tabulation. Total percents are occasionally used instead, but only when neither the row nor the column variable can be identified as independent. Graphic presentations are often employed to enhance the readability and appeal of research findings. Pie charts have limited utility, being most appropriate for providing a simple illustration of nominal-level data that can be divided into only a few categories. Bar graphs and histograms are more widely used, because they can accommodate any number of categories. Stressing continuity along a scale, frequency polygons are especially useful for depicting ordinal and interval data. Among their many applications, line charts are particularly useful for tracing trends over time. Finally, maps provide a method for displaying the geographical patterns in a set of data.

Terms to Remember

Frequency distribution	Median
Percentage distribution	Cross-tabulation
Proportion	Total percent
Percentage	Row percent
Ratio	Column percent
Rate	Pie chart
Grouped frequency distribution	Bar graph
Class interval	Histogram
Class limit	Frequency polygon
Midpoint	Cumulative frequency polygon
Cumulative frequency	Kurtosis
Cumulative percentage	Skewness
Percentile rank	Negatively skewed distribution
Deciles	Positively skewed distribution
Quartiles	Line chart

Questions and Problems

1. A cross-tabulation of serious illnesses is a table in which the distribution of illnesses is
   a. presented separately for the categories of a second variable, such as gender, age, or race.
   b. presented in a table.
   c. presented in a graph.
   d. presented in a pie chart.

2. Frequency distributions can be used to
   a. compare gender differences in violent criminal behavior.
   b. display the grades on a midterm examination for all students in a sociology course.
   c. compare attitudes of college students and their parents regarding war.

      d. show attitudes of all students on a campus regarding war.

      e. All of the above

3. Which of the following is *not* used to make comparisons between distributions when their total frequencies differ?

      a. Proportions

      b. Rates

      c. Ratios

      d. Class limits

4. By definition, class intervals contain more than one

      a. score value.

      b. score.

      c. respondent.

      d. category.

5. Which of the following is employed when comparing a score on a final examination against the entire distribution of grades in a class?

      a. Midpoint

      b. Class interval

      c. Class limits

      d. Percentile rank

6. A cross-tabulation would be useful for comparing

      a. the amount of attention—a little, a moderate amount, or a great deal—a particular teacher gives to two different students.

      b. the amount of attention—a little, a moderate amount, or a great deal—a particular teacher gives to 30 black and 30 white students.

      c. the amount of attention—a little, a moderate amount, or a great deal—a particular teacher gives 30 students.

      d. the amount of attention—a little, a moderate amount, or a great deal—a particular teacher gives to honor students.

7. A frequency distribution of the number of defendants sentenced to death in each of the 50 states during the year 2001 would be depicted best in the form of a

      a. histogram.

      b. bar graph.

      c. frequency polygon.

      d. line chart.

8. The direction of skewness is determined by the relative position of the

      a. peak of the distribution.

      b. midpoint of the distribution.

      c. tail of the distribution.

      d. class limits of the distribution.

9. To show changes in birth rate from 1980 to the present, by year, a researcher would probably use a

      a. pie chart.

      b. bar graph.

      c. line chart.

      d. frequency polygon.

10. From the following table representing achievement for 173 television viewers and 183 non-viewers, find (a) the percent of nonviewers who are high achievers, (b) the percent of viewers who are high achievers, (c) the proportion of nonviewers who are high achievers, and (d) the proportion of viewers who are high achievers.

Achievements for Television Viewers and Nonviewers

	Viewing Status	
Achievement	*Nonviewers*	*Viewers*
High achievers	93	46
Low achievers	90	127
Total	183	173

11. From the following table representing family structure for black and white children in a particular community, find (a) the percent of black children having two-parent families, (b) the percent of white children having two-parent families, (c) the proportion of black children having two-parent families, and (d) the proportion of white children having two-parent families.

Family Structure for Black and White Children

	Race of Child	
Family Structure	*Black*	*White*
One parent	53	59
Two parents	60	167
Total	113	226

12. From the following table illustrating the handedness of a random sample of men and women, find (a) the percent of men who are left-handed, (b) the percent of women who are left-handed, (c) the proportion of men who are left-handed, and (d) the proportion of women who are left-handed. (e) What can you conclude about gender and the prevalence of left-handedness?

Handedness of Men and Women

	Gender	
Handedness	*Male*	*Female*
Left-handed	15	8
Right-handed	86	114
Total	101	122

13. In a group of 125 males and 80 females, what is the gender ratio (number of males per 100 females)?

14. In a group of 15 black children and 20 white children, what is the ratio of blacks to whites?

15. If 300 live births occur among 3,500 women of childbearing age, what is the birth rate (per 1,000 women of childbearing age)?

16. What is the rate of change for a population increase from 15,000 in 1960 to 25,000 in 2000?

17. What is the rate of change for a tax increase from $32 billion per year to $37 billion per year?

18. A researcher studying the prevalence of alcohol use among seniors in a particular high school asked 45 of these youths how many drinks they had consumed in the last week. Convert the following frequency distribution of responses (number of drinks) into a grouped frequency distribution containing four class intervals, and (a) determine the size of the class intervals, (b) indicate the upper and lower limits of each class interval, (c) identify the midpoint of each class interval, (d) find the percentage for each class interval, (e) find the cumulative frequency of each class interval, and (f) find the cumulative percentage for each class interval.

Number of Drinks	f
7	5
6	9
5	6
4	11
3	4
2	3
1	3
0	4
	$N = 45$

19. The Psychopathy Checklist—Revised (PCL—R) is an assessment tool used to identify psychopaths, with scores ranging from 0 to 40 (a score of 30 or higher being indicative of psychopathy). A forensic psychologist interested in the prevalence of psychopaths in a prison administered the PCL—R to 74 random prison inmates and obtained the following distribution of scores. Convert this into a grouped frequency distribution containing five class intervals, and (a) determine the size of the class intervals, (b) indicate the upper and lower limits of each class interval, (c) identify the midpoint of each class interval, (d) find the percentage for each class interval, (e) find the cumulative frequency for each class interval, and (f) find the cumulative percentage for each class interval.

Score Value	f
39	4
38	4
35	2
32	3
31	4
27	9
26	7
25	6
21	13
20	10
17	5
15	7
	$N = 74$

20. In the following distribution of scores on a promotional exam, find the percentile rank for (a) a score of 75 (passing) and (b) a score of 52.

Class Interval	f	cf
90–99	6	48
80–89	9	42
70–79	10	33
60–69	10	23
50–59	8	13
40–49	5	5
	N = 48	

21. A sports sociologist collected data on the number of points scored by football teams over a two-week period. In the following distribution of scores, find the percentile rank for (a) a score of 36 and (b) a score of 18.

Class Interval	f
40–44	5
35–39	5
30–34	8
25–29	9
20–24	10
15–19	8
10–14	6
5–9	5
	N = 56

22. The following is a cross-tabulation of whether respondents rent or own their home by social class for a sample of 240 heads of households:

	Housing Status		
Social Class	*Rent*	*Own*	**Total**
Lower class	62	18	80
Middle class	47	63	110
Upper class	11	39	50
Total	120	120	240

    a. Which is the independent variable and which is the dependent variable?
    b. Compute row percents for the cross-tabulation.
    c. What percent of the sample owns their home?
    d. What percent of the sample rents?
    e. What percent of the lower-class respondents owns?
    f. What percent of the middle-class respondents rents?
    g. Which social class has the greatest tendency to rent?
    h. Which social class has the greatest tendency to own?
    i. What can be concluded about the relationship between social class and housing status?

23. The following is a cross-tabulation of votes in the 2008 Democratic presidential primary by age for a local sample of respondents aged 18 and over:

| | Age | | | | |
Vote	*18–29*	*30–44*	*45–59*	*60+*	**Total**
Clinton	27	25	34	11	97
Edwards	23	35	10	10	78
Obama	30	20	10	15	75
Total	80	80	54	36	250

    a. Which is the independent variable and which is the dependent variable?
    b. Compute column percents for the cross-tabulation.
    c. What percent of the sample voted for Clinton?
    d. What percent of the 18–29 age group voted for Edwards?
    e. Among which age group was Clinton most popular?
    f. Among which age group was Edwards most popular?
    g. Among which age group was Obama most popular?

24. A sample of respondents was asked their opinions of the death penalty for convicted murderers and of mercy killing for the terminally ill. The responses are given in the following cross-tabulation:

| | **Death Penalty** | | |
Mercy Killing	*Favor*	*Oppose*	**Total**
Favor	63	29	92
Oppose	70	18	88
Total	133	47	180

    a. Why is there no independent or dependent variable?
    b. Compute total percents for the cross-tabulation.
    c. What percent of the sample favors the use of the death penalty?
    d. What percent of the sample favors mercy killing?
    e. What percent of the sample favors both types of killing?
    f. What percent of the sample opposes both types of killing?
    g. What percent of the sample favors one type of killing but not the other?
    h. What can be concluded about the relationship between the variables?

25. From the following table representing the gender of demonstrators outside a local Planned Parenthood, find (a) the percent of pro-choice demonstrators who are female, (b) the percent of pro-life demonstrators who are female, (c) the proportion of pro-choice demonstrators who are female, and (d) the proportion of pro-life demonstrators who are female.

**Gender of Pro-Choice and
Pro-Life Demonstrators**

	Stance	
Gender	*Pro-Choice*	*Pro-Life*
Male	23	32
Female	37	39
Total	60	71

26. The following is a cross-tabulation of sexual orientation by gender for a random sample of respondents aged 18 and over living in Chicago:

	Gender		
Sexual Orientation	*Male*	*Female*	**Total**
Heterosexual	87	106	193
Homosexual	14	9	23
Bisexual	6	3	9
Total	107	118	225

    a. Are there independent and dependent variables in this case? If so, what are they? If not, explain why not.
    b. Compute column percents for the cross-tabulation.
    c. What percent of the sample is heterosexual?
    d. What percent of the sample is comprised of female homosexuals?
    e. What percent of the sample is bisexual?
    f. What percent of the sample is comprised of male heterosexuals?
    g. What can we conclude about gender differences in sexual orientation?

27. A random sample of women over the age of 18 was asked if they considered themselves to be depressed. Their responses are given next, cross-tabulated with their marital status:

State of	**Marital Status**				
Depression	*Single*	*Married*	*Divorced*	*Widowed*	**Total**
Depressed	24	37	11	3	75
Nondepressed	113	82	68	14	277
Total	137	119	79	17	352

    a. Compute total percents for the cross-tabulation.
    b. What percent of the sample considered themselves to be depressed?
    c. What percent of the sample did not consider themselves to be depressed?
    d. What percent of the sample is divorced women who are not depressed?
    e. What percent of the sample is single women who are depressed?
    f. Which marital status is associated with the highest percentage of depressed women?

28. A sample of respondents was asked their opinions of bilingual education and affirmative action. The responses are given in the following cross-tabulation:

Affirmative Action	Bilingual Education		Total
	Favor	*Oppose*	
Favor	56	33	89
Oppose	25	97	122
Total	81	130	211

a. Why is there no independent variable?
b. Compute total percents for the cross-tabulation.
c. What percent of the sample favors affirmative action?
d. What percent of the sample favors bilingual education?
e. What percent of the sample favors both affirmative action and bilingual education?
f. What percent of the sample opposes them both?
g. What percent of the sample opposes one but not the other?
h. What can be concluded about the relationship between the variables?

29. Use a pie chart to depict the following information about the religious affiliation of students in a class:

Religion	*f*	%
Protestant	84	56
Catholic	45	30
Jewish	12	8
Muslim	6	4
Other	3	2
	150	100

30. Depict the following data in a bar graph:

Country of Origin of International Students	*f*
Canada	5
China	7
England	2
Germany	5
Greece	3
Other	4
	N = 26

31. On graph paper, draw both a histogram and a frequency polygon to illustrate the following distribution of IQ scores:

Class Interval	*f*
130–144	3
115–129	11
100–114	37
85–99	34
70–84	16
65–69	2
	N = 103

32. Display the following suicide rates (per 100,000) both as a histogram and as a line chart:

Age	Suicide Rate
15–24	13.1
25–34	15.7
35–44	15.2
45–54	16.4
55–64	17.0
65–74	19.7
75–84	25.2
85+	20.8

33. The distribution of Scholastic Assessment Test (SAT) scores for 38 high school seniors who graduated in the top third of their class is as follows:

SAT Scores	*f*
750–800	1
700–740	2
650–690	3
600–640	5
550–590	10
500–540	8
450–490	4
400–440	3
350–390	2
	N = 38

   a. For each class interval, find the size, midpoint, upper and lower limits, the cumulative frequency, the percentage, and the cumulative percentage.
   b. To depict the distribution of SAT scores for the 38 students, draw a histogram and a frequency polygon.
   c. To depict the cumulative distribution of these SAT scores, draw a cumulative frequency polygon.

34. Using a blank map of the United States, show the unemployment rates from Table 2.10 with the following key: (a) shade in states with unemployment rates of 6.0 and over; (b) draw right diagonal lines in states with unemployment rates of 5.0 through 5.9; and (c) leave blank the states with unemployment rates under 5.0.

SPSS Exercises

1. Using SPSS to analyze the Monitoring the Future Study data, find the valid percentage of American high school seniors who say that there is little or no risk in trying marijuana once or twice (V1767). Find the percentage of seniors who say that there is little or no risk in trying cocaine once or twice (V1770). What do you think may explain why some students see no risk while other students perceive great risk in trying these drugs? Hint: The frequencies procedure in SPSS can be located by clicking on ANALYZE, then DESCRIPTIVE STATISTICS, and finally FREQUENCIES. Also, check to ensure that the WEIGHT is on by selecting DATA, then WEIGHT CASES, and select V5 as the weighting variable.

2. Using SPSS to analyze the General Social Survey data, find the valid percentage of Americans who favor the death penalty for murder (CAPPUN). Find the percentage of people who say that they are in excellent health (HEALTH). Choose another variable and comment on some aspect of it. Hints: The data set is a portable file. Change the file type to "por." Weight cases by selecting DATA, then WEIGHT CASES, and select WTSSALL. The frequencies procedure in SPSS can be located by clicking on ANALYZE, then DESCRIPTIVE STATISTICS, and finally FREQUENCIES.

3. Using SPSS, generate a pie chart from the Monitoring the Future Study for students' race (V151). Be sure to provide a title, source, and show percentages on the pie chart.

4. Using SPSS, generate a bar graph from the Monitoring the Future Study for the number of days in the past four weeks that students have skipped a class (V178).

5. Use SPSS to produce a line chart for the violent crime rate in the United States from 1987 to 2006 based on the following table of data values:

Year	Violent Crime Rate
1987	612.5
1988	640.6
1989	666.9
1990	729.6
1991	758.2
1992	757.7
1993	747.1
1994	713.6
1995	684.5
1996	636.6
1997	611.0
1998	567.6
1999	523.0
2000	506.5
2001	504.5
2002	494.4
2003	475.8
2004	463.2
2005	469.0
2006	473.5

3

Measures of Central Tendency

Researchers in many fields have used the term *average* to ask such questions as these: What is the *average* income earned by high school and college graduates? How many cigarettes are smoked by the *average* teenager? What is the grade-point *average* of the female college student? On *average*, how many automobile accidents happen as the direct result of alcohol or drug use?

A useful way to describe a group as a whole is to find a single number that represents what is average or typical of that set of data. In social research, such a value is known as a measure of *central tendency,* because it is generally located toward the middle or center of a distribution where most of the data tend to be concentrated.

What the layperson means by the term *average* is often vague and even confusing. The social researcher's conception is much more precise; it is expressed numerically as one of several different kinds of measures of average or central tendency that may take on quite different numerical values in the same set of data. Only the three best known measures of central tendency are discussed here: the mode, the median, and the mean.

The Mode

The *mode* (Mo) is the most frequent, most typical, or most common value in a distribution. For example, there are more Protestants in the United States than people of any other religion; and so we refer to this religion as the mode. Similarly, if at a given university, engineering is the most popular major, this, too, would represent the mode. The mode is the only measure of central tendency available for nominal-level variables, such as religion and college major. It can be used, however, to describe the most common score in any distribution, regardless of the level of measurement.

To find the mode, find the score or category that occurs most often in a distribution. The mode can be easily found by inspection, rather than by computation. For instance, in the set of scores ①, 2, 3, ①, ①, 6, 5, 4, ①, 4, 4, 3, the mode is 1, because it is the number that occurs more than any other score in the set (it occurs four times). Make no mistake: The mode is *not* the frequency of the most frequent score ($f = 4$), but the value of the most frequent score (Mo $= 1$).

Some frequency distributions contain two or more modes. In the following set of data, for example, the scores 2 and 6 *both* occur most often: 6, 6, 7, 2, 6, 1, 2, 3, 2, 4. Graphically, such distributions have two points of maximum frequency, suggestive of the humps on a camel's back. These distributions are referred to as being *bimodal* in contrast to the more common *unimodal* variety, which has only a single hump or point of maximum frequency.

Figure 3.1, for example, shows the test scores on an English and a Spanish final. The English scores are unimodal; that is, the achievement levels cluster around a single mode. The Spanish scores, however, are bimodal; that is, the achievement levels cluster around two modes. In the Spanish class, apparently, there were many students who caught on, yet many students who did not.

The Median

When ordinal or interval data are arranged in order of size, it becomes possible to locate the *median* (Mdn), the *middlemost* point in a distribution. Therefore, the median is regarded as the measure of central tendency that cuts the distribution into two equal parts, just as the median strip of a highway cuts it in two.

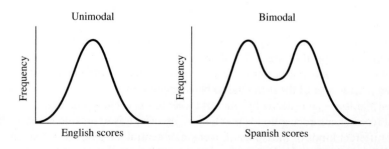

FIGURE 3.1   Graphic presentations of unimodal and bimodal distributions of test scores

The position of the median value can be located by inspection or by the formula

$$\text{Position of median} = \frac{N + 1}{2}$$

If we have an odd number of cases, then the median will be the case that falls exactly in the middle of the distribution. Thus, 16 is the median value for the scores 11, 12, 13, ⑯, 17, 20, 25; this is the case that divides the distribution of numbers, so that there are three scores on either side of it. According to the formula $(7 + 1)/2$, we see that the median 16 is the fourth score in the distribution counting from either end.

If the number of cases is even, the median is always that *point* above which 50% of the cases fall and below which 50% of the cases fall. For an even number of cases, there will be two middle cases. To illustrate, the numbers 16 and 17 represent the middle cases for the following data: 11, 12, 13, ⑯, ⑰, 20, 25, 26. By the formula $(8 + 1)/2 = 4.5$, the median will fall midway between the fourth and fifth cases; the middlemost point in this distribution turns out to be 16.5, because it lies halfway between 16 and 17, the fourth and fifth scores in the set. Likewise, the median is 9 in the data 2, 5, ⑧, ⑩, 11, 12, again because it is located exactly midway between the two middle cases $(6 + 1)/2 = 3.5$.

Another circumstance must be explained and illustrated—we may be asked to find the median from data containing several middle scores having identical numerical values. The solution is simple—that numerical value becomes the median. Therefore, in the data 11, 12, 13, ⑯, ⑯, ⑯, 25, 26, 27, the median case is 16, although it occurs more than once.

Finally, if the data are not in order from low to high (or high to low), you should put them in order before trying to locate the median. Thus, in the data 3, 2, 7, the median is 3, the middle score after arranging the numbers 2, ③, 7.

The Mean

By far the most commonly used measure of central tendency, the arithmetic mean X, is obtained by adding up a set of scores and dividing by the number of scores. Therefore, we define the *mean* more formally as *the sum of a set of scores divided by the total number of scores in the set.* By formula,

$$\boxed{\bar{X} = \frac{\Sigma X}{N}}$$

where $\bar{X}$ = mean (read as X bar)
   Σ = sum (expressed as the Greek capital letter sigma)[1]
   X = raw score in a set of scores
   N = total number of scores in a set

---

[1]The Greek capital letter sigma (Σ), called the summation sign, will be encountered many times throughout the text. It simply indicates that we must *sum* or add up what follows. In the present example, ΣX indicates adding up the raw scores. See Appendix B for a discussion of the summation sign.

TABLE 3.1   *Calculating the Mean: An Illustration*

Respondent	X (IQ)
Gene	125
Steve	92
Bob	72
Michael	126
Joan	120
Jim	99
Jane	130
Mary	<u>100</u>
	$\Sigma X = 864$

$$\bar{X} = \frac{\Sigma X}{N}$$

$$= \frac{864}{8}$$

$$= 108$$

Using the previous formula, we learn that the mean IQ for the eight respondents listed in Table 3.1 is 108.

Unlike the mode, the mean is not always the score that occurs most often. Unlike the median, it is not necessarily the middlemost point in a distribution. Then, what does the *mean* mean? How can it be interpreted?

As we shall see, the mean can be regarded as the "center of gravity" of a distribution. It is similar to the notion of a seesaw or a fulcrum and lever (see Figure 3.2). Four blocks of weight are placed on the lever. The block marked *11* is 7 units (inches, feet, or whatever) to the right of the fulcrum. It balances with the blocks marked *1, 2,* and *2,* which are 3, 2, and 2 units to the left of the fulcrum, respectively. In a distribution of data, the mean acts as a fulcrum: It is the point in a distribution around which the scores above it balance with those below it.

To understand this characteristic of the mean, we must first understand the concept of *deviation*. The deviation indicates the distance and direction of any raw score from the mean, just as we noted that a particular block is 7 units to the right of the fulcrum.

To find the deviation for a particular raw score, we simply subtract the mean from that score:

$$\text{Deviation} = X - \bar{X}$$

where X = any raw score in the distribution
$\bar{X}$ = mean of the distribution

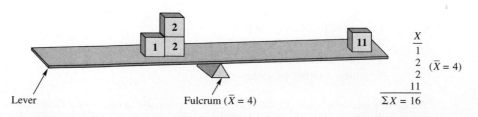

Lever   Fulcrum ($\bar{X} = 4$)

$$\begin{array}{c} X \\ \hline 1 \\ 2 \\ 2 \\ \underline{11} \\ \Sigma X = 16 \end{array} \quad (\bar{X} = 4)$$

FIGURE 3.2   Lever and fulcrum analogy to the mean

TABLE 3.2 *Deviations of a Set of Raw Scores from $\overline{X}$*

X	$X - \overline{X}$
9	$+3$ $\left.\right\}$
8	$+2$ $\left.\right\}+5$
6	0 $\qquad \overline{X} = 6$
5	-1 $\left.\right\}$
2	-4 $\left.\right\}-5$

For the set of raw scores 9, 8, 6, 5, and 2 in Table 3.2, $\overline{X} = 6$. The raw score 9 lies exactly three raw score units above the mean of 6 (or $X - \overline{X} = 9 - 6 = +3$). Similarly, the raw score 2 lies four raw-score units below the mean (or $X - \overline{X} = 2 - 6 = -4$). Thus, the greater the deviation $(X - \overline{X})$, the greater is the distance of that raw score from the mean of the distribution.

Considering the mean as a point of balance in the distribution, we can now say that the sum of the deviations that fall above the mean is equal in absolute value (ignoring the minus signs) to the sum of the deviations that fall below the mean. Let us return to the set of scores 9, 8, 6, 5, 2 in which $\overline{X} = 6$. If the mean for this distribution is the "center of gravity," then disregarding minus signs and adding together the positive deviations (deviations of raw scores 8 and 9) should equal adding together the negative deviations (deviations of raw scores 5 and 2). As shown in Table 3.2, this turns out to be the case because the sum of deviations below $\overline{X}(-5)$ equals the sum of deviations above $\overline{X}(+5)$.

Taking another example, 4 is the mean for the numbers 1, 2, 3, 5, 6, and 7. We see that the sum of deviations below this score is -6, whereas the sum of deviations above it is $+6$. We shall return to the concept of deviation in Chapters 4 and 5.

The Weighted Mean

Researchers sometimes find it useful to obtain a "mean of means"—that is, to calculate a total mean for a number of different groups. Suppose, for example, that the students in three different sections of introductory sociology received the following mean scores on their final exams for the course:

Section 1: $\overline{X}_1 = 85$ $N_1 = 28$
Section 2: $\overline{X}_2 = 72$ $N_2 = 28$
Section 3: $\overline{X}_3 = 79$ $N_3 = 28$

Because exactly the same number of students were enrolled in each section of the course, it becomes quite simple to calculate a total mean score:

$$\frac{\overline{X}_1 + \overline{X}_2 + \overline{X}_3}{3} = \frac{85 + 72 + 79}{3} = \frac{236}{3} = 78.67$$

In most cases, groups differ in size. By looking again at sections of introductory sociology, for example, it is probably unusual to find precisely the same number of students enrolling in different sections of a course. More likely, the number of students who take a final exam in three different sections of introductory sociology would differ. In this case,

Section 1: $\overline{X}_1 = 85$ $N_1 = 95$
Section 2: $\overline{X}_2 = 72$ $N_2 = 25$
Section 3: $\overline{X}_3 = 79$ $N_3 = 18$

When groups differ in size, you cannot just sum their means and divide by 3 to obtain a total mean for all groups combined (if we had followed the "unweighted" procedure for the present example, we would have erroneously concluded that the mean of means was almost 79). Instead, you must *weight* each group mean by its size *(N)*. The *weighted mean* may be calculated by first multiplying each group mean by its respective N before summing the products, and then dividing by the total number in all groups:

$$\overline{X}_w = \frac{\Sigma N_{\text{group}} \overline{X}_{\text{group}}}{N_{\text{total}}}$$

where $\overline{X}_{\text{group}}$ = mean of a particular group
        N_{group} = number in a particular group
        N_{total} = number in all groups combined
        $\overline{X}_w$ = weighted mean

In the preceding example,

$$\overline{X}_w = \frac{N_1\overline{X}_1 + N_2\overline{X}_2 + N_3\overline{X}_3}{N_{\text{total}}}$$

$$= \frac{95(85) + 25(72) + 18(79)}{138}$$

$$= \frac{8,075 + 1,800 + 1,422}{138}$$

$$= \frac{11,297}{138}$$

$$= 81.86$$

We have determined that the average or mean final exam grade for all sections combined was 81.86, more heavily weighted toward the mean of the largest section.

The weighted mean is particularly useful for averaging values across geographic units of varying size. For example, the following table displays the percentage of black population and overall resident population (in thousands) for the six New England states in

2002. The usual mean for %black across the six states is 4.10%. In this calculation, however, each state contributes an equal share, as if all of them had the same number of residents. Thus, the mean percentage black population gives high-population states like Massachusetts and Connecticut too little emphasis and low-population states like New Hampshire and Vermont too much influence.

State	%Black	Population (in 1,000s)	Population × %Black
Connecticut	10.0%	3,461	34,610.0
Maine	0.6%	1,294	776.4
Massachusetts	6.7%	6,428	43,067.6
New Hampshire	0.9%	1,275	1,147.5
Rhode Island	5.8%	1,070	6,206.0
Vermont	0.6%	617	370.2
Sum	24.6%	14,145	86,177.7
Mean	4.10%		6.09%

We can use the populations as weights in the weighted mean formula to adjust for lack of comparability in population. The weighted mean (treating the state values for %black as group means) multiplies each state's value by its population, and then the sum of these products is divided by the total population of the six states. Using the data from the given table,

$$\bar{X}_w = \frac{\Sigma \text{Population} \times \%\text{Black}}{\Sigma \text{Population}} = \frac{86,177.7}{14,145} = 6.09$$

The weighted mean of 6.09% gives a better sense of the population diversity of the New England region than the unweighted mean. More generally, whenever using data for states, cities, or other units that vary considerably in size, one would be wise to consider weighting the scores as we did here before combining units of dissimilar size.

Taking One Step at a Time

When you open a cookbook to find a method for making a chocolate cake, the recipe can at first appear overwhelming. But when you approach the cake "formula" step by step, you often find that it was easier than it looked. In a similar way, some of the statistical "recipes" that you will encounter later in this book can also look overwhelming, or at least very complicated. Our advice is to confront formulas step by step—that is, to perform a series of small mathematical tasks to achieve the eventual solution. Throughout this book, we will often demonstrate calculations through step-by-step illustrations. Try not to focus so much on whether there are four, six, or seven steps, but on the progression from one to another. Now let's review the steps to calculate the mode, median, and mean.

BOX 3.1 • *Step-by-Step Illustration: Mode, Median, and Mean*

Suppose that a volunteer canvasses houses in her neighborhood collecting for a local charity. She receives the following donations (in dollars):

5	10	25	15	18	2	5

Step 1   Arrange the scores from highest to lowest.

$$25$$
$$18$$
$$15$$
$$10$$
$$5$$
$$5$$
$$2$$

Step 2   Find the most frequent score.

$$Mo = \$5$$

Step 3   Find the middlemost score. Because there are seven scores (an odd number), the fourth from either end is the median.

$$Mdn = \$10$$

Step 4   Determine the sum of the scores.

$$25$$
$$18$$
$$15$$
$$10$$
$$5$$
$$5$$
$$\underline{2}$$
$$\Sigma X = \$80$$

Step 5   Determine the mean by dividing the sum by the number of scores.

$$\bar{X} = \frac{\Sigma X}{N} = \frac{\$80}{7} = \$11.43$$

Thus, the mode, median, and mean provide very different pictures of the average level of charitable giving in the neighborhood. The mode suggests that the donations were typically small, whereas the median and the mean suggest greater generosity overall.

Obtaining the Mode, Median, and Mean from a Simple Frequency Distribution

In the last chapter, we saw how a set of raw scores could be rearranged in the form of a simple frequency distribution—that is, a table of the frequency of occurrence of each score value. It is important to note that a simple frequency distribution does not change the data; it only shows them in a different way. Therefore, the mode, median, and mean obtained from a simple frequency distribution also do not change, but they are calculated differently.

Let's consider the following raw scores representing the age at first marriage of a sample of 25 adults:

18 18 19 19 19 19 20 20 20 21 21 22
22 23 23 24 25 26 26 26 27 27 29 30 31

There are more respondents first married at age 19; thus, Mo = 19. The middlemost score (the thirteenth from either end) is 22; thus, Mdn = 22. Finally, the scores sum to 575; thus, $\overline{X}$ = 575/25 = 23.

These data can be rearranged as a simple frequency distribution as follows:

	X	f
	31	1
	30	1
	29	1
	28	0
	27	2
	26	3
	25	1
	24	1
	23	2
	22	2
	21	2
	20	3
Mo →	19	4
	18	2

In the case of a simple frequency distribution in which the score values and frequencies are presented in separate columns, the mode is the score value that appears most often in the frequency column of the table. Therefore, Mo = 19 in the simple frequency distribution shown previously. This agrees, of course, with the mode obtained from the raw scores.

To find the median for this simple frequency distribution, we start by identifying the position of the median. There are 25 age at first marriage scores (as opposed to 14 score values, 18 through 31). With $N = 25$,

$$\text{Position of median} = \frac{25 + 1}{2}$$
$$= \frac{26}{2}$$
$$= 13$$

The median turns out to be the thirteenth score in this frequency distribution. To help locate the thirteenth score, we construct a cumulative frequency distribution, as shown in the third column of the following table (for a small number of scores, this can be done in your head):

	X	f	cf
	31	1	25
	30	1	24
	29	1	23
	28	0	22
	27	2	22
	26	3	20
	25	1	17
	24	1	16
	23	2	15
Mdn →	22	2	13
	21	2	11
	20	3	9
	19	4	6
	18	2	2

Beginning with the lowest score value (18), we add frequencies until we reach a score value that represents the thirteenth score in the distribution. This is accomplished by searching for the smallest score value having a cumulative frequency that is at least 13.

In this distribution of age at first marriage, the cumulative frequency for the score value 18 is 2, meaning that the two youngest ages were 18. The cumulative frequency for the score value 19 is 6, indicating that 6 respondents were first married by the age of 19. Continuing, we eventually see that the score value of 22 has a cumulative frequency of at least 13. Here, 13 respondents were married by the age of 22. Thus, the median, the thirteenth score, is 22, which agrees with the result obtained from raw scores. Note, finally, that the median is *not* the middlemost score *value* (there are 14 different values, and 22 is not middlemost in the column).

To obtain the mean, we first need to calculate the sum of the scores. In a simple frequency distribution, this can be done efficiently by noting that there are, for example, two scores of 18, four scores of 19, and so on. Thus, rather than adding 18 twice and 19 four times, we can first multiply 18 by 2 and 19 by 4 *before* adding. That is, we can multiply the score values by their respective frequencies and then add the products in order to obtain the sum of scores. Just as in the raw-score formula, we divide the sum by the number of scores to determine the mean.

Thus, a more practical and less time-consuming way to compute the mean from a simple frequency distribution is provided by the following formula:

$$\bar{X} = \frac{\Sigma fX}{N}$$

where $\bar{X}$ = mean

X = a score value in the distribution

f = frequency of occurrence of X

N = total number of scores

In the following table, the third column (headed fX) contains the products of the score values multiplied by their frequencies of occurrence. Summing the fX column, we obtain $\Sigma fX = 575$.

X	f	fX
31	1	31
30	1	30
29	1	29
28	0	0
27	2	54
26	3	78
25	1	25
24	1	24
23	2	46
22	2	44
21	2	42
20	3	60
19	4	76
18	2	36
		$\Sigma fX = 575$

Thus,

$$\bar{X} = \frac{\Sigma fX}{N} = \frac{575}{25} = 23$$

This result also agrees with the mean obtained from the raw scores themselves.

Comparing the Mode, Median, and Mean

The time comes when the social researcher chooses a measure of central tendency for a particular research situation. Will he or she employ the mode, the median, or the mean? The decision involves several factors, including the following:

1. Level of measurement
2. Shape or form of the distribution of data
3. Research objective

Level of Measurement

Because the mode requires only a frequency count, it can be applied to any set of data at the nominal, ordinal, or interval level of measurement. For instance, we might determine that the modal category in a nominal-level measure of religious affiliation (Protestant, Catholic, Muslim, or Jewish) is Protestant, because the largest number of our respondents identify themselves as such. Similarly, we might learn that the largest number of students attending a particular university have a 2.5 grade-point average (Mo = 2.5).

The median requires an ordering of categories from highest to lowest. For this reason, it can only be obtained from ordinal or interval data, *not* from nominal data. To illustrate, we might find the median annual income is $77,000 among dentists in a small town. The result gives us a meaningful way to examine the central tendency in our data. By contrast, it would make little sense if we were to compute the median for religious affiliation, gender, or country of origin, when ranking or scaling cannot be carried out.

The use of the mean is exclusively restricted to interval data. Applying it to ordinal or nominal data yields a meaningless result, generally not at all indicative of central tendency. What sense would it make to compute the mean for a distribution of religious affiliation or gender? Although less obvious, it is equally inappropriate to calculate a mean for data that can be ranked but not scored.

Shape of the Distribution

The shape, or form, of a distribution is another factor that can influence the researcher's choice of a measure of central tendency. In a perfectly symmetrical unimodal distribution, the mode, median, and mean will be identical, because the point of maximum frequency (Mo) is also the middlemost score (Mdn), as well as the "center of gravity" ($\overline{X}$). As shown in Figure 3.3, the measures of central tendency will coincide at the most central point, the "peak" of the symmetrical distribution.

When social researchers work with a symmetrical distribution, their choice of measure of central tendency is chiefly based on their particular research objectives and the level at

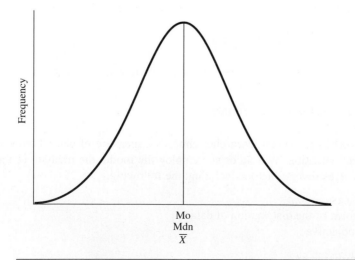

FIGURE 3.3   A unimodal, symmetrical distribution showing that the mode, median, and mean have identical values

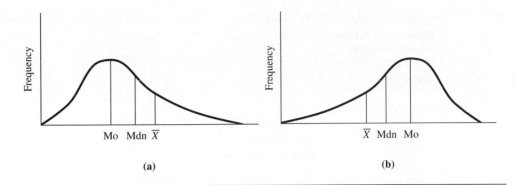

FIGURE 3.4   Relative positions of measures of central tendency in (a) a positively skewed distribution and (b) a negatively skewed distribution

which their data are measured. When they work with a skewed distribution, however, this decision is very much influenced by the shape, or form, of their data.

As shown in Figure 3.4, the mode, median, and mean do not coincide in skewed distributions. The mode is at the peak of the curve, because this is the point where the most frequent scores occur. The mean, by contrast, is found closest to the tail, where the relatively few extreme scores are located. For this reason, the mean score in the positively skewed distribution in Figure 3.4(a) lies toward the high-score values; the mean in the negatively skewed distribution in Figure 3.4(b) falls close to the low-score values.

Whereas the mean is very much influenced by extreme scores in either direction, the median is modified little, if at all, by changes in extreme values. This is because the mean considers all of the scores in any distribution, whereas the median (by definition) is concerned only with the numerical value of the score that falls at the middlemost position in a distribution. As illustrated, changing an extreme score value from 10 in distribution A to 95 in distribution B does not at all modify the median value (Mdn = 7.5), whereas the mean shifts from 7.63 to 18.25:

Distribution A: 5   6   6   7   8   9   10   10   Mdn = 7.5 $\overline{X}$ = 7.63

Distribution B: 5   6   6   7   8   9   10   95   Mdn = 7.5 $\overline{X}$ = 18.25

In a skewed distribution, the median usually falls somewhere between the mean and the mode. It is this characteristic that makes the median the most desirable measure of central tendency for describing a skewed distribution of scores. To illustrate this advantage of the median, let us turn to Table 3.3 and examine the average annual salary among secretaries working in a public-interest law firm. If we were public relations practitioners hired by the firm to give it a favorable public image, we would probably want to calculate the mean to show that the average employee makes $50,000 and is relatively well paid. On the other hand, if we were union representatives seeking to upgrade salary levels, we would probably want to employ the mode to demonstrate that the most common salary is only $30,000. Finally, if we were social researchers seeking to report accurately the average salary among the secretaries in the firm, we would wisely employ the median ($40,000), because it falls between the other

TABLE 3.3   *Measures of Central Tendency in a Skewed Distribution of Annual Salaries*

Salary	
$120,000	
$60,000	$\bar{X} = \$50,000$
$40,000	$\text{Mdn} = \$40,000$
$40,000	$\text{Mo} = \$30,000$
$30,000	
$30,000	
$30,000	

measures of central tendency and, therefore, gives a more balanced picture of the salary structure. The most acceptable method would be to report all three measures of central tendency and permit the audience to interpret the results. It is unfortunately true that few researchers—let alone public relations practitioners or union representatives—report more than a single measure of central tendency. Even more unfortunate is that some reports of research fail to specify exactly which measure of central tendency—the mode, median, or mean—was used to calculate the average amount or position in a group of scores. As shown in Table 3.3, a reasonable interpretation of findings may be impossible without such information.

It was noted earlier that some frequency distributions can be characterized as being bimodal, because they contain two points of maximum frequency. To describe bimodal distributions properly, it is generally useful to identify *both* modes; important aspects of such distributions can be obscured using either the median or the mean.

Consider the situation of a social researcher who conducted personal interviews with 20 lower-income respondents to determine their ideal conceptions of family size. Each respondent was asked, "Suppose you could decide exactly how large your family should be. Including all children and adults, how many people would you like to see in your ideal family?"

1   2   2   2   3   3   3   3   4   4   5   6   6   7   7   7   7   8   8   9
                 └──Mode──┘   └──Mode──┘

As shown in the previous responses, there is a wide range of family size preferences, from living alone (1) to having a big family (9). Using either the mean ($\bar{X} = 4.85$) or the median (Mdn $= 4.5$), we might conclude that the average respondent's ideal family contained between four and five members. Knowing that the distribution is bimodal, however, we see that there were actually *two* ideal preferences for family size in this group of respondents—one for a small family (Mo $= 3$) and the other for a large family (Mo $= 7$).[2]

[2]Sometimes distributions can be "almost" bimodal. For example, suppose that one more respondent reported 3 rather than 2 as the ideal family size. There would only be one real mode (Mo $= 3$). But because there would be almost as many seven-person ideal families, 7 would almost, but not quite, be a mode as well. In such a case, it is wise to report that the distribution has two peaks, although one is somewhat smaller than the other.

Research Objective

Until this point we have discussed the choice of a measure of central tendency in terms of the level of measurement and the shape of a distribution of scores. We now ask what the social researcher expects to do with this measure of central tendency. If he or she seeks a fast, simple, yet crude descriptive measure or is working with a bimodal distribution, the researcher generally will employ the mode. In most situations encountered by the researcher, however, the mode is useful only as a preliminary indicator of central tendency, which can be quickly obtained from briefly scanning the data. If he or she seeks a precise measure of central tendency, the decision is usually between the median and the mean.

To describe a skewed distribution, the social researcher generally chooses the median, since it tends to give a balanced picture of extreme scores. (The mean, in contrast, tends to be distorted by very extreme scores.) In addition, the median is sometimes used as a point in the distribution where the scores can be divided into two categories containing the same number of respondents. For instance, we might divide respondents at the median into two categories according to their family size preferences—those who like a small family versus those who like a large family.

For a precise measure of distributions that are at least roughly symmetrical, the mean tends to be preferred over the median since the mean can be easily used in more advanced statistical analyses, such as those introduced in subsequent chapters of the text. Moreover, the mean is more stable than the median, in that it varies less across different samples drawn from any given population. This advantage of the mean—although perhaps not yet understood or appreciated by the student—will become more apparent in subsequent discussions of the decision-making function of statistics.

BOX 3.2 • *Practical and Statistical: Many Happy Returns*

One of the most commonplace uses of descriptive statistics is in sports. For every sport there is an endless array of percentages and averages both about players and teams. In baseball, for example, the batting average is the number of hits divided by the number of at bats; the slugging average is the mean number of bases per at bat; and the earned run average (ERA) is the number of earned runs divided by the number of full-length games pitched. Basketball has average points, rebounds, and assists per game, as well as won–loss percentages. Football also promotes a variety of "stats" on passing, running, kicking, and tackling.

Whenever you hear about an average in sports, you can be sure that that average is the mean. We can think of no average used in sports involving the mode or the median. Although this consistent use of the mean may simplify matters (since they do not have to say what kind of average), in a number of instances skewness in the data greatly distorts the mean, and the sports statistician would be better off breaking tradition by using a median or two.

Consider the table displaying kickoff return statistics for the National Football League. The players are ranked by their average (mean) returns, which are obtained by dividing the total yards in kickoff returns by the number of returns. The average performance of some players tends to be exaggerated by breakaway runs when they manage to find a hole in the kickoff coverage and

(continued)

BOX 3.2   Continued

breeze down the field for a touchdown. Kevin Faulk of the New England Patriots, for instance, broke two runs for touchdowns, pushing him up near the top of the list.

Most of the hard work happens in the first 20 to 40 yards. It might give us a better sense of the usual or typical kickoff return by reducing the distorting effects of the touchdown runs. If we trim the breakaways so as to count only the first 50 yards and then recalculate, the modified rankings are quite different. The new leader is Albert Johnson of the Miami Dolphins. Although he hadn't managed to break one for a touchdown,

Johnson tended to do comparatively well. Faulk, without the excessive boost from his breakaway touchdowns, falls to thirteenth in the new rankings.

By now the lesson should be clear: the mean is not a particularly good measure of central tendency if there is skewness or extreme cases (like the touchdown runs). For an improved measure of average in such situations, one might modify the atypical or extreme scores or, better yet, use the median instead. At least in the world of sports, however, it is doubtful that the median will ever get much play.

2002 NFL Kick Return Statistics

Player	Team	Returns	Average	Rank	Long	TDs	Modified Average	Modified Rank
MarTay Jenkins	ARI	20	28.0	1	95	1	25.4	5
Kevin Faulk	NE	26	27.9	2	87	2	24.0	13
Albert Johnson	MIA	12	27.5	3	49	0	27.5	1
Brian Mitchell	PHI	43	27.0	4	57	0	27.0	2
Kevin Kasper	DEN	15	26.2	5	56	0	26.2	3
Chad Morton	NYJ	58	26.0	6	98	2	24.3	11
Eddie Drummond	DET	40	26.0	7	91	0	26.0	4
Michael Lewis	NO	70	25.8	8	97	0	24.4	9
Aaron Stecker	TB	37	25.2	9	67	0	25.2	6
Brandon Bennett	CIN	49	25.1	10	94	1	24.1	12
Reggie Swinton	DAL	28	24.9	11	100	1	23.1	15
Scottie Montgomery	DEN	15	24.7	12	40	0	24.7	7
Ladell Betts	WAS	28	24.6	13	60	0	24.6	8
Marcus Knight	OAK	29	24.3	14	65	0	24.3	10
Maurice Morris	SEA	34	24.1	15	97	1	22.7	16
Deion Branch	NE	36	24.0	16	63	0	24.0	14

Summary

In this chapter, we introduced the three best known measures of central tendency—indicators of what is average or typical of a set of data. The mode is the category or score that occurs most often; the median is the middlemost point in a distribution; and the mean is the sum of a set of scores divided by the total number of scores in a set. Which of these three measures of central tendency is appropriate to employ in any research project can be determined with reference to three criteria: level of measurement, shape or form of distribution, and research objective. The mode can be used with nominal, ordinal, or interval data, is especially useful for displaying a bimodal distribution, and provides a fast, simple, but rough measure of central tendency. The median can be used with ordinal or interval data, is most appropriate for displaying a highly skewed distribution, and provides a precise measure of central tendency. Moreover, the median can sometimes be used for more advanced statistical operations or for splitting distributions into two categories (for example, high versus low). The mean can be employed only with interval data, is most appropriate for displaying a unimodal symmetrical distribution, and provides a precise measure of central tendency. In addition, the mean can often be used for more advanced statistical operations including the decision-making tests we discuss in subsequent chapters of the text.

Terms to Remember

Central tendency
Mode
Unimodal distribution
Bimodal distribution

Median
Mean
Deviation
Weighted mean

Questions and Problems

1. The measures in this chapter are known as *measures of central tendency* because they tend to
   a. fall toward the center of a distribution, where most of the scores are located.
   b. be central to our understanding of statistics.
   c. be located at the midpoint of a class interval.
   d. All of the above

2. Which measure of central tendency represents the point of maximum frequency in a distribution?
   a. Mode   b. Median   c. Mean

3. Which measure of central tendency is considered the point of balance in a distribution?
   a. Mode   b. Median   c. Mean

4. Which measure of central tendency cuts a distribution in half when the scores are arranged from high to low?
   a. Mode   b. Median   c. Mean

5. Deviation indicates the _____ of any raw score from the mean.
   a. distance
   b. direction
   c. distance and direction
   d. frequency

6. A distribution of income is highly skewed. Which measure of central tendency are you likely to employ for the purpose of characterizing income?
   a. Mode   **b.** Median   **c.** Mean

7. A distribution of the strength of attitudes toward legalized abortion has two points of maximum frequency, indicating that many people strongly oppose it and many people strongly support it. Which measure of central tendency are you likely to employ for the purpose of characterizing the strength of attitudes toward legalized abortion?
   a. Mode
   b. Median
   c. Mean

8. You have a distribution of children's empathy scores that approximate a normal curve. Which measure of central tendency are you likely to use for the purpose of characterizing empathy?
   a. Mode
   b. Median
   c. Mean

9. The following is a list of seven movies and their ratings:

March of the Penguins	G
National Treasure: Book of Secrets	PG
Mamma Mia	PG-13
Sex and the City	R
There Will Be Blood	R
The 40-Year-Old Virgin	R
Showgirls	NC-17

   a. Find the modal film rating.
   b. Find the median film rating.
   c. Explain why it is inappropriate to calculate a mean film rating.

10. Anita suffers from multiple personality disorder (now known as dissociative identity disorder). The following table provides data on each of her six alternative personalities:

Name	IQ	Gender	Ethnicity	Age	Frequency of Appearance
Betty	104	Female	White	32	Very often
Rosa	98	Female	Hispanic	37	Rarely
John	76	Male	White	16	Sometimes
Charles	112	Male	Black	44	Often
Ann	137	Female	White	33	Very rarely
Colleen	106	Female	White	24	Rarely

    Calculate the most appropriate measure of central tendency for each of the variables (IQ, gender, ethnicity, age, and frequency of appearance).

11. A caregiver at a local nursing home has been accused of abusing some elderly people living there. The following table provides information about the nine victims who have come forward:

Name	Age	Health Status	Type of Abuse	Duration of Abuse (month)
Balfour, S.	68	Fair	Physical abuse	1
Enger, R.	79	Fair	Active neglect	4
Bradshaw, C.	73	Poor	Financial abuse	15
Marcus, L.	82	Good	Verbal and emotional abuse	8
McCarthy, K.	87	Poor	Financial abuse	2
Conley, R.	74	Fair	Active neglect	1
Quinn, D.	91	Poor	Financial abuse	7
Stein, J.	70	Good	Passive neglect	6
Martinez, M.	84	Poor	Financial abuse	5

Calculate the most appropriate measure of central tendency for each of the variables (age, health status, type of abuse, and duration of abuse).

12. A researcher interested in the effectiveness of organized diet groups on weight loss weighed five clients after several weeks on the program. The weight-loss scores (in pounds) were as follows:

    13   12   6   9   10

Calculate (a) the median and (b) the mean for these weight-loss scores.

13. A group of high school students was surveyed about its use of various drugs, including alcohol. Asked how frequently they had engaged in binge drinking during the previous six months, the students responded as follows:

    4   2   0   2   1   3   0   1   7   5   3

Calculate (a) the median and (b) the mean for these self-report scores.

14. Five convicts were given the following five prison sentences (in years):

    4   5   3   3   40

Find (a) the mode, (b) the median, and (c) the mean. Which measure provides the most accurate indication of central tendency for these scores?

15. The hourly wages (in dollars) of seven employees in a small company are as follows:

    18   16   20   12   14   12   10

Find (a) the mode, (b) the median, and (c) the mean.

16. Suppose that the small company in Problem 15 hired another employee at an hourly wage of $24, resulting in the following hourly wages (in dollars):

    18   16   20   12   14   12   10   24

    Find (a) the mode, (b) the median, and (c) the mean.

17. Ten long-term psychiatric patients were questioned about how many times they were visited by family members during the previous month. Their responses were as follows:

    4   0   6   1   0   0   3   5   4   2

    Find (a) the mode, (b) the median, and (c) the mean.

18. The following scores represent the number of households per city block in which a resident owns an unregistered handgun:

    2   7   3   6   5   7   0   9   6   1   4   7   5   2   9   3   4   1   8

    a. Calculate the mode, median, and mean from the scores.
    b. Rearrange the scores into a simple frequency distribution and recalculate the mode, median, and mean.

19. Referring to the sentence lengths given in Problem 14, calculate the deviations (from the mean) for each of the five convicts. What do these deviations indicate about the sentence lengths received by the convicts?

20. Referring to the hourly wage data given in Problem 15, calculate the deviations (from the mean) for each of the seven employees. What do these deviations indicate about the wages earned by the employees?

21. The following scores represent the number of children in a group of 20 households:

    2   0   2   1   5   3   4   0   1   1   2   1   3   2   4   6   3   2   0   2

    a. Calculate the mode, median, and mean from the scores.
    b. Rearrange the scores into a simple frequency distribution and recalculate the mode, median, and mean.

22. A particular company that produced Christmas toys organized its factory workers into four different groups to compare the productivity of four different manufacturing protocols. As shown in what follows, measures of productivity over a period of one month indicated that the workers in Protocol 4 were the most productive of workers in any of the four work groups (higher scores indicate the number of Christmas toys assembled). Using the weighted mean, determine the overall mean productivity for all four work groups combined.

    $$\text{Protocol 1: } \overline{X}_1 = 20; \quad N_1 = 10$$
    $$\text{Protocol 2: } \overline{X}_2 = 15; \quad N_2 = 14$$
    $$\text{Protocol 3: } \overline{X}_3 = 18; \quad N_3 = 15$$
    $$\text{Protocol 4: } \overline{X}_4 = 22; \quad N_4 = 8$$

23. The scores of attitudes toward older people for 30 students were arranged in the following simple frequency distribution (higher scores indicate more favorable attitudes toward older people):

Attitude Score Value	f
7	3
6	4
5	6
4	7
3	5
2	4
1	1
	$N = 30$

Find (a) the mode, (b) the median, and (c) the mean.

24. The following frequency distribution contains the social class of 46 people who voted Republican in a recent election.

Social Class	f
Upper	17
Upper middle	12
Middle middle	6
Lower middle	7
Lower	4
	$N - 46$

Given the level at which social class was measured (nominal, ordinal, or interval), calculate the appropriate measures of central tendency for describing the distribution given.

25. A team of psychologists studying narcolepsy (a sleep disorder characterized by sudden nodding off during the day) decided to follow 20 narcoleptic volunteers throughout the course of one day and record how many times they each fell asleep. The data they collected have been arranged in the following simple frequency distribution:

Number of Times Asleep	f
7	2
6	1
5	3
4	5
3	3
2	4
1	2
	$N = 20$

Find (a) the mode, (b) the median, and (c) the mean.

26. The simple frequency distribution below shows the scores from 40 people who were asked to rate on a scale from 1 to 7 their attitudes toward stem cell research (with 1 being the most favorable attitude toward stem cell research and 7 being the least favorable).

Attitude toward Stem Cell Research	f
7	8
6	5
5	7
4	6
3	4
2	3
1	7
	$N = 40$

Find (a) the mode, (b) the median, and (c) the mean. What must you assume about the scale for the mean to be meaningful?

27. The scores on a religiosity scale (higher scores indicate greater commitment to religious expression) were obtained for 46 adults. For the following simple frequency distribution, calculate (a) the mode, (b) the median, and (c) the mean:

Score Value	f
10	3
9	4
8	6
7	8
6	9
5	7
4	5
3	2
2	1
1	1
	$N = 46$

28. A focus group of 10 adults was chosen to evaluate the performance of a candidate during a presidential debate. They were instructed to rate the candidate on two characteristics, knowledge and delivery, using a scale of 1 (poor) to 10 (superior). The ratings given the candidate were as follows:

Rater	Knowledge Rating	Delivery Rating
A	7	5
B	8	6

C	9	5
D	3	5
E	5	6
F	8	7
G	9	6
H	4	5
I	8	6
J	7	5

a. Find the mode, the median, and the mean for the knowledge ratings.
b. Find the mode, the median, and the mean for the delivery ratings.
c. On which characteristic was the candidate rated more favorably?

29. Find the mean, median, and mode for the following set of suicide rates (number of suicides per 100,000 population), representing six large metropolitan areas:

Metropolitan Area	Suicide Rate
A	15.2
B	13.7
C	13.0
D	18.5
E	20.6
F	13.9

30. The following frequency distribution contains the educational levels attained by the 39 office personnel of a small company:

Educational Level	f
Graduate school	3
Completed college	5
Some college	11
Completed high school	14
Some high school	6
	$N = 39$

Given the level at which educational level was measured (nominal, ordinal, or interval/ratio), calculate the measures of central tendency appropriate for describing the distribution given.

SPSS Exercises

1. How often does an average high school senior get a speeding ticket or get stopped and warned for a moving violation? Using SPSS to analyze the Monitoring the Future Survey, find the mode, median, and mean for traffic violations (V197). Hint: Open the STATISTICS option within the FREQUENCIES procedure to obtain the mode, median, and mean for selected variables. Verify that the data set is weighted by V5 by checking DATA and WEIGHT CASES.

2. How often do high school seniors get hurt so badly in a fight, assault, or auto accident that they have to go to the doctor? Use the Monitoring the Future Study to find the median and mode for V1734. Should you calculate a mean for this variable? Why or why not?

3. How much money does a high school senior earn during an average week from a job or other work? Use SPSS to analyze self-reported income from a job (V192). What is the level of measurement for this variable? Which measures of central tendency are appropriate for this kind of variable? Use SPSS to find appropriate measures. (Hint: You should get SPSS to calculate only two of these measures of central tendency.)

4. What is the average commute time to work in American metropolitan statistical areas? Using SPSS and the Best Places data set, analyze commute time (COMMUTE). What is the level of measurement for this variable? Which measures of central tendency are appropriate for this kind of variable? Use SPSS to find them.

5. Using SPSS to analyze the Best Places data set, find the quartiles for the suicide rate per 100,000 persons in American metropolitan statistical areas (SUICIDE).

4

Measures of Variability

In Chapter 3, we saw that the mode, median, and mean could be used to summarize in a single number what is average or typical of a distribution. When employed alone, however, any measure of central tendency yields only an incomplete picture of a set of data and, therefore, can mislead or distort, as well as clarify.

To illustrate this possibility, consider that Honolulu, Hawaii, and Phoenix, Arizona, have almost the same mean daily temperature of 75°F. Can we therefore assume that the temperature is basically alike in both localities? Or is it not possible that one city is better suited for year-round swimming and other outdoor activities?

As shown in Figure 4.1, Honolulu's temperature varies only slightly throughout the year, usually ranging between 70°F and 80°F. By contrast, the temperature in Phoenix can

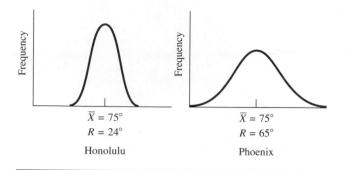

$\overline{X} = 75°$ $\overline{X} = 75°$
$R = 24°$ $R = 65°$

Honolulu   Phoenix

FIGURE 4.1   *Differences in variability:* The distribution of temperature in Honolulu and Phoenix (approximate figures)

differ seasonally from a low of about 40°F in January to a high of over 100°F in July and August. Needless to say, Phoenix's swimming pools are not overcrowded year-round.

Consider another example. Suppose that Judge A and Judge B both average (mean) 24 months in the prison sentences that they hand down to defendants convicted of assault. One could easily be misled by this statistic into thinking that the two judges agree in their philosophies about proper sentencing. Suppose we learn further that Judge A, believing in complete equality in sentencing, gives all defendants convicted of assault 24 months, whereas Judge B gives anywhere from 6 months to 6 years, depending on her assessment of both the defendants' demeanor in court and the nature of the prior criminal record. If you were an attorney maneuvering to have your client's case heard by a particular judge, which judge would you choose?

It can be seen that we need, in addition to a measure of central tendency, an index of how the scores are scattered around the center of the distribution. In a word, we need a measure of what is commonly referred to as *variability* (also known as *spread, width,* or *dispersion*). Returning to an earlier example, we might say that the distribution of temperature in Phoenix has greater variability than the distribution of temperature in Honolulu. In the same way, we can say that the distribution of sentences given by Judge A has less variability than the distribution of sentences given by Judge B. This chapter discusses only the best known measures of variability: the range, the inter-quartile range, the variance, and the standard deviation.

The Range

To get a quick but rough measure of variability, we might find what is known as the *range (R)*, the difference between the highest and lowest scores in a distribution. For instance, if Honolulu's hottest temperature of the year was 89°F and its coldest temperature of the year was 65°F then the range of temperature in Honolulu would be 24°F ($89 - 65 = 24$). If Phoenix's hottest day was 106°F and its coldest day 41°F, the range of temperature in Phoenix would be 65°F ($106 - 41 = 65$). By formula,

$$R = H - L$$

where R = range
        H = highest score in a distribution
        L = lowest score in a distribution

The advantage of the range—its quick and easy computation—is also its most important disadvantage. That is, the range is totally dependent on only two score values: the largest and smallest cases in a given set of data. As a result, the range usually gives merely a crude index of the variability of a distribution.

For example, Professor A has a seminar with eight students ages 18, 18, 19, 19, 20, 20, 22, and 23; Professor B also has a seminar with eight students ages 18, 18, 19, 19, 20, 21, 22, and 43. Professor A comments then that his students range in age from 18 to 23 (or a 5-year range). Professor B boasts, however, that his students range in age from 18 to 43 (or a 25-year range). With the exception of one student, however, the two classes are similar in age distribution. The range is clearly influenced or even distorted by just one student. Any measure that is so much affected by the score of only one case may not give a precise idea as to variability and, for most purposes, should be considered only a preliminary or very rough index.

The Inter-Quartile Range

As we have seen, the range *(R)* is completely dependent on only the two most extreme scores in a distribution, the highest and the lowest. In a very small sample (say, 10 or 15 scores), this fact may not reduce the usefulness of the range as an indicator of variability. But as the sample increases in size into the hundreds or even thousands of cases, the value or size of the range can only remain the same or grow larger. To describe the spread of a large number of scores, then, the range loses much of its reliability and becomes merely a crude measure of variability.

In Chapter 2, we introduced the concept of quartiles, defined as points along a distribution of scores, arranged in order of size, that divide it into quarters. Thus, if a score is located at the first quartile ($Q1$), we know that 25% of the cases fall at or below it; if a score is at the second quartile ($Q2$), 50% of the cases fall at or below it; and if a score is located at the third quartile ($Q3$), 75% of the cases fall at or below it.

Using quartiles rather than the highest and lowest individual score values, we can calculate a measure of variability known as the *inter-quartile range (IQR)* which—unlike R which is dependent on the most extreme scores—is sensitive to the way in which scores are concentrated around the center of the distribution. Like the range, the larger the size of IQR, the greater the spread or variability.

By definition, the inter-quartile range includes the middle 50% of score values in the distribution, when they are arranged in order of size. IQR is obtained by calculating the distance between the first and third quartiles,

$$IQR = Q3 - Q1$$

where $Q1$ is the score value at or below which 25% of the cases fall and $Q3$ is the score value at or below which 75% of the cases fall.

Consider the following set of 20 scores, arranged in order of size, on a college history midterm exam,

94 92 91 88 85 | 84 80 79 77 76 74 74 71 69 65 | 62 56 53 48 40

Q3   Q1

In order to determine the inter-quartile range, you must first locate the *position of Q1* and *Q3*: For the 20 scores above, it is possible simply to arrange the scores from high to low and then count through the array to the points where 25% (after the fifth score) and 75% (after the fifteenth score) of the cases fall.

In more complicated situations, however, it may be helpful to apply the following formulas:

Position of $Q3 = .75(N + 1) = 15.75$

Position of $Q1 = .25(N + 1) = 5.25$

Thus, the third quartile, indicating the point at or below which three-quarters (or 75%) fall, lies between the 15th score (84) and the 16th score (85). *To simplify, any quartile found to be located between score values will be treated as falling half-way or a distance of one-half from adjacent score values.* Thus, $Q3 = 84.50$. At the other extreme, the first quartile, indicating the point at or below which one-quarter (or 25%) falls, lies between the 5th and 6th scores, respectively 62 and 65. Thus, $Q1 = 63.5$. The inter-quartile range then is 21.

$IQR = Q3 - Q1$

$= 84.5 - 63.5$

$= 21$

Let us say that we want to compare the variability of the 20 midterm scores above with the 19 final exam scores for the same history class (one of the 20 students—he got a 49 on the midterm—dropped the course prior to the final, leaving only 19 scores). We ask: Were the students' scores on the final exam concentrated more than the midterm grades around the center of the distribution? Did the students' exam scores spread more around the center on the midterm than on the final? To find out, we calculate the inter-quartile range for the 19 final exam scores,

94 90 87 85 | 82 80 80 79 78 78 77 76 74 73 72 | 71 67 56 40

Q3   Q1

In this case,

Position of $Q3 = .75(19 + 1) = 15$

Position of $Q1 = .25(19 + 1) = 5$

$Q3$ and $Q1$ are located at the following scores. Therefore,

$IQR = Q3 - Q1$

$= 82 - 72$

$= 10$

Because the highest and lowest scores on both exams—midterm and final—do not differ (they are 94 and 40 on both exams), the range ($R = 54$) is the same for the midterm and the final scores, falsely indicating no difference in variability or spread. The inter-quartile range tells a different story. It indicates that the middle 50% of scores on the midterm ($IQR = 21$) is more widely spread than the middle 50% of scores on the final ($IQR = 10$). We can conclude from this comparison of inter-quartile ranges that the variability was greater on the midterm than on the final exam. Given its lack of reliability, colleges rarely, if ever, report the range (R) of SAT scores for students in their freshman class. They (along with *U.S. News and World Report*) are more likely to report the middle 50%.

The Variance and the Standard Deviation

In the previous chapter, the concept of deviation was defined as the distance of any given raw score from its mean. To find a deviation, we were told to subtract the mean from any raw score ($X - \overline{X}$). If we now wish to obtain a measure of variability that takes into account every score in a distribution (rather than only two score values), we might be tempted to add together all the deviations. However, the sum of actual deviations, $\Sigma(X - \overline{X})$, is always zero. Plus and minus deviations cancel themselves out and therefore cannot be used to describe or compare the variability of distributions.

To overcome this problem, we might square the actual deviations from the mean and add them together $\Sigma(X - \overline{X})^2$. As illustrated in Table 4.1, using the number of weeks that six jobless people have received unemployment benefits, this procedure would get rid of minus signs, because squared numbers are always positive.

Having added the squared deviations from the mean, we might divide this sum by N in order to control for the number of scores involved. This is the mean of the squared deviations, but it is better known as the *variance*. Symbolized by s^2, the variance is

$$s^2 = \frac{\Sigma(X - \overline{X})^2}{N}$$

where
$$s^2 = \text{variance}$$
$$\Sigma(X - \overline{X})^2 = \text{sum of the squared deviations from the mean}$$
$$N = \text{total number of scores}$$

TABLE 4.1 *Squaring Deviations* (with $\overline{X} = 5$)

X	$X - \overline{X}$	$(X - \overline{X})^2$
9	+4	16
8	+3	9
6	+1	1
4	−1	1
2	−3	9
1	−4	16
	0	$\Sigma(X - \overline{X})^2 = 52$

Continuing with the illustration in Table 4.1, we see that the variance is

$$s^2 = \frac{\Sigma(X - \overline{X})^2}{N}$$

$$= \frac{52}{6}$$

$$= 8.67$$

One further problem arises, however. As a direct result of having squared the deviations, the unit of measurement is altered, making the variance rather difficult to interpret. The variance is 8.67, but 8.67 of what? The variance is expressed as the square of whatever unit expresses our data. In this case, we would have squared weeks!

To put the measure of variability into the right perspective, that is, to return to our original unit of measurement, we take the square root of the variance. This gives us the *standard deviation*, a measure of variability that we obtain by summing the squared deviations from the mean, dividing by N and then taking the square root. Symbolized by s, the standard deviation is

$$s = \sqrt{\frac{\Sigma(X - \overline{X})^2}{N}}$$

where

$$s = \text{standard deviation}$$
$$\Sigma(X - \overline{X})^2 = \text{sum of the squared deviations from the mean}$$
$$N = \text{the total number of scores}$$

BOX 4.1 • *Step-by-Step Illustration: Standard Deviation*

With reference to the weeks-of-unemployment data, the following steps are carried out to calculate the standard deviation:

Step 1   Find the mean for the distribution.

X	
9	$\overline{X} = \dfrac{\Sigma X}{N}$
8	
6	
4	$= \dfrac{30}{6}$
2	
1	$= 5$
$\Sigma X = \overline{30}$	

Step 2   Subtract the mean from each raw score to get the deviation.

X	$X - \bar{X}$
9	+4
8	+3
6	+1
4	−1
2	−3
1	−4

Step 3   Square each deviation before adding together the squared deviations.

X	$X - \bar{X}$	$(X - \bar{X})^2$
9	+4	16
8	+3	9
6	+1	1
4	−1	1
2	−3	9
1	−4	16
		$\Sigma(X - \bar{X})^2 = 52$

Step 4   Divide by N to get the square root of the result.

$$s = \sqrt{\frac{\Sigma(X - \bar{X})^2}{N}}$$

$$= \sqrt{\frac{52}{6}}$$

$$= \sqrt{8.67}$$

$$= 2.94$$

We can now say that the standard deviation is 2.94 weeks for the six unemployment recipients. On average, that is, the scores in this distribution deviate from the mean by nearly 3 weeks. For example, the 2 in this distribution is below the mean, but only by an average amount.

The Raw-Score Formula for Variance and Standard Deviation

Until now, we have used deviations $(X - \bar{X})$ to obtain the variance and standard deviation. There is an easier method for computing these statistics, especially with the help of a calculator. This method works directly with the raw scores. The raw-score formulas for variance and standard deviation are

$$s^2 = \frac{\Sigma X^2}{N} - \bar{X}^2$$

$$s = \sqrt{\frac{\Sigma X^2}{N} - \bar{X}^2}$$

where ΣX^2 = sum of the squared raw scores. (*Important:* Each raw score is *first* squared and then these squared raw scores are summed.)
N = total number of scores
$\bar{X}^2$ = mean squared

BOX 4.2 • *Step-by-Step Illustration: Variance and Standard Deviation Using Raw Scores*

The step-by-step procedure for computing s^2 and s by the raw-score method can be illustrated by returning to the weeks on unemployment data: 9, 8, 6, 4, 2, and 1.

Step 1   Square each raw score before adding together the squared raw scores.

X	X^2
9	81
8	64
6	36
4	16
2	4
1	1
	$\Sigma X^2 = 202$

Step 2   Obtain the mean and square it.

X
9
8
6
4
2
1
$\Sigma X = 30$

$\bar{X} = \dfrac{\Sigma X}{N} = \dfrac{30}{6} = 5$

$\bar{X}^2 = 25$

Step 3   Insert the results from Steps 1 and 2 into the formulas.

$$s^2 = \frac{\Sigma X^2}{N} - \overline{X}^2 \qquad s = \sqrt{\frac{\Sigma X^2}{N} - \overline{X}^2}$$

$$= \frac{202}{6} - 25 \qquad = \sqrt{\frac{202}{6} - 25}$$

$$= 33.67 - 25 \qquad = \sqrt{8.67}$$

$$= 8.67 \qquad = 2.94$$

As Step 3 shows, applying the raw-score formulas to the weeks-on-unemployment data yields exactly the same results as the original method, which worked with deviations.

Obtaining the Variance and Standard Deviation from a Simple Frequency Distribution

In the last chapter, we saw how measures of central tendency could be calculated from a set of scores rearranged in the form of a simple frequency distribution. The variance and standard deviation can be obtained in a similar fashion.

Let's return to the following raw scores representing the age at first marriage of a sample of 25 adults:

18 18 19 19 19 19 20 20 20 21 21 22
22 23 23 24 25 26 26 26 27 27 29 30 31

Calculated from these raw scores, the variance $s^2 = 14.56$ and the standard deviation $s = 3.82$.

These data can be rearranged as a simple frequency distribution as follows:

X	f	X	f
31	1	24	1
30	1	23	2
29	1	22	2
28	0	21	2
27	2	20	3
26	3	19	4
25	1	18	2

To obtain the variance and standard deviation from a simple frequency distribution, we apply the following formulas:

$$s^2 = \frac{\Sigma f X^2}{N} - \overline{X}^2$$

$$s = \sqrt{\frac{\Sigma f X^2}{N} - \overline{X}^2}$$

where X = a score value
 f = a score frequency
 N = number of cases
 $\overline{X}$ = mean of the simple frequency distribution

BOX 4.3 • *Step-by-Step Illustration: Variance and Standard Deviation of a Simple Frequency Distribution*

To obtain the variance and standard deviation, we must first calculate the mean using the steps outlined in the previous chapter.

Step 1   Multiply each score value *(X)* by its frequency *(f)* to obtain the *fX* products, and then sum the *fX* column.

X	f	fX
31	1	31
30	1	30
29	1	29
28	0	0
27	2	54
26	3	78
25	1	25
24	1	24
23	2	46
22	2	44
21	2	42
20	3	60
19	4	76
18	2	36
		$\Sigma fX = 575$

Step 2 Square each score value (X^2) and multiply by its frequency (f) to obtain the fX^2 products, and then sum the fX^2 column.

X	f	fX	fX^2
31	1	31	961
30	1	30	900
29	1	29	841
28	0	0	0
27	2	54	1,458
26	3	78	2,028
25	1	25	625
24	1	24	576
23	2	46	1,058
22	2	44	968
21	2	42	882
20	3	60	1,200
19	4	76	1,444
18	2	36	648
			$\Sigma fX^2 = 13{,}589$

Step 3 Obtain the mean and square it.

$$X = \frac{\Sigma fX}{N} = \frac{575}{25} = 23$$
$$\overline{X}^2 - (23)^2 = 529$$

Step 4 Calculate the variance using the results from the previous steps.

$$s^2 = \frac{\Sigma fX^2}{N} - \overline{X}^2$$
$$= \frac{13{,}589}{25} - 529$$
$$= 543.56 - 529$$
$$= 14.56$$

(continued)

BOX 4.3   Continued

Step 5   Calculate the standard deviation (the square root of the variance).

$$s = \sqrt{\frac{\Sigma f X^2}{N} - \bar{X}^2}$$

$$= \sqrt{\frac{13{,}589}{25} - 529}$$

$$= \sqrt{543.56 - 529}$$

$$= \sqrt{14.56}$$

$$= 3.82$$

Finally, note that the variance and standard deviation calculated from the simple frequency distribution are identical to the values obtained from the raw scores.

The Meaning of the Standard Deviation

We noted earlier that the standard deviation is more interpretable than the variance because it is in the correct unit of measurement. Even so, the series of steps required to compute the standard deviation can leave the reader with an uneasy feeling as to the meaning of his or her result. For example, suppose we learn that $s = 4$ in a particular distribution of scores. What is indicated by this number? Exactly what can we say now about that distribution that we could not have said before?

Chapter 5 will seek to clarify the full meaning of the standard deviation. For now, we note briefly that the standard deviation represents the "average" variability in a distribution, because it measures the average of deviations from the mean. The procedures of squaring and taking the square root also enter the picture, but chiefly to eliminate minus signs and return to the more convenient unit of measurement, the raw-score unit.

We note also that the greater the variability around the mean of a distribution, the larger the standard deviation. Thus, $s = 4.5$ indicates greater variability than $s = 2.5$ For instance, the distribution of daily temperatures in Phoenix, Arizona, has a larger standard deviation than does the distribution of temperatures for the same period in Honolulu, Hawaii.

Let's also reconsider the case of prison sentencing that we encountered earlier to see the importance of variance and standard deviation for understanding and interpreting distributions. Table 4.2 displays the sentences given two sets of six defendants in robbery trials by the respective judges. Note first the advantage of the standard deviation over the variance. Even though they are equal in their abilities to measure variability or dispersion, the standard deviation has a more tangible interpretation. In this case, the standard deviation is expressed in terms of months—something that has meaning to us. The variance, however, is stated in terms of months squared, which renders the variance more difficult to understand.

Returning to a comparison of the two judges, we see that Judge A has a larger mean yet a smaller variance and standard deviation than Judge B. One might say, at least on the

TABLE 4.2 *Sentences in Months for Robbery Given by Two Judges*

Judge A	Judge B
34	26
30	43
31	22
33	35
36	20
34	34
$\overline{X} = 33.0$	$\overline{X} = 30.0$
$s^2 = 4.0$	$s^2 = 65.0$
$s = 2.0$	$s = 8.1$

basis of these data alone, that Judge A is harsher but fairer, and Judge B is more lenient but inconsistent. For an attorney, your best bet might be Judge A. Even though you can expect a longer sentence (because of the higher mean), you may not want to risk the severe sentences that Judge B has been known to give.

Now let's add another piece to the puzzle. The highly variable sentences of Judge B may not be so unreasonable as they may seem. The long sentences were given to offenders with long criminal records and the short sentences to first- and second-time offenders. (We will consider techniques for measuring the sources of variability in later chapters.) As an attorney, your preference for a judge would depend, therefore, on the criminal history of your client. If he had a history of minor offenses or no criminal history at all, you would prefer Judge B, because you would expect a shorter sentence from her than from Judge A, who now seems rather inflexible. On the other hand, if representing a repeat offender, you would prefer Judge A, because she seems to focus less on the background of the offender than on the current charge.

Thus, the standard deviation is a useful device for measuring the degree of variability in a distribution or for comparing the variability in different distributions. It is also employed, and quite frequently, for calibrating the relative standing of individual scores within a distribution. The standard deviation in this sense is the standard against which we assess the placement of one score (such as your examination score) within the whole distribution (such as the examination scores of the entire class).

To understand this meaning of the standard deviation, consider first an analogy to the placement of a plant in a room. If we wish to discuss the distance of a plant from a living room wall, we might think in terms of feet as a unit of measurement of distance (for example, "The plant in the living room is located a distance of 5 feet from this wall"). But how do we measure the width of a base line of a frequency polygon that contains the scores of a group of respondents arranged from low to high (in ascending order)? As a related matter, how do we come up with a method to find the distance between any raw score and the mean—a standardized method that permits comparisons between raw scores in the same distribution as well as between different distributions? If we were talking about plants, we might find that one plant is 5 feet from the living room wall, and another plant is 10 feet from the kitchen wall. In the concept of feet, we have a standard unit of measurement and, therefore, we can make such comparisons in a meaningful way. But how about comparing raw scores? For instance, can we always compare 85 on an English exam with 80 in German? Which grade is really higher? A little thought will show that it depends on how the other students in each class performed.

One method for giving a rough indicator of the width of a base line is the range, because it gives the distance between the highest and lowest scores along the base line. But the range cannot be effectively used to locate a score relative to the mean, because—aside from its other weaknesses—the range covers the entire width of the base line. By contrast, the size of the standard deviation is smaller than that of the range and covers far less than the entire width of the base line.

Just as we "lay off" a carpet in feet or yards, so we might lay off the base line in units of standard deviation. For instance, we might add the standard deviation to the value of the mean to find which raw score is located exactly one standard deviation above the mean. As shown in Figure 4.2, therefore, if $\overline{X} = 80$ and $s = 5$, then the raw score 85 lies exactly one standard deviation *above* the mean $(80 + 5 = 85)$, a distance of $+1s$. This direction is *plus* because all deviations above the mean are positive; all deviations below the mean are *minus,* or negative.

We continue laying off the base line by adding the value of the standard deviation to the raw score 85. This procedure gives us the raw score 90, which lies exactly two standard deviations above the mean $(85 + 5 = 90)$. Likewise, we add the standard deviation to the raw score 90 and obtain 95, which represents the raw score falling exactly three standard deviations from the mean: We subtract 5 from 80, 5 from 75, and 5 from 70 to obtain $-1s$, $-2s$, and $-3s$, respectively.

The process of laying off the base line in units of standard deviation is in many respects similar to measuring the distance between a plant and the wall in units of feet. However, the

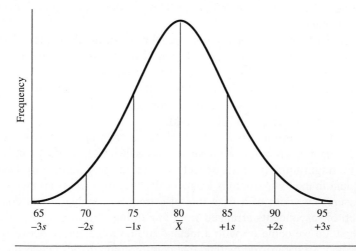

FIGURE 4.2 Measuring the base line in units of standard deviation when the standard deviation (*s*) is 5 and the mean $(\overline{X})$ is 80

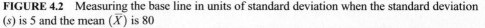

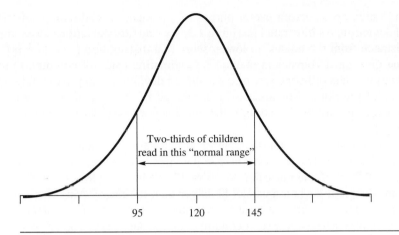

FIGURE 4.3   Distribution of reading speed

analogy breaks down in at least one important respect: Whereas feet and yards are of constant size (1 ft always equals 12 in; 1 yd always equals 3 ft), the value of the standard deviation varies from distribution to distribution. Otherwise, we could not use the standard deviation as previously illustrated to compare the variability of distributions (for example, the judges in Table 4.2). For this reason, we must calculate the size of the standard deviation for any distribution with which we happen to be working. Also as a result, it is usually more difficult to understand the standard deviation as opposed to feet or yards as a unit of measurement.

As we will see in detail in the next chapter, the standard deviation has a very important meaning for interpreting scores in what we call the *normal distribution*. Actually, unless a distribution is highly skewed, approximately two-thirds of its scores fall within one standard deviation above and below the mean. Sometimes this range is called the *normal range* because it contains cases that are considered close to the norm. For example, the child whose reading score falls precisely at the mean is, in a strict sense, average; but children whose reading speeds are close to the mean in either direction, more specifically within one standard deviation of the mean, also are regarded as being within normal range. Thus, if for a given age group, the mean reading speed is 120 words per minute with a standard deviation of 25 words per minute, then the normal range may be defined as 95–145 words per minute, and approximately two-thirds of children in this age group read at a speed that lies within the normal range (see Figure 4.3). We will return to this concept of the standard deviation in Chapter 5.

Comparing Measures of Variability

The range is regarded generally as a preliminary or rough index of the variability of a distribution. It is quick and simple to obtain but not very reliable, and it can be applied to interval or ordinal data. The inter-quartile range is similarly determined by only two points in the distribution, but has the advantage of representing the middle rather than the extreme scores in a data set.

The range does serve a useful purpose in connection with computations of the standard deviation. As illustrated in Figure 4.3, six standard deviations cover almost the entire distance from the highest to lowest score in a distribution ($-3s$ to $+3s$). to This fact alone gives us a convenient method for estimating (but not computing) the standard deviation. Generally, the size of the standard deviation is approximately one-sixth of the size of the range. For instance, if the range is 36, then the standard deviation might be expected to fall close to 6; if the range is 6, the standard deviation will likely be close to 1.

This rule can take on considerable importance for the reader who wishes to find out whether her or his result is anywhere in the vicinity of being correct. To take an extreme case, if $R = 10$ and our calculated $s = 12$, we have made some sort of an error, because the standard deviation cannot be larger than the range. A note of caution: The one-sixth rule applies when we have a large number of scores. For a small number of cases, there will generally be a smaller number of standard deviations to cover the range of a distribution.

Whereas the range and inter-quartile range are calculated from only two score values, both the variance and the standard deviation take into account every score value in a distribution. The standard deviation has become the initial step for obtaining certain other statistical measures, especially in the context of statistical decision making. We shall be exploring this characteristic of the standard deviation in detail in subsequent chapters.

Despite its usefulness as a reliable measure of variability, the standard deviation also has its drawbacks. As compared with other measures of variability, the standard deviation tends to be difficult and time consuming to calculate. However, this disadvantage is being more and more overcome by the use of calculators and computers to perform statistical analyses. The variance and the standard deviation also have the characteristic of being interval-level measures and, therefore, cannot be used with nominal or ordinal data—the type of data with which many social researchers often work.

Visualizing Distributions

When we attempt to describe the appearance of an acquaintance, we tend to focus on such features as height, weight, age, and hair length. Similarly, when describing distributions of data, social researchers tend to indicate their central tendency, dispersion, skewness, or other characteristics. Yet, it is said that a picture is worth a thousand words. A photograph of a person is far more useful in making an impression concerning appearance than is a list of attributes. In the same way, a graphic representation of a distribution serves far better than a list of statistics.

In recent years, the box plot has become a popular graphic technique for simultaneously displaying a number of aspects about a distribution. Consider, for example, the following collection of "hold times" (in minutes) in response to an experiment designed to measure patience. In the study, subjects call a toll-free number given in a newspaper advertisement for a too good to be true bargain on a new high-definition television.

Those inquiring about the "sale" are placed on hold indefinitely by the switchboard operator, until they hang up in frustration.

3   7   2   1   4   3   8   3   5   7   4   5   6   7   6   5   4   3   9   5

Using the techniques and formulas learned up to this point, we can calculate that the mean waiting time is 4.85 minutes, whereas the median is 5.0 minutes. The closeness of these two central tendency measures would suggest a distribution that is fairly symmetrical. In terms of variability, the holding times range from a low of 1 minute to a high of 9 minutes. The standard deviation is 2.03 minutes, indicating that approximately two-thirds of the callers held for 2 minutes of the mean (or 2.82 to 6.88).

All these details can be displayed in the box plot shown in Figure 4.4. Although the exact design is a matter of taste and preference (as is true of most artwork), this plot shows the range with a line segment extending from the minimum to the maximum. The rectangular box within the plot represents the interval between one standard deviation above and one standard deviation below the mean. Finally, the horizontal line through the box shows the median, and the circle in the middle of the box is the mean. (Other measures, such as the first and third quartiles can also be used to form the box, rather than the standard deviations.)

When we first encountered graphs in Chapter 2, bar graphs were employed to compare two distributions (for example, seat belt usage by males and females). That was fine for an ordinal variable such as seat belt usage, but with an interval measure such as holding times, we can do better. Just as we might compare the photos of two people side by side, we can display the box plots of two groups side by side to obtain a fuller understanding of group differences. Figure 4.5 shows the following breakdown of holding times for males and females:

Males: 5   2   7   9   3   4   3   1   3   8
Females: 3   5   7   4   5   6   7   6   5   4

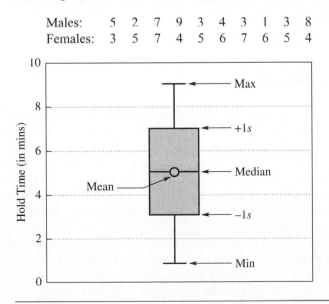

FIGURE 4.4   Box plot of hold time distribution

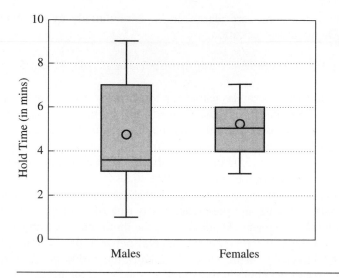

FIGURE 4.5   Box plot of holding times for males and females

We see that, on average, the women were more patient in holding, in terms of both mean and median times. The men, however, were far more diverse in their levels of patience. That is, the box for the males, representing the interval between the first standard deviation on either side of the mean, is far longer than that for the females. In addition, the range of hold times for the males is wider than that for the females. We can also see, based on the relative positions of the mean and median, that the hold time distribution for males is somewhat skewed, whereas that for the females is not.

Obtaining the Mean and Standard Deviation from a Grouped Frequency Distribution

This and the previous chapter place heavy emphasis on calculating the mean and standard deviation for interval-level data, whether the scores are presented in raw form or arranged conveniently and compactly in a simple frequency distribution. Calculations for important measures of central tendency and dispersion could be performed by working directly with the collection of N cases (in the instance of using raw data) or by working with the score values and their associated frequencies (in the instance of a simple frequency distribution). Regardless of which approach is used, the calculated values for $\overline{X}$ and s are the same.

Recall from Chapter 2 that often times social researchers present data in the form of a grouped frequency distribution in order to summarize and simplify a large collection of score values. It is wise in such instances to calculate the mean and standard deviation based on the original raw-score data, as the grouping process blurs the distinction among score values contained in the same class interval.

There are occasions, however, when a social researcher only has a grouped frequency distribution available so that it is not possible to resort to the precise raw scores for calculating the mean and standard deviation. For example, surveys involving such sensitive questions as respondent age and income typically offer a set of categorical response choices (that is, various age and income ranges), so as not to ask the respondent to reveal a precise age or income. Or, you might encounter a grouped distribution presented in a journal article or government report. In such cases for which the actual score values are not available, how can the mean and standard deviation be determined with only a grouped frequency distribution available?

The mean and standard deviation can be approximated from a grouped frequency distribution by adapting the formulas for a simple frequency distribution in replacing the score values (X) with the class-interval midpoints (m). Specifically, for grouped data,

$$\bar{X} = \frac{\Sigma fm}{N}$$

$$s = \sqrt{\frac{\Sigma fm^2}{N} - \bar{X}^2}$$

To illustrate, suppose that a survey of the ages of 100 expectant mothers uses a set of age-range responses and yields the following sample data:

Age of mother	f
15 up to 20	7
20 up to 25	36
25 up to 30	23
30 up to 35	18
35 up to 40	10
40 up to 45	6
	100

The necessary calculations for finding the mean and standard deviation for age of the respondents can be obtained using the following table

Age of mother	m	f	fm	fm^2
15 up to 20	17.5	7	122.5	2,143.75
20 up to 25	22.5	36	810	18,225.00
25 up to 30	27.5	23	632.5	17,393.75
30 up to 35	32.5	18	585	19,012.50
35 up to 40	37.5	10	375	14,062.50
40 up to 45	42.5	6	255	10,837.50
		100	2,780	81,675.00

By formula,

$$\overline{X} = \frac{\Sigma fm}{N}$$

$$= \frac{2{,}780}{100}$$

$$= 27.8$$

and

$$s = \sqrt{\frac{\Sigma fm^2}{N} - \overline{X}^2}$$

$$= \sqrt{\frac{81{,}675}{100} - (27.8)^2}$$

$$= \sqrt{816.75 - 772.84}$$

$$= \sqrt{43.91}$$

$$= 6.63$$

Finally, it is important to remember that these are only approximations, because the midpoints serve as proxies or substitutes for all the score values in their respective class intervals. Had the researcher had available the precise ages of the women in the study, it would have been preferable to use the usual formulas for raw scores. However, in most cases, this approximation based on a group frequency distribution produces rather good results.

Summary

In Chapter 3, we discussed measures of how scores cluster around the center of a distribution. Notwithstanding the immense value of the mode, median, and mean, there is much more to learn about a distribution than just central tendency. In this chapter, we introduced three measures of variability—the range (including the inter-quartile range), variance, and standard deviation—for indicating how the scores are scattered around the center of a distribution. Just as was true of indicators of central tendency, each measure of variability has its particular weaknesses and strengths. The range is a quick but very rough indicator of variability, which can be easily determined by taking the difference between the highest and lowest scores in a distribution. The inter-quartile range, however, measures the spread of the middle half of the distribution, that is, the range between the first and third quartiles. A more precise measure of the spread of scores can be found in the variance, the mean of the squared deviations from the mean. The variance and its square root known as the standard deviation are two reliable, interval-level measures of variability which can be employed for more advanced descriptive and decision-making operations. In addition to its ability to compare the variability in different distributions, for example, the standard deviation is useful for calibrating the relative standing of individual scores within a distribution. We actually lay off the base line of a distribution in units of standard deviation in order to determine how far any particular score falls from the mean.

The full meaning of the standard deviation will be explored in subsequent discussions of generalizing from a sample to a population.

Terms to Remember

Variability   Variance
Range   Standard deviation
Inter-quartile range   Box plot

Questions and Problems

1. A measure of how scores scatter around the center of a distribution is
   a. variance.
   b. standard deviation.
   c. range.
   d. All of the above

2. Which of the following statements is most true about the relative size of measures of variability?
   a. The standard deviation is usually larger than the variance.
   b. The variance is usually larger than the standard deviation.
   c. The variance and the standard deviation are equal in size.
   d. The inter-quartile range is usually larger than the range.

3. The greater the variability around the mean of a distribution, the larger the
   a. range.
   b. inter-quartile range.
   c. variance.
   d. All of the above

4. How many standard deviations tend to cover the entire range of scores in a distribution?
   a. 2   b. 4   c. 6   d. 8

5. The so-called normal range within which approximately two-thirds of all scores fall is located
   a. within one standard deviation above and below the mean.
   b. between the highest and lowest scores in a distribution.
   c. two standard deviations above the mean.
   d. close to the value of the inter-quartile range.

6. The variance and standard deviation assume
   a. nominal data.
   b. ordinal data.
   c. interval data.
   d. a normal distribution.

7. Two students in a math class compared their scores on a series of five quizzes:

Student A	Student B
4	6
9	5
3	7
8	5
9	6

Considering the concepts of both central tendency and variability, find (a) which student tended to perform better on the quizzes and (b) which student tended to perform more consistently on the quizzes.

8. On a scale designed to measure attitude toward immigration, two college classes scored as follows:

Class A	Class B
4	4
6	3
2	2
1	1
1	4
1	2

Compare the variability of attitudes toward immigration among the members of the two classes by calculating for each class (a) the range, (b) the inter-quartile range, and (c) the standard deviation. Which class has greater variability of attitude scores?

9. A researcher interested in the effectiveness of organized diet groups on weight loss weighed five clients after several weeks on the program. The weight-loss scores (in pounds) were as follows:

   13   12   6   9   10

Calculate the (a) range, (b) inter-quartile range, and (c) variance and standard deviation for these weight-loss scores.

10. A focus group of 10 adults was chosen to evaluate the performance of a candidate during a presidential debate. Each member of the group rated the overall performance on a scale of 1 (poor) to 10 (superior). The ratings given the candidate were as follows:

    4   5   8   7   9   8   7   3   6   7

Calculate the (a) range, (b) inter-quartile range, and (c) variance and standard deviation for these rating scores.

11. A group of high school students was surveyed about their use of various drugs, including alcohol. Asked how frequently they had been drunk in the previous six months, the students responded:

    4   2   0   2   1   3   0   1   7   5   3

Calculate the (a) range, (b) inter-quartile range, and (c) variance and standard deviation for these self-report scores.

12. On a measure of authoritarianism (higher scores reflect greater tendency toward prejudice, ethnocentrism, and submission to authority), seven students scored as follows:

    1   6   6   3   7   4   10

Calculate the (a) range, (b) inter-quartile range, and (c) variance and standard deviation for these authoritarianism scores.

13. On a 20-item measure of self-esteem (higher scores reflect greater self-esteem), five teenagers scored as follows:

    16   5   18   9   11

    Calculate the (a) range, (b) inter-quartile range, and (c) variance and standard deviation for these self-esteem scores.

14. The following are the numbers of hours that 10 police officers have spent being trained in how to handle encounters with people who are mentally ill:

    4   17   12   9   6   10   1   5   9   3

    Calculate the (a) range, (b) inter-quartile range, (c) variance, and (d) standard deviation.

15. A local city is considering building a new train to alleviate the heavy traffic on its highways. To find out how many people would make use of this new train, a government worker went out and interviewed a random sample of commuters. The following are the number of miles driven each week by 10 people who said that they would use the train instead of driving to work:

    150   750   300   425   175   600   450   250   900   275

    Calculate the (a) range, (b) inter-quartile range, and (c) variance and standard deviation.

16. A psychologist interested in nonverbal communication decided to do an informal study on eye contact. Knowing that eye contact tends to be associated with an emotional connection, she believed that most people would be uncomfortable maintaining eye contact for more than a few moments with a total stranger. To test this, she took a walk on a crowded street and counted how many seconds she could maintain eye contact with various strangers before they looked away. Her findings, in seconds, were:

    3.2   2.1   1.7   1.1   2.6   2.2   3.1   1.9   1.7   4.7   2.3   1.6   2.4

    Calculate the (a) range, (b) inter-quartile range, and (c) variance and standard deviation for these lengths of eye contact.

17. In the previous chapter, you were asked to find the mode, median, and mean for the following set of suicide rates (number of suicides per 100,000 population), rounded to the nearest whole number, representing six large metropolitan areas. Now, determine the range, inter-quartile range, variance and standard deviation:

Metropolitan Area	Suicide Rate
A	15
B	13
C	13
D	18
E	20
F	13

18. Find the variance and standard deviation for the following frequency distribution of hours studied during a particular week by the 40 students in a statistics course:

X	f
10	1
9	0
8	2
7	4
6	7
5	11
4	5
3	3
2	4
1	2
0	1
	N = 40

19. Find the standard deviation for the following frequency distribution of the hours of television watched on one Saturday night by a sample of 18 junior high boys:

X	f
5	3
4	5
3	6
2	2
1	2
	N = 18

20. Find the variance and standard deviation for the following frequency distribution of attitudes toward capital punishment held by 25 college students (seven-point scale; higher score indicates more favorable attitude toward capital punishment):

X	f
7	2
6	3
5	5
4	7
3	4
2	3
1	1
	N = 25

21. Find the variance and the standard deviation for the following frequency distribution of ages of juvenile offenders tried and sentenced in adult criminal courts:

X	f
17	12
16	7
15	5
14	4
13	3
12	1
11	2
10	1
	N = 35

22. Find the variance and standard deviation for the following frequency distribution of attitudes toward Internet censorship held by 40 high school teachers (seven-point scale; higher score indicates more favorable attitude toward Internet censorship):

X	f
7	4
6	4
5	7
4	8
3	6
2	5
1	6
	N = 40

SPSS Exercises

1. Using SPSS to analyze the Monitoring the Future Study, examine the dispersion of satisfaction with personal safety (V1643). Hint: Open the STATISTICS option within the FREQUENCIES procedure and check the range and standard deviation for selected variables. Remember to weight cases by V5. For your information, SPSS uses *N-1* rather than *N* when it calculates variance and standard deviation. This slight difference in computation will be explained in more detail in the text in Chapter 6.

2. In the previous chapter, you were asked to find the mode, median, and mean for high school seniors getting a speeding ticket or getting stopped and warned for a moving violation. Now, use SPSS and the Monitoring the Future Survey to find the range, variance, and standard deviation for traffic violations (V197).

3. Using SPSS to analyze the General Social Survey, which measures of central tendency and variability are most appropriate for socioeconomic index (SEI)? Using SPSS, weight the data set by WTSSALL to find the information.

4. Which measures of central tendency and variability are most appropriate for an analysis of respondent's income (RINCOME98)? Use SPSS to analyze the General Social Survey to calculate this information.

5. Use SPSS to analyze a variable of your choice for which you find the most appropriate measures of central tendency and variability.

Looking at the Larger Picture: Describing Data

At the close of Chapter 1, we presented a hypothetical survey of 250 students, grades 9 through 12, in a typical urban high school. The survey questionnaire asked about the use of cigarettes and alcohol, as well as about a variety of other variables. Before we can attempt to understand the factors underlying smoking and drinking, it is necessary to describe the variables in our survey—the extent of smoking and drinking as well as some of the academic, social, and background characteristics. Let's examine, in particular, frequency distributions and bar charts of the primary variables of interest—smoking and drinking.

The extent of smoking was assessed with two questions—whether the respondent smoked during the past month and, if so, how many cigarettes on a typical day. Suppose that 95 of the 250 respondents reported that they had not smoked; and among those who did smoke, the responses were wide ranging, from many who reported one, two, or three cigarettes per day to a few who reported smoking as many as two packs or more per day. Because the number of cigarettes smoked per day has so many possible values, from zero for the nonsmokers on up, it is convenient to use a grouped frequency distribution to present these data.

Daily Cigarette Consumption

Number	f	%
0	95	38.0
1–9	45	18.0
10–19	58	23.2
20–29	33	13.2
30–39	15	6.0
40+	4	1.6
Total	250	100.0

As shown in Figure 4.6, 38% of the students could be considered nonsmokers, not having smoked at all during the past month; 18% could be considered light smokers, consuming less than a half a pack per day; another 23.2% could be called moderate smokers, between a half a pack and

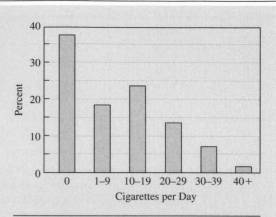

FIGURE 4.6   Smoking among high school students

under a pack per day. The remaining 20.8% (combining the last three categories) could be labeled heavy smokers, having at least a pack per day.

Alcohol use (occasions using alcohol during the month) can be shown in a simple frequency distribution, with values ranging from 0 through 5. Barely more than half the respondents drank on no more than one occasion (20.8% reporting no drinking and 30.8% reporting drinking on one occasion; see Figure 4.7). At the other extreme, only about 5% reported drinking on four or more occasions during the month (or at least once per week).

Alcohol Consumption during Month

Occasions	f	%
0	52	20.8
1	77	30.8
2	82	32.8
3	26	10.4
4	9	3.6
5	4	1.6
Total	250	100.0

The distributions of smoking and drinking can be displayed in the form of a bar graph. The smoking distribution peaks at zero, representing a relatively large group of nonsmokers. As a result, the distribution is extremely skewed to the right.

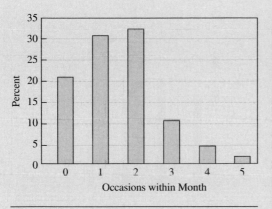

FIGURE 4.7 Drinking among high school students

Unlike the situation with smoking, there does not appear to be such a clearly defined group of non-drinkers to dominate the distribution. Although not perfectly so, the drinking variable is therefore more symmetrical in shape than the smoking distribution. In future analyses, we will look at whether or not a respondent smokes separately from how many cigarettes the smokers consume daily. In other words, we will examine two distinct measures of smoking: smokers versus nonsmokers and the extent of smoking among those who do use cigarettes. By contrast, the analysis of drinking will include all respondents, even those who did not drink during the month.

We can further describe these variables in terms of central tendency and variability. Looking just at the 155 smokers, we see that the mean is 16.9 cigarettes per day, slightly above the mode and median. There is still some slight skewness— a few really heavy smokers at the high end. For the drinking variable, the mean is 1.58 occasions per month, with a median at 1 (once in the month) and mode at 2 (twice in the month).

The standard deviations for daily smoking (for smokers) and drinking occasions are quite dissimilar reflecting very different kinds of distributions. Because of the wider variability in smoking (from just a few cigarettes to over 40 cigarettes) compared to drinking (from a low of 0 occasions to a high of 5), the standard deviation for cigarette consumption is many times greater than that for alcohol. We can also say that roughly two-thirds of the smokers are within one standard deviation of the mean (16.9 ± 10.4 indicating that about two-thirds smoke between 7 and 27 cigarettes daily). For drinking, about two-thirds are included in the range 1.58 ± 1.16. In the next part of the text, we will say a lot more about intervals like these.

We might also attempt to describe differences in smoking and drinking between male and female students. Shown in Figure 4.8, a much higher percentage of the males (47.3%) than the females (28.5%) are nonsmokers, and a higher percentage of the females could be considered heavy smokers. For drinking, the gender differences are reversed. A larger percentage of the females had not had alcohol during the month (26.8% versus 14.2%), whereas the males tended more toward moderate or frequent drinking (see Figure 4.9). In the third part of the text, we will return to this issue, and try to determine if these gender differences are sizable enough to be truly meaningful or indicative of male–female differences among urban high school students in general.

Finally, it is useful to examine descriptive statistics for some of the background variables as well. Because age is measured at the interval level, not only can we create a simply frequency distribution, but we can also calculate all three measures of central tendency. There are more 15-year-olds in the survey than any other age (Mo = 15), but both the median and mean are a bit higher. For sports–exercise participation, both

Central Tendency and Variability Measures

Variable	N	Mean	Mdn	Mo	s^2	s
Smoking (smokers only)	155	16.99	15	15	108.5	10.4
Drinking occasions	250	1.58	1	2	1.35	1.16

(continued)

Continued

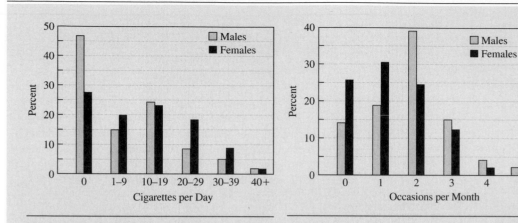

FIGURE 4.8   Smoking among students by gender

FIGURE 4.9   Drinking among students by gender

the mode and median are "Sometimes," whereas the mean should not be calculated since this variable is at the ordinal level. The variable "race" is measured at the nominal level; we can see that the mode is "white," but there can be no median or mean. Last, parental smoking is nominal, with most students not having a parent who smokes.

At this juncture, there is little more that we can do to analyze these and other background characteristics. In the final two parts of the book, however, we shall examine differences in smoking and drinking by race, as well as such things as age and athleticism in their relationship with the smoking and drinking measures.

Descriptive Measures for Background Characteristics

Variable	Group	%	Mean	Median	Mode
Age	19	2.4			
	18	15.6			
	17	20.0			
	16	26.0	16.1	16	16
	15	21.2			
	14	14.8			
Sports/exercise participation	Very freq.	17.2			
	Sometimes	34.8	—	Sometimes	Sometimes
	Seldom	32.4			
	Never	15.6			
Race	White	62.4			
	Black	17.6			
	Latino	12.8	—	—	White
	Asian	4.0			
	Other	3.2			
Parental smoking	Yes	26%			
	No	74%	—	—	No

Part II

From Description to Decision Making

Chapter 5
Probability and the Normal Curve

Chapter 6
Samples and Populations

5

Probability and the Normal Curve

In Part I, we focused on ways to describe variables. In particular, we began to explore variables by concentrating on their distributions—by categorizing data and graphing frequencies. This permitted us to see patterns and trends, to see the most frequent occurrences and the most extreme. We further summarized these distributions by computing measures of central tendency and variability.

Up until now, our interpretations and conclusions about variables have come solely from what we observed. We collected information about a variable and then described what we obtained using a variety of statistical measures, such as percentages and means. From this point on, our approach will be somewhat different. We will first suggest certain theories, propositions, or hypotheses about variables, which will then be tested using the data we observe.

The cornerstone of *decision making*—the process of testing hypotheses through analysis of data—is probability. Probability is a difficult concept to grasp, yet we use it quite frequently. We ask such questions as, "How likely is it that I will get an A on this exam?" "How likely is it that this marriage will last?" "If I draw a card, what is the chance that it will be smaller than a 5?" "What is the chance that this team will win the series?" In everyday conversation, we answer these questions with vague and subjective hunches, such as "probably," "pretty good chance," or "unlikely."

Researchers attempt to answer such questions in a far more precise and scientific way, using the concept of *probability*. There are two types of probability, one based on theoretical mathematics and the other based on systematic observation.

Theoretical probabilities reflect the operation of chance or randomness, along with certain assumptions we make about the events. In the simplest case, we know that the probability of getting a head on a coin flip is .5 under the very reasonable assumption that the coin is fairly weighted so that both sides are equally likely to land face up. Similarly, you may guess the answer to a five-item multiple-choice question for which you haven't a clue with a .20 probability of getting the correct answer. This assumes, of course, that the five responses are equally likely to be the correct answer.

Empirical probabilities are those for which we depend on observation to determine or estimate their values. The probability of giving birth to a boy is about .51 based on long-term demographic data. Although there are two possible outcomes (leaving aside the possibility of twins or triplets), there are slightly more male than female births.

Empirical probabilities are essentially percentages based on a large number of observations. The probability that the home team wins a pro football game is about .6 (6 out of 10 games won by the home team), a "fact" we know from observing hundreds of games over the years. If our percentages are based on large numbers of observations, we can safely treat empirical probabilities as fairly good estimates of the truth.

In both forms, probability *(P)* varies from 0 to 1.0, although sometimes a percentage rather than a decimal is used to express the level of probability. For example, a .50 probability (or 5 chances out of 10) is sometimes called a 50% chance. Although percentages may be used more in everyday language, the decimal form is more appropriate for statistical use.

A zero probability implies that something is impossible; probabilities near zero, such as .01, .05, or .10, imply very unlikely occurrences. At the other extreme, a probability of 1.0 constitutes certainty, and high probabilities such as .90, .95, or .99 signify very probable or likely outcomes.

Some probabilities are easy to calculate: Most people know that the probability of getting heads on a coin flip is .50. However, more complex situations involve the application of various basic rules of probability. Just as we had to learn basic arithmetic operations, so we must learn a few basic operations that will permit us to calculate more complex and interesting probabilities.

Rules of Probability

The term *probability* refers to the relative likelihood of occurrence of any given outcome or event—that is, the probability associated with an event is the number of times that event can occur relative to the total number of times any event can occur:

$$Probability = \frac{Number\ of\ times\ the\ outcome\ or\ event\ can\ occur}{Total\ number\ of\ times\ any\ outcome\ or\ event\ can\ occur}$$

For example, suppose a particular jury consists of five men and seven women. Furthermore, suppose that the court clerk selects the jury foreperson by randomly drawing an index card from a pile of 12, each card with the name of a juror printed on it. Denoting $P(F)$ as the probability that the foreperson will be female,

$$P(F) = \frac{\text{Number of female jurors}}{\text{Total number of jurors}} = \frac{7}{12} = .58$$

The probability of an event not occurring, known as the *converse rule* of probability, is 1 minus the probability of that event occurring. Thus, the probability that the foreperson of the jury is not female, denoted $P(\overline{F})$ where the bar over the F symbolizes "not," or the converse, is

$$P(\overline{F}) = 1 - P(F) = 1 - .58 = .42$$

For another example, suppose that a particular city police department is able to clear (or solve) 60% of its homicides. Thus, the probability of homicide clearance is $P(C) = .60$ (or simply .6). For any particular homicide, say the first one of the year, the probability that it will not be cleared, $P(\overline{C}) = .4$.

An important charactcristic of probability is found in the *addition rule*, which states that the probability of obtaining any one of several different and distinct outcomes equals the sum of their separate probabilities. That is, the probability that either event A or event B occurs is

$$P(A \text{ or } B) = P(A) + P(B)$$

For example, suppose that a defendant in a first-degree murder trial has a .52 probability of being convicted as charged, a .26 probability of being convicted of a lesser charge, and a .22 chance of being found not guilty. The chance of a conviction on any charge is the probability of a conviction on the first-degree murder charge plus the probability of a conviction on a lesser charge, or $.52 + .26 = .78$. Note also that this answer agrees with the converse rule by which the probability of being found guilty of any charge (the converse of not guilty) is $1 - .22 = .78$.

The addition rule always assumes that the outcomes being considered are *mutually exclusive*—that is, no two outcomes can occur simultaneously. More precisely, the occurrence of any particular outcome (say, conviction on the first-degree murder charge) excludes the possibility of any other outcome, and vice versa.

Assuming mutually exclusive outcomes, we can say that the probability associated with all possible outcomes of an event always equals 1. Here, adding the probabilities for the three possible verdicts that the jury may return, we find:

$$P(\text{Guilty as charged}) + P(\text{Guilty lesser charge}) + P(\text{Not guilty})$$
$$= .52 + .26 + .22 = 1$$

This indicates that some outcome must occur: If not a conviction on the charge of murder in the first degree, then either conviction on a lesser crime or an acquittal.

Another important characteristic of probability becomes evident in the *multiplication rule*, which focuses on the possibility of obtaining two or more outcomes in combination. The multiplication rule states that the probability of obtaining a combination of independent outcomes equals the product of their separate probabilities. Thus, the probability that both event A and event B occur is

$$P(\text{A and B}) = P(\text{A}) \times P(\text{B})$$

The assumption of *independent outcomes* means that the occurrence of one outcome does not change the likelihood of the other. Rather than considering the probability of outcome 1 or outcome 2 occurring as in the addition rule, the multiplication rule concerns the probability that both outcome 1 and outcome 2 will occur.

Returning to the city police department with a 60% clearance rate, the probability that any two particular homicides during the year are both cleared is

$$
\begin{aligned}
P(\text{Homicides A and B are both cleared}) &= P(\text{A is cleared}) \times P(\text{B is cleared}) \\
&= (.6)(.6) \\
&= .36
\end{aligned}
$$

Extending it just a bit further, the probability that any three particular homicides are all solved is the product of the three respective probabilities:

$$
\begin{aligned}
P(\text{Homicides A, B, and C are all cleared}) &= P(\text{A is cleared}) \times P(\text{B is cleared}) \\
&\quad \times P(\text{C is cleared}) \\
&= (.6)(.6)(.6) \\
&= .216
\end{aligned}
$$

For this calculation to be valid, we must assume that the three homicides considered are independent of each other. If the three homicides were part of a string of killings believed to have been committed by an unidentified serial killer, then neither the independence assumption nor the calculation itself would be valid.

Finally, the multiplication rule applies more broadly than just to repetitions of similar events such as solving homicides. Rather, the multiplication rule allows us to calculate the joint probability of any number of outcomes so long as they are independent. Suppose that a prosecutor is working on two cases, a robbery trial and a kidnapping trial. From previous experience, she feels that she has a .80 chance of getting a conviction on the robbery and a .70 chance of a conviction on the kidnapping. Thus, the probability that she will get convictions on both cases is $.7 \times .8 = .56$ (slightly better than half).

Probability Distributions

In Part I, we encountered distributions of data in which frequencies and percentages associated with particular values were determined. For example, Table 2.8 shows the frequency distribution of PSAT scores of 336 students. The possible values of the scores are represented

BOX 5.1 • *Practical and Statistical: Overtime Parking*

Hans Zeisel and Harry Kalven Jr. reported on a fascinating use (or rather misuse) of probability in a Swedish trial of a man charged with overtime parking.[1] It seems that a police officer had marked down the directions in which the valves on the curb-side tires of the defendant's car were pointed, using the aviation method of hours on a clock face (e.g., one o'clock, two o'clock, three o'clock, etc.). Let's say that the front tire valve pointed to one o'clock and the rear tire valve to six o'clock. Returning later, the officer found the car still parked, in excess of the posted time limit, and to verify this he noted that the valves still pointed to one and six o'clock. He then issued a ticket for overtime parking.

The owner of the vehicle requested a trial, claiming that he had left the space for awhile and had returned to it later. Confronted with the evidence against him concerning tire position, he argued that it was mere coincidence.

An expert testified in court as to the probability of the two tires being repositioned in the same way after reparking. Reasoning that each tire had a $\frac{1}{12}$ probability of returning to the same position, he calculated that the probability that both tires returned to their former positions by chance would have been $\frac{1}{12} \times \frac{1}{12} = \frac{1}{144}$ (because of the multiplication rule).

The judge apparently felt that a 1 in 144 probability of a coincidence left a reasonable doubt and held in the defendant's favor. He also noted that the outcome may have been different had the officer noted all four tires, since a $\frac{1}{12} \times \frac{1}{12} \times \frac{1}{12} \times \frac{1}{12} = \frac{1}{20,736}$ left little reasonable likelihood of coincidence.

Actually, the judge made the right decision but used the wrong statistical reasoning. To apply the multiplication rule to find the probability of both tires matching their former positions, one must assume that the tires are independent—that is, that they rotate independently. This is not the case, however: the front and back wheels turn in unison. When the front tire moves, the back tire follows (it is only the left and right sides that spin at different rates when the car goes around a corner).

The true probability that both the front and rear tires return to their previous positions is simply the probability that one of them does. In other words, if one tire matches its former position (with a probability of $\frac{1}{12}$), the other must also. Therefore, the probability that both tires would be positioned the same way if the defendant had left and returned to the space is as much as $\frac{1}{12}$ or .0833.

[1] Hans Zeisel and Harry Kalven, Jr., Parking Tickets and Missing Women: Statistics and the Law," in Judith M. Tanur et al., *Statistics: A Guide to the Unknown* (San Francisco, CA: Holden Day, 1978).

by the categories, and the frequencies and percents represent the relative occurrences of the scores among the group.

A *probability distribution* is directly analogous to a frequency distribution, except that it is based on theory (probability theory) rather than on what is observed in the real world (empirical data). In a probability distribution, we specify the possible values of a variable and calculate the probabilities associated with each value. The probabilities represent the likelihood of each value, directly analogous to the percentages in a frequency distribution.

Suppose we were to flip two coins, and let X represent the number of heads we obtain. The variable X has three possible values, 0, 1, and 2, according to whether we obtain zero, one, or two heads. Zero heads ($X = 0$) has a probability of

$$P(0 \text{ heads}) = P(\text{tails on flip 1})P(\text{tails on flip 2})$$
$$= (.50)(.50) = .25$$

We multiply the probability of getting tails on the coins because the flips of the two coins are independent.

Similarly, for two heads ($X = 2$),

$$P(2 \text{ heads}) = P(\text{heads on flip 1})P(\text{heads on flip 2})$$
$$= (.50)(.50) = .25$$

Determining the probability of heads on one of two flips ($X = 1$) requires an additional consideration. There are two ways to obtain one flip of heads: (1) heads on flip 1 and tails on flip 2 (HT) or (2) tails on flip 1 and heads on flip 2 (TH). Because these two possible outcomes are mutually exclusive (they cannot both occur), we may add their respective probabilities. That is,

$$P(1 \text{ heads}) = P(\text{HT or TH})$$
$$= P(\text{HT}) + P(\text{TH})$$

As before, the individual coin flips are independent. Thus, we can multiply the probability of a head times the probability of a tail to get the probability of first a head then a tail, $P(\text{HT})$; we can also multiply the probability of a tail times the probability of a head to obtain the probability of first a tail and then a head, $P(\text{TH})$. That is,

$$P(\text{HT}) + P(\text{TH}) = P(\text{H})P(\text{T}) + P(\text{T})P(\text{H})$$
$$= (.50)(.50) + (.50)(.50)$$
$$= .25 + .25$$
$$= .50$$

The complete probability distribution for variable X is summarized in Table 5.1. Note that the probabilities sum to 1.00. The distribution can be plotted much the way we did with frequency distributions in Chapter 2. A bar graph of a probability distribution places the values of the variable along the horizontal base line, and the probabilities along the vertical axis (see Figure 5.1). We see that the distribution is symmetric, as one would expect of coins that favor neither of their sides.

We can construct a probability distribution for the number of clearances among a group of homicides in a manner similar to the simple case of two coins, except that the

TABLE 5.1 *Probability Distribution for Number of Heads in Two Flips*

X	Probability (P)
0	.25
1	.50
2	.25
Total	1.00

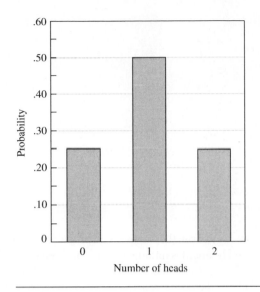

FIGURE 5.1 Probability distribution of two coins

shape will not be symmetric because clearance and nonclearance are not equally likely as are the two sides of a coin. Determining the probability distribution for the number of clearances among three homicides is manageable. Beyond three, the logic is unchanged, but the steps become rather exhausting. There are alternative methods for handling larger problems, but they are outside the scope of this introductory taste of probability.

Shown in the following table are all the possible combinations of clearance and nonclearance, denoted by C and N, respectively, for the group of three homicide cases. Because of the multiplication rule for independent outcomes, the probability of any one of the sequences of Cs and Ns can be expressed as the product of their respective probabilities. Finally, the eight different possible combinations of outcomes are mutually exclusive. That is, whatever combination does occur for a particular group of three homicides rules out the other seven alternative combinations. As a consequence of the combinations being mutually exclusive, the probabilities for the alternative combinations can be added, and the probabilities of all eight possible combinations sum to 1.

Combination	Outcomes	Probability
A	NNN	$P(NNN) = P(N)P(N)P(N) = (.4)(.4)(.4) = .064$
B	CNN	$P(CNN) = P(C)P(N)P(N) = (.6)(.4)(.4) = .096$
C	NCN	$P(NCN) = P(N)P(C)P(N) = (.4)(.6)(.4) = .096$
D	NNC	$P(NNC) = P(N)P(N)P(C) = (.4)(.4)(.6) = .096$
E	CCN	$P(CCN) = P(C)P(C)P(N) = (.6)(.6)(.4) = .144$
F	CNC	$P(CNC) = P(C)P(N)P(C) = (.6)(.4)(.6) = .144$
G	NCC	$P(NCC) = P(N)P(C)P(C) = (.4)(.6)(.6) = .144$
H	CCC	$P(CCC) = P(C)P(C)P(C) - (.6)(.6)(.6) = \underline{.216}$
		1.0

TABLE 5.2 *Probability Distribution of the Number of Clearances for Three Homicides*

X	P(X)
0	.064
1	.288
2	.432
3	.216
	1.0

Given these probabilities, we can form the theoretical probability distribution in Table 5.2 for X, the number of cleared homicides among a group of three cases. Of course, X has four possible values, ranging from 0 (none cleared) to 3 (all cleared). In forming the probability distribution, we can add the three different ways (combinations B through D) that one case out of three can be solved, as well as adding the probabilities for the three ways (combinations E through G) that two out of three homicides can be cleared.

The Difference between Probability Distributions and Frequency Distributions

It is important to be clear on the difference between frequency distributions, like those we saw in Chapter 2, and probability distributions. Look again at Figure 5.1, which shows the probability distribution of the number of heads in two flips of a coin. This is a perfectly symmetrical distribution, with the probability of two heads equal to .25, identical to the probability of no heads. Furthermore, it shows that the probability of one head out of two is .50. This distribution is based on probability theory. It describes what *should* happen when we flip two coins.

Now let's observe some data. Flip two coins and record the number of heads. Repeat this nine more times. How many times did you get zero heads, one head, and two heads? Our own results of a sample of 10 flips of a pair of coins are shown in Table 5.3.

TABLE 5.3 *Frequency Distribution of 10 Flips of Two Coins*

Number of Heads	f	%
0	3	30.0
1	6	60.0
2	1	10.0
Total	10	100.0

This is a frequency distribution, not a probability distribution. It is based on actual observations of flipping two coins 10 times. Although the percentages (30%, 60%, and 10%) may seem like probabilities, they are not. The percentage distribution does not equal the probability distribution given earlier in Table 5.1. A probability distribution is theoretical or ideal: It portrays what the percentages should be in a perfect world. Unfortunately, we did not get a perfect outcome—there were more zero-head outcomes than two-head outcomes, for instance. The problem is that just about anything can happen in only 10 pairs of flips. In fact, we could have obtained even more skewed results than these.

Imagine if we were to repeat our flips of two coins many more times. The results of our flipping the two coins 1,000 times are shown in the frequency distribution in Table 5.4.

This frequency distribution (with $N = 1,000$) looks a lot better. Why is that? Was our luck just better this time than when we flipped the coins 10 times? In a sense, it is a matter of luck, but not completely. As we said previously, with only 10 sets of flips almost anything can happen—you could even get a streak of zero heads. But when we run our experiment for 1,000 pairs of flips, things even out over the long run. There will have been streaks of zero heads, but so too will there have been streaks of one head and two heads. As we approach an infinite number of flips of the two coins, the laws of probability become apparent. Our luck, if you want to call it that, evens out.

A probability distribution is essentially a frequency distribution for an infinite number of flips. Thus, we may never observe this distribution of infinite flips, but we know it looks like Figure 5.1.

Mean and Standard Deviation of a Probability Distribution

Returning to the frequency distribution in Table 5.3 for 10 flips of two coins, let's compute the mean number of heads:

$$\bar{X} = \frac{\Sigma X}{N}$$
$$= \frac{0 + 0 + 0 + 1 + 1 + 1 + 1 + 1 + 1 + 2}{10}$$
$$= .8$$

TABLE 5.4 *Frequency Distribution of 1,000 Flips of Two Coins*

Number of Heads	f	%
0	253	25.3
1	499	49.9
2	248	24.8
Total	1,000	100.0

This is low. The probability distribution of the two coin flips shown in Figure 5.1 clearly suggests that the average should be 1.00. That is, for the flip of two coins, in the long run you should expect to average one head. Note, however, that $N = 1,000$ frequency distribution seems to be more in line with this expectation. For Table 5.4 (collecting values),

$$\bar{X} = \frac{\Sigma X}{N}$$

$$= \frac{(253)(0) + (499)(1) + (248)(2)}{1,000}$$

$$= .995$$

Again, our "luck" averages out in the long run.

As you might suspect, a probability distribution has a mean. Because the mean of a probability distribution is the value we expect to average in the long run, it is sometimes called an *expected value*. For our case of the number of heads in two coin flips, the mean is 1. We use the Greek letter μ (mu) for the mean of a probability distribution (here $\mu = 1$) to distinguish it from $\bar{X}$, the mean of a frequency distribution. $\bar{X}$ is something we calculate from a set of observed data and their frequencies. On the other hand, the mean of a probability distribution (μ) is a quantity that comes from our theory of what the distribution should look like.

A probability distribution also has a standard deviation, symbolized by σ (sigma), the Greek letter equivalent of *s*. Up to this point, we have used *s* to signify standard deviation generally. But, from here on, *s* will represent the standard deviation of a set of observed data obtained from research, and σ shall represent the standard deviation of a theoretical distribution that is not observed directly. Similarly, s^2 will denote the variance of a set of observed data, and σ^2 will be the variance of a theoretical distribution. We will encounter μ, σ and σ^2 often in chapters to come, and it is important to keep in mind the difference between $\bar{X}$, *s*, and s^2 (summary measures of observed data) on the one hand, and μ, σ, and σ^2 (characteristics of theoretical distributions) on the other.

The mean, variance, and standard deviation of a probability distribution are calculated from the possible values of *X* and their associated probabilities $P(X)$.

$$\mu = \Sigma X P(X)$$
$$\sigma^2 = \Sigma (X - \mu)^2 P(X)$$
$$\sigma = \sqrt{\Sigma (X - \mu)^2 P(X)}$$

where the summation extends over the possible values of *X*. Using these formulas, for example, we can determine that the probability distribution for the number of cleared homicides (*X*) among a group of three cases has the following mean, variance, and standard deviation:

$$\mu = \Sigma X P(X)$$
$$= 0(.064) + 1(.288) + 2(.432) + 3(.216)$$
$$= 0 + .288 + .864 + .648$$
$$= 1.8$$

$$\sigma^2 = \Sigma(X - \mu)^2 P(X)$$
$$= (0 - 1.8)^2(.064) + (1 - 1.8)^2(.288) + (2 - 1.8)^2(.432) + (3 - 1.8)^2(.216)$$
$$= (-1.8)^2(.064) + (-.8)^2(.288) + (.2)^2(.432) + (1.2)^2(.216)$$
$$= 3.24(.064) + .64(.288) + .04(.432) + 1.44(.216)$$
$$= .207 + .184 + .017 + .311$$
$$= .72$$

$$\sigma = \sqrt{\Sigma(X - \mu)^2 P(X)}$$
$$= \sqrt{.72}$$
$$= .8485$$

Thus, although for any group of three homicides, the police could solve $X = 0$, 1, 2, or 3 of them, in the long run we would expect an average of 1.8 clearances for every three homicide cases and a standard deviation of about .85.

The Normal Curve as a Probability Distribution

Previously, we saw that frequency distributions can take a variety of shapes or forms. Some are perfectly symmetrical or free of skewness, others are skewed either negatively or positively, still others have more than one hump, and so on. This is true as well for probability distributions.

Within this great diversity, there is one probability distribution with which many students are already familiar, if only from being graded on "the curve." This distribution, commonly known as the *normal curve*, is a theoretical or ideal model that was obtained from a mathematical equation, rather than from actually conducting research and gathering data. However, the usefulness of the normal curve for the researcher can be seen in its applications to actual research situations.

As we will see, for example, the normal curve can be used for describing distributions of scores, interpreting the standard deviation, and making statements of probability. In subsequent chapters, we will see that the normal curve is an essential ingredient of statistical decision making, whereby the researcher generalizes her or his results from samples to populations. Before proceeding to a discussion of techniques of decision making, it is first necessary to gain an understanding of the properties of the normal curve.

Characteristics of the Normal Curve

How can the normal curve be characterized? What are the properties that distinguish it from other distributions? As indicated in Figure 5.2, the normal curve is a type of smooth, symmetrical curve whose shape reminds many individuals of a bell and is thus widely known as the bell-shaped curve. Perhaps the most outstanding feature of the normal curve is its *symmetry:* If we were to fold the curve at its highest point at the center, we would create halves, each the mirror image of the other.

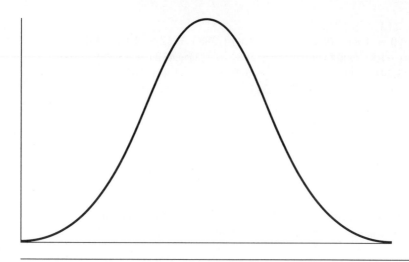

FIGURE 5.2   The shape of the normal curve

In addition, the normal curve is unimodal, having only one peak or point of maximum likelihood—that point in the middle of the curve at which the mean, median, and mode coincide (the student may recall that the mean, median, and mode occur at different points in a skewed distribution—see Chapter 3). From the rounded central peak of the normal distribution, the curve falls off gradually at both tails, extending indefinitely in either direction and getting closer and closer to the base line without actually reaching it.

The Model and the Reality of the Normal Curve

Because it is a probability distribution, the normal curve is a theoretical ideal. We might then ask: To what extent do distributions of actual data (i.e., the data collected by researchers in the course of doing research) closely resemble or approximate the form of the normal curve? For illustrative purposes, let us imagine that all social, psychological, and physical phenomena were normally distributed. What would this hypothetical world be like?

So far as physical human characteristics are concerned, most adults would fall within the 5 to 6 ft range of height, with far fewer being either very short (less than 5 ft) or very tall (more than 6 ft). As shown in Figure 5.3, IQ would be equally predictable—the greatest proportion of IQ scores would fall between 85 and 115. We would see a gradual falling off of scores at either end with few "geniuses" who score higher than 145 and equally few who score lower than 55. Likewise, relatively few individuals would be regarded as political extremists, either of the right or of the left, whereas most would be considered politically moderate or middle-of-the-roaders. Finally, even the pattern of wear resulting from the flow of traffic in doorways would resemble the normal distribution: most wear would take place in the center of the doorway, whereas gradually decreasing amounts of wear would occur at either side.

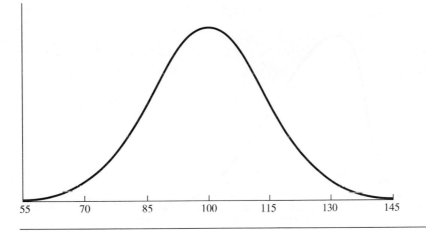

55   70   85   100   115   130   145

FIGURE 5.3   The distribution of IQ

Some readers have, by this time, noticed that the hypothetical world of the normal curve does not differ radically from the real world. Characteristics such as height, IQ, political orientation, and wear in doorways do, in fact, seem to approximate the theoretical normal distribution. Because so many phenomena have this characteristic—because it occurs so frequently in nature (and for other reasons that will soon become apparent)—researchers in many fields have made extensive use of the normal curve by applying it to the data that they collect and analyze.

But it should also be noted that some variables in social science as elsewhere, simply do not conform to the theoretical notion of the normal distribution. Many distributions are skewed, others have more than one peak, and some are symmetrical but not bell-shaped. As a concrete example, let us consider the distribution of wealth throughout the world. It is well known that the "have-nots" greatly outnumber the "haves." Thus, as shown in Figure 5.4, the distribution of wealth (as indicated by per capita income) is extremely skewed, so that a small proportion of the world's population receives a large proportion of the world's income. Likewise, population specialists tell us that the United States is soon to become a land of the young and the old. From an economic standpoint, this distribution of age represents a burden for a relatively small labor force made up of middle-aged citizens providing for a disproportionately large number of retired as well as school-age dependents.

Where we have good reason to expect radical departures from normality, as in the case of age and income, the normal curve cannot be used as a model of the data we have obtained. Thus, it cannot be applied at will to all of the distributions encountered by the researcher but must be used with a good deal of discretion. Fortunately, statisticians know that many phenomena of interest to the social researcher take the form of the normal curve.

The Area under the Normal Curve

It is important to keep in mind that the normal curve is an ideal or theoretical distribution (that is, a probability distribution). Therefore, we denote its mean by μ and its standard

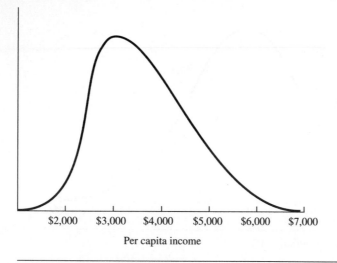

FIGURE 5.4 The distribution of per capita income among the nations of the world (in U.S. dollars)

deviation by σ. The mean of the normal distribution is at its exact center (see Figure 5.5). The standard deviation (σ) is the distance between the mean (μ) and the point on the base line just below where the reversed S-shaped portion of the curve shifts direction.

To employ the normal distribution in solving problems, we must acquaint ourselves with the area under the normal curve: the area that lies between the curve and the base line containing 100% or all of the cases in any given normal distribution. Figure 5.5 illustrates this characteristic.

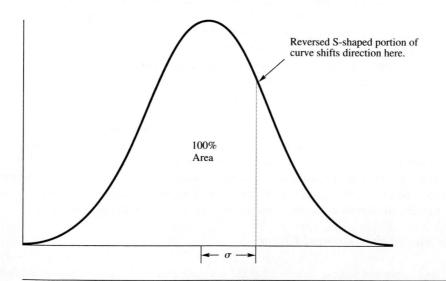

FIGURE 5.5 The total area under the normal curve

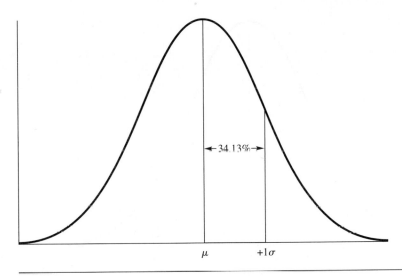

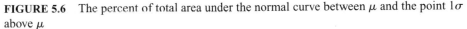

FIGURE 5.6   The percent of total area under the normal curve between μ and the point 1σ above μ

We could enclose a portion of this total area by drawing lines from any two points on the baseline up to the curve. For instance, using the mean as a point of departure, we could draw one line at the mean (μ) and another line at the point that is 1σ (1 standard deviation distance) above the mean. As illustrated in Figure 5.6, this enclosed portion of the normal curve includes 34.13% of the total frequency.

In the same way, we can say that 47.72% of the cases under the normal curve lie between the mean and 2σ above the mean and that 49.87% lie between the mean and 3σ above the mean (see Figure 5.7).

As we shall see, a constant proportion of the total area under the normal curve will lie between the mean and any given distance from the mean as measured in sigma σ units. This is true regardless of the mean and standard deviation of the particular distribution and applies universally to all data that are normally distributed. Thus, the area under the normal curve between the mean and the point 1σ above the mean *always* turns out to include 34.13% of the total cases, whether we are discussing the distribution of height, intelligence, political orientation, or the pattern of wear in a doorway. The basic requirement, in each case, is that we are working with a *normal* distribution of scores.

The symmetrical nature of the normal curve leads us to make another important point: Any given sigma distance above the mean contains the identical proportion of cases as the same sigma distance below the mean. Thus, if 34.13% of the total area lies between the mean and 1σ *above* the mean, then 34.13% of the total area also lies between the mean and 1σ *below* the mean; if 47.72% lies between the mean and 2σ *above* the mean, then 47.72% lies between the mean and 2σ *below* the mean; if 49.87% lies between the mean and 3σ *above* the mean, then 49.87% also lies between the mean and 3σ *below* the mean. In other words, as illustrated in Figure 5.8, 68.26% of the total area of the normal curve

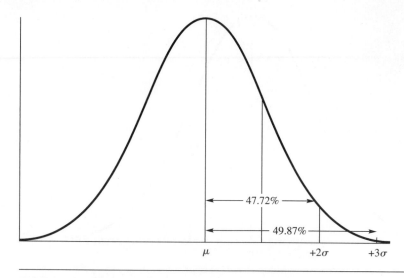

FIGURE 5.7   The percent of total area under the normal curve between μ and the points that are 2σ and 3σ from μ

(34.13% + 34.13%) falls between -1σ and $+1\sigma$ from the mean; 95.44% of the area (47.72% + 47.72%) falls between -2σ and $+2\sigma$ from the mean; and 99.74%, or almost all, of the cases (49.87% + 49.87%) falls between -3σ and $+3\sigma$ from the mean. It can be said, then, that six standard deviations include practically all of the cases (more than 99%) under any normal distribution.

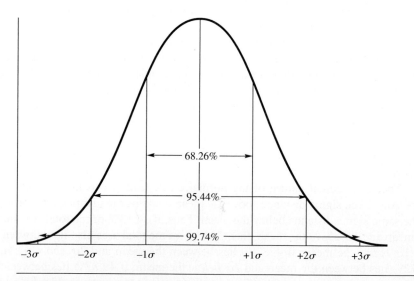

FIGURE 5.8   The percent of total area under the normal curve between -1σ and $+1\sigma$, -2σ and $+2\sigma$, and -3σ and $+3\sigma$

Clarifying the Standard Deviation

An important function of the normal curve is to help interpret and clarify the meaning of the standard deviation. To understand how this function is carried out, let us examine what some researchers tell us about gender differences in IQ. Despite the claims of male supremacists, there is evidence that both males and females have mean IQ scores of approximately 100. Let us also say these IQ scores differ markedly in terms of variability around the mean. In particular, let us suppose hypothetically that male IQs have greater *heterogeneity* than female IQs; that is, the distribution of male IQs contains a much larger percent of extreme scores representing very bright as well as very dull individuals, whereas the distribution of female IQs has a larger percent of scores located closer to the average, the point of maximum frequency at the center.

Because the standard deviation is a measure of variation, these gender differences in variability should be reflected in the sigma value of each distribution of IQ scores. Thus, we might find that the standard deviation is 15 for male IQs, but only 10 for female IQs. Knowing the standard deviation of each set of IQ scores and assuming that each set is normally distributed, we could then estimate and compare the percent of males and females having any given range of IQ scores.

For instance, measuring the baseline of the distribution of male IQ in standard deviation units, we would know that 68.26% of male IQ scores fall between -1σ and $+1\sigma$ from the mean. Because the standard deviation is always given in raw-score units and $\sigma = 15$, we would also know that these are points on the distribution at which IQ scores of 115 and 85 are located ($\mu - \sigma = 100 - 15 = 85$ and $\mu + \sigma = 100 + 15 = 115$). Thus, 68.26% of the males would have IQ scores that fall between 85 and 115.

Moving away from the mean and farther out from these points, we would find, as illustrated in Figure 5.9, that 99.74%, or practically all of the males have IQ scores between 55 and 145 (between -3σ and $+3\sigma$).

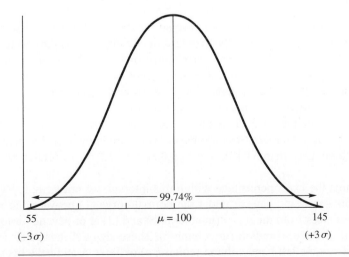

FIGURE 5.9 A distribution of male IQ scores

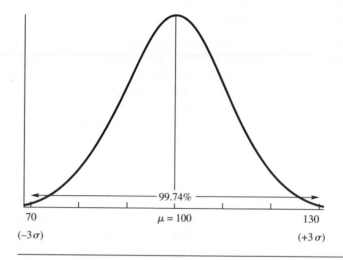

FIGURE 5.10   A distribution of female IQ scores

In the same manner, looking next at the distribution of female IQ scores as depicted in Figure 5.10, we see that 99.74% of these cases would fall between scores of 70 and 130 (between -3σ and $+3\sigma$). In contrast to males, then, the distribution of female IQ scores could be regarded as being relatively homogeneous, having a smaller range of extreme scores in either direction. This difference is reflected in the comparative size of each standard deviation and in the range of IQ scores falling between -3σ and $+3\sigma$ from the mean.

Using Table A

In discussing the normal distribution, we have so far treated only those distances from the mean that are exact multiples of the standard deviation. That is, they were precisely one, two, or three standard deviations either above or below the mean. This question now arises: What must we do to determine the percent of cases for distances lying between any two score values? For instance, suppose we wish to determine the percent of total area that falls between the mean and, say, a raw score located 1.40σ above the mean. As illustrated in Figure 5.11, a raw score 1.40σ above the mean is obviously greater than 1σ but less than 2σ from the mean. Thus, we know this distance from the mean would include more than 34.13% but less than 47.72% of the total area under the normal curve.

To determine the exact percentage within this interval, we must employ Table A in Appendix C. This shows the percent under the normal curve (1) between the mean and various sigma distances from the mean (in column b) and (2) at or beyond various scores toward either tail of the distribution (in column c). These sigma distances (from 0.00 to 4.00) are labeled z in the left-hand column (column a) of Table A and have been given to two decimal places.

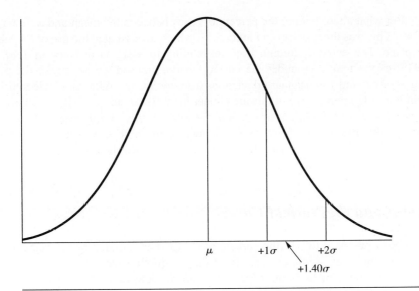

FIGURE 5.11 The position of a raw score that lies 1.40σ above μ

Notice that the symmetry of the normal curve makes it possible to give percentages for only one side of the mean, that is, only one-half of the curve (50%). Values in Table A represent either side. The following is a portion of Table A:

(a)	(b)	(c)
z	**Area between Mean and z**	**Area beyond z**
.00	.00	50.00
.01	.40	49.60
.02	.80	49.20
.03	1.20	48.80
.04	1.60	48.40
.05	1.99	48.01
.06	2.39	47.61
.07	2.79	47.21
.08	3.19	46.81
.09	3.59	46.41

When learning to use and understand Table A, we might first attempt to locate the percent of cases between a sigma distance of 1.00 and the mean (the reason being that we already know that 34.13% of the total area falls between these points on the base line). Looking at column b of Table A, we see it indeed indicates that exactly 34.13% of the total frequency falls between the mean and a sigma distance of 1.00. Likewise, we see that the area between the mean and the sigma distance 2.00 includes exactly 47.72% of the total area under the curve.

But what about finding the percent of cases between the mean and a sigma distance of 1.40? This was the problem in Figure 5.11, which necessitated the use of the table in the first place. The entry in column b corresponding to a sigma distance of 1.40 includes 41.92% of the total area under the curve. Finally, how do we determine the percent of cases at or beyond 1.40 standard deviations from the mean? Without a table to help us, we might locate the percentage in this area under the normal curve by simply subtracting our earlier answer from 50%, because this is the total area lying on either side of the mean. However, this has already been done in column c of Table A, where we see that exactly 8.08% ($50 - 41.92 = 8.08$) of the cases fall at or above the score that is 1.40 standard deviations from the mean.

Standard Scores and the Normal Curve

We are now prepared to find the percent of the total area under the normal curve associated with any given sigma distance from the mean. However, at least one more important question remains to be answered: How do we determine the sigma distance of any given raw score? That is, how do we translate our raw score—the score that we originally collected from our respondents—into units of standard deviation? If we wished to translate feet into yards, we would simply divide the number of feet by 3, because there are 3 ft in a yard. Likewise, if we were translating minutes into hours, we would divide the number of minutes by 60, because there are 60 min in every hour. In precisely the same manner, we can translate any given raw score into sigma units by dividing the distance of the raw score from the mean by the standard deviation. To illustrate, let us imagine a raw score of 16 from a distribution in which μ is 13 and σ is 2. Taking the difference between the raw score and the mean and obtaining a deviation ($16 - 13$), we see that a raw score of 16 is 3 raw-score units above the mean. Dividing this raw-score distance by $\sigma = 2$, we see that this raw score is 1.5 (one and one-half) standard deviations above the mean. In other words, the sigma distance of a raw score of 16 *in this particular distribution* is 1.5 standard deviations above the mean. We should note that regardless of the measurement situation, there are always 3 ft in a yard and 60 min in an hour. The constancy that marks these other standard measures is not shared by the standard deviation. It changes from one distribution to another. For this reason, we must know the standard deviation of a distribution by calculating it, estimating it, or being given it by someone else before we are able to translate any particular raw score into units of standard deviation.

The process that we have just illustrated—that of finding sigma distance from the mean, yields a value called a *z score* or *standard score*, which indicates the *direction and degree that any given raw score deviates from the mean of a distribution on a scale of sigma units* (notice that the left-hand column of Table A in Appendix C is labeled z). Thus, a z score of +1.4 indicates that the raw score lies 1.4σ (or almost $1\frac{1}{2}\sigma$) *above* the mean, whereas a z score of −2.1 means that the raw score falls slightly more than 2σ *below* the mean (see Figure 5.12).

We obtain a z score by finding the deviation ($X - \mu$), which gives the distance of the raw score from the mean, and then dividing this raw-score deviation by the standard deviation.

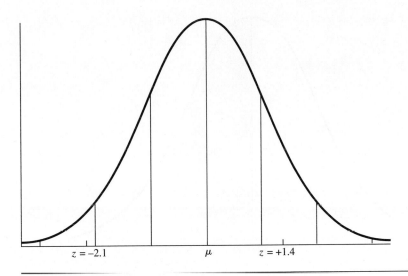

FIGURE 5.12   The position of $z = -2.1$ and $z = +1.4$ in a normal distribution

Computed by formula,

$$z = \frac{X - \mu}{\sigma}$$

where μ = mean of a distribution
σ = standard deviation of a distribution
z = standard score

As an example, suppose we are studying the distribution of annual income for home healthcare workers in a large agency in which the mean annual income is \$20,000 and the standard deviation is \$1,500. Assuming that the distribution of annual income is normally distributed we can translate the raw score from this distribution, \$22,000, into a standard score in the following manner:

$$z = \frac{22,000 - 20,000}{1,500} = +1.33$$

Thus, an annual income of \$22,000 is 1.33 standard deviations above the mean annual income of \$20,000 (see Figure 5.13).

As another example, suppose that we are working with a normal distribution of scores representing job satisfaction among a group of city workers. The scale ranges from 0 to 20, with higher scores representing greater satisfaction with the job.

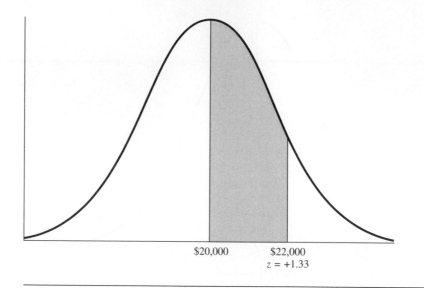

$20,000 $22,000
                z = +1.33

FIGURE 5.13   The position of $z = +1.33$ for the raw score \$22,000

Let us say this distribution has a mean of 10 and a standard deviation of 3. To determine how many standard deviations a score of 3 lies from the mean of 10, we obtain the difference between this score and the mean, that is,

$$X - \mu = 3 - 10$$
$$= -7$$

We then divide by the standard deviation:

$$z = \frac{X - \mu}{\sigma}$$

$$= \frac{-7}{3}$$

$$= -2.33$$

Thus, as shown in Figure 5.14, a raw score of 3 in this distribution of scores falls 2.33 standard deviations below the mean.

Finding Probability under the Normal Curve

As we shall now see, the normal curve can be used in conjunction with z scores and Table A to determine the probability of obtaining any raw score in a distribution. In the present context, the normal curve is a distribution in which it is possible to determine probabilities associated with various points along its baseline. As noted earlier, the normal curve is

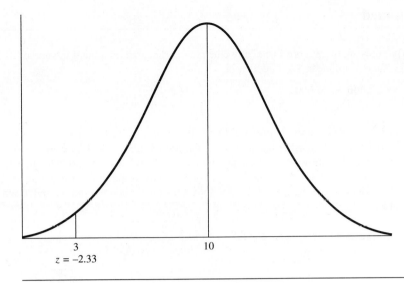

3
z = −2.33

10

FIGURE 5.14   The position of $z = -2.33$ for the raw score 3

a *probability distribution* in which the total area under the curve equals 100%; it contains a central area surrounding the mean, where scores occur most frequently, and smaller areas toward either end, where there is a gradual flattening out and thus a smaller proportion of extremely high and low scores. In probability terms, then, we can say that probability decreases as we travel along the baseline away from the mean in either direction. Thus, to say that 68.26% of the total frequency under the normal curve falls between -1σ and $+1\sigma$ from the mean is to say that the probability is approximately 68 in 100 that any given raw score will fall within this interval. Similarly, to say that 95.44% of the total frequency under the normal curve falls between -2σ and $+2\sigma$ from the mean is also to say that the probability is approximately 95 in 100 that any raw score will fall within this interval, and so on.

This is precisely the same concept of probability or *relative frequency* that we saw in operation when flipping pairs of coins. Note, however, that the probabilities associated with areas under the normal curve are always given relative to 100%, which is the entire area under the curve (for example, 68 in 100, 95 in 100, 99 in 100).

BOX 5.2 • *Step-by-Step Illustration: Probability under the Normal Curve*

To apply the concept of probability in relation to the normal distribution, let us return to an earlier example. We were then asked to translate into its *z*-score equivalent a raw score from an agency's distribution of annual salary for healthcare workers, which we assumed approximated a normal curve. This distribution of income had a mean of $20,000 with a standard deviation of $1,500.

(continued)

BOX 5.2   Continued

By applying the *z*-score formula, we learned earlier that an annual income of $22,000 was 1.33σ above the mean $20,000, that is,

$$z = \frac{22,000 - 20,000}{1,500} = +1.33$$

Let us now determine the probability of obtaining a score that lies between $20,000, the mean, and $22,000. In other words, what is the probability of randomly choosing, in just one attempt, a healthcare worker whose annual income falls between $20,000 and $22,000? The problem is graphically illustrated in Figure 5.15 (we are solving for the shaded area under the curve) and can be solved in two steps using the *z*-score formula and Table A in Appendix C.

Step 1   Translate the raw score ($22,000) into a *z* score.

$$z = \frac{X - \mu}{\sigma}$$

$$= \frac{22,000 - 20,000}{1,500}$$

$$= +1.33$$

Thus, a raw score, $22,000 is located 1.33σ above the mean.

Step 2   Using Table A, find the percent of total area under the curve falling between the *z* score (*z* = +1.33) and the mean.

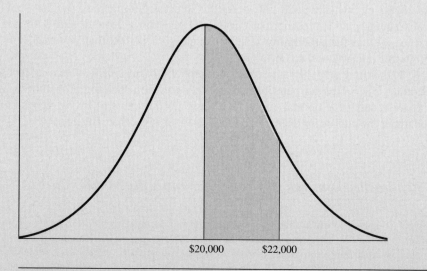

$20,000 $22,000

FIGURE 5.15   The portion of area under the normal curve for which we seek to determine the probability of occurrence

In column b of Table A, we find that 40.82% of the healthcare workers earn between $20,000 and $22,000 (see Figure 5.16). Thus, by moving the decimal two places to the left, we see that the probability (rounded off) is 41 in 100. More precisely, $P = .4082$ is the probability that we would obtain an individual whose annual income lies between $20,000 and $22,000.

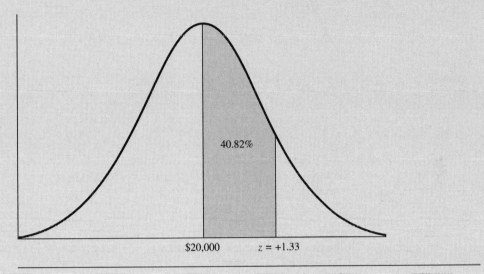

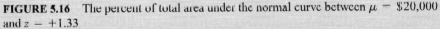

FIGURE 5.16 The percent of total area under the normal curve between $\mu - $20,000$ and $z - +1.33$

In the previous example, we were asked to determine the probability associated with the distance between the mean and a particular sigma distance from it. Many times however, we may wish to find the percent of area that lies at or *beyond* a particular raw score toward either tail of the distribution or to find the probability obtaining these scores. For instance, in the present case, we might wish to learn the probability of obtaining an annual income of $22,000 or *greater.*

This problem can be illustrated graphically, as shown in Figure 5.17 (we are solving for the shaded area under the curve). In this case we would follow Steps 1 and 2, thus obtaining the z score and finding the percent under the normal curve between $20,000 and a $z - 1.33$ (from Table A). In the present case, however, we must go a step beyond and *subtract* the percentage obtained in Table A from 50%—that percent of the total area lying on either side of the mean. Fortunately, column c of Table A has already done this for us.

Therefore, subtracting 40.82% from 50% or simply looking in column c of Table A, we learn that slightly more than 9% (9.18%) fall *at* or *beyond* $22,000. In probability terms, we can say (by moving the decimal two places to the left) there are only slightly more than 9 chances in $100(P = .0918)$ that we would find a healthcare worker in this agency whose income was $20,000 or greater.

It was earlier noted that any given sigma distance above the mean contains the identical proportion of cases as the same sigma distance below the mean. For this reason, our

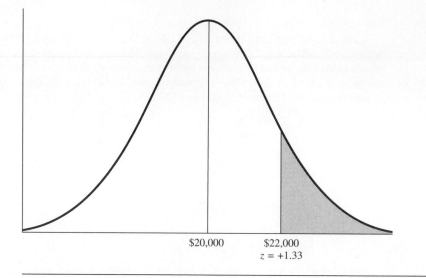

$20,000 $22,000
                 z = +1.33

FIGURE 5.17   The portion of total area under the normal curve for which we seek to determine the probability of occurrence

procedure in finding probabilities associated with points below the mean is identical to that followed in the previous examples.

For instance, the percent of total area between the z score -1.33 ($18,000) and the mean is identical to the percent between the z score $+1.33$ ($22,000) and the mean. Therefore, we know that $P = .4082$ of obtaining a worker whose income falls between $18,000 and $20,000. Likewise, the percent of total frequency at or beyond -1.33 ($18,000 or less) equals that at or beyond $+1.33$ ($22,000 or more). Thus, we know $P = .0918$ that we shall obtain a home healthcare worker from the agency with an annual income of $18,000 or less.

We can use Table A to find the probability of obtaining more than a single portion of the area under the normal curve. For instance, we have already determined that $P = .09$ for incomes of $18,000 or less and for incomes of $22,000 or more. To find the probability of obtaining *either* $18,000 or less *or* $22,000 or more, we simply add their separate probabilities as follows:

$$P = .0918 + .0918$$
$$= .1836$$

In a similar way, we can find the probability of obtaining a home healthcare worker whose income falls between $18,000 and $22,000 by adding the probabilities associated with z scores of 1.33 on either side of the mean. Therefore,

$$P = .4082 + .4082$$
$$= .8164$$

Notice that .8164 + .1836 equals 1.00, representing all possible outcomes under the normal curve.

The application of the multiplication rule to the normal curve can be illustrated by finding the probability of obtaining four home healthcare workers whose incomes are $22,000 greater. We already know that $P = .0918$ associated with finding a single cashier whose income is at least $22,000 Therefore,

$$P = (.0918)(.0918)(.0918)(.0918)$$
$$= (.0918)^4$$
$$= .00007$$

Applying the multiplication rule, we see that the probability of obtaining four healthcare workers at random with incomes of $22,000 or more is only 7 chances in 100,000.

Finding Scores from Probability Based on the Normal Curve

In the previous section, we used knowledge about the mean and standard deviation of a normally distributed variable (such as starting salaries of healthcare workers) to determine particular areas under the normal curve. The portions of area were used in turn to calculate various percentages or probabilities about the distribution, such as determining the percentage of healthcare workers who earn over $22,000 in salary.

The process of using Table A to translate z-scores into portions of the total area can be reversed to calculate score values from particular portions of area or percentages. We know without having to consult Table A that the salary above which 50% of the workers earn is $20,000. Because of symmetry of shape, the median salary dividing the distribution in half is the same as the mean. But what if we wanted to know the salary level that defines the top 10% of earners, the shaded portion in Figure 5.18?

The threshold for the top 10% can be determined by entering Table A for a particular portion of area (such as the highest 10% tail) to determine the z-score values and then

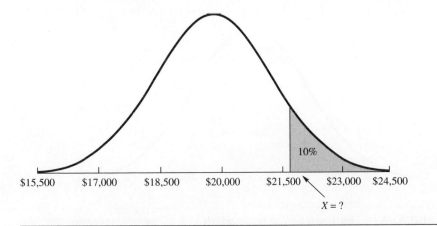

FIGURE 5.18   Finding the salary above which 10% of home healthcare workers are paid

translating that z score into its raw score (salary) equivalent. Specifically, we scan through column c to locate the area beyond (in this case, above) z that is closest to 10%. Apparently, 10.03% falls beyond a z-score of 1.28. We can then use a modified form of the usual z-score formula to solve for X given a particular value of z.

$$X = \mu + z\sigma$$

To find the top 10% threshold with $\mu = \$20{,}000$ and $\sigma = \$1{,}500$, and as we found from Table A, $z = 1.28$,

$$\begin{aligned} X &= 20{,}000 + 1.28(1{,}500) \\ &= 20{,}000 + 1{,}920 \\ &= 21{,}920 \end{aligned}$$

Thus, on the basis of the normal shape of the salary distribution of salaries of home healthcare workers with a mean of $20,000 and a standard deviation of $1,500, we can determine that the top 10% earners have salaries above $21,920.

BOX 5.3 • *Step-by-Step Illustration: Finding Scores from Probability Based on the Normal Curve*

Suppose that a particular ambulance company has data showing that the 911 response time from receiving the call to arrival is normally distributed with a mean of 5.6 min and a standard deviation of 1.8. The boss wants to know how much time is required for three-quarters (75%) of all calls to be handled, that is, the value below which 75% of response times fall.

Step 1   Locate in Table A the z-score that cuts off the area closest to 75% below it. As shown in Figure 5.19, the portion of area representing 75% of response times has 25% of the area between the mean and the z score. Scanning column b of Table A for an area closest to 25, we see that $z = .67$ cuts off 24.86 of the total area.

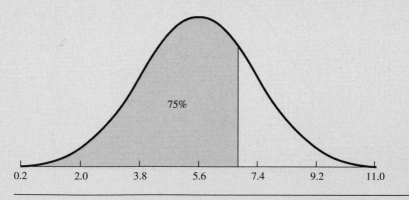

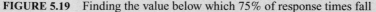

FIGURE 5.19   Finding the value below which 75% of response times fall

Step 2   Convert the z value to its raw score equivalent

$$X = \mu + z\sigma$$
$$= 5.6 + .67(1.8)$$
$$= 5.6 + 1.2$$
$$= 6.8$$

Thus, we can say that about three-quarters of the 911 calls for medical assistance are handled within 6.8 min. Conversely, one-quarter of the calls require more than 6.8 min.

Summary

In this chapter, we introduced the concept of probability—the relative likelihood of occurrence of any given outcome or event—as the foundation for decision making in statistics. Indicated by the number of times an event can occur relative to the total number of times any event can occur, probabilities range from 0 (an impossible event) to 1 (a certainty). Probabilities can be summed as in the addition rule, so as to establish the likelihood of obtaining any one of several different outcomes; they can also be multiplied as in the multiplication rule, to determine the likelihood of obtaining a combination of independent outcomes. As we saw in earlier chapters, a frequency distribution is based on actual observations in the real world. By contrast, a probability distribution is theoretical or ideal; it specifies what the percentages should be in a perfect world. Thus, we specify the possible values of a variable and calculate the probabilities associated with each. Analogous to the percentages in a frequency distribution, the probabilities we obtain in a probability distribution represent the likelihood (rather than the frequency of occurrence) of each value.

The theoretical model known as the normal curve is a particularly useful probability distribution for its applications to actual research situations. Known for its symmetrical bell shape, the normal curve can be used for describing distributions of scores, interpreting the standard deviation, and making statements of probability. In conjunction with standard scores, we can determine the percent of the total area under the normal curve associated with any given sigma distance from the mean and the probability of obtaining any raw score in a distribution. In subsequent chapters, we shall see that the normal curve is an essential ingredient of statistical decision making.

Terms to Remember

Probability

Converse rule

Addition rule

Mutually exclusive outcomes

Multiplication rule

Independent outcomes

Normal curve

Area under the normal curve

z score (standard score)

Questions and Problems

1. The relative likelihood of occurrence of any given event is known as that event's
   a. standard deviation.
   b. area under the normal curve.
   c. probability.
   d. All of the above

2. Probability varies from
   a. zero to infinity.
   b. zero to 1.0.
   c. 1.0 to 100.0.
   d. −1.0 to +1.0.

3. A criminal investigator is working to bring two separate and distinct serial killers to justice—one who preys on prostitutes and another who targets college students. From previous experience, the investigator is led to believe that he has a .50 chance of apprehending the prostitute slayer and a .65 chance of apprehending the student slayer. To calculate the probability that he will capture both killers, you must _____ the two probabilities.
   a. add   b. subtract   c. multiply   d. divide

4. A probability distribution is based on
   a. actual observations.
   b. probability theory.
   c. the addition rule.
   d. the multiplication rule.

5. Which of the following is *not* true of a normal curve?
   a. It is skewed.
   b. It is a probability distribution.
   c. Its total area contains 100% of the cases.
   d. The mode, median, and mean are identical.

6. A _____ indicates how far an individual raw score falls from the mean of a distribution; the _____ indicates how the scores in general scatter around the mean.
   a. standard deviation; z score
   b. z score; standard deviation
   c. probability; z score
   d. standard deviation; probability

7. The equation $z = -1.33$ indicates that a particular raw score lies
   a. 1.33 standard deviations below the mean.
   b. 1.33 percent below the mean.
   c. 1.33 percentile ranks below the mean.
   d. 1.33 z scores below the standard deviation.

8. The equation $P = .33$ for obtaining an income between $40,000 and $50,000 represents
   a. a percentage.
   b. a probability expressed as a proportion.
   c. a frequency of occurrence.
   d. a z score.

9. The following students are enrolled in a course in Introductory Sociology. They are listed along their year in school and whether they are majoring in sociology.

Student	Year	Sociology Major
1	Sophomore	Yes
2	Senior	No
3	Junior	Yes
4	Freshman	No
5	Freshman	Yes
6	Sophomore	Yes
7	Sophomore	Yes
8	Junior	No
9	Sophomore	Yes
10	Sophomore	No

What is the probability of selecting at random from this class
a. a sophomore?
b. a student majoring in sociology?
c. a freshman or a sophomore?
d. not a freshman?

10. Ten politicians were asked about their views of euthanasia. They are listed along with their political party.

Politician	Political Party	Supports Euthanasia
1	Republican	No
2	Democrat	No
3	Democrat	Yes
4	Republican	No
5	Republican	No
6	Democrat	No
7	Democrat	Yes
8	Democrat	Yes
9	Republican	No
10	Democrat	Yes

What is the probability of selecting at random from this group of politicians
a. a Republican?
b. a Democrat who supports euthanasia?
c. a politician who does not support euthanasia?
d. a Republican who does not support euthanasia?

11. Suppose that 16% of inmates in state prisons have psychiatric illnesses. What is the probability that
a. a particular prisoner does not suffer from a psychiatric illness?
b. that three prisoners do not suffer from a psychiatric illness?

12. Research has shown that nearly 25% of homeless adults are veterans. What is the probability of selecting
a. a particular homeless person who is not a veteran?
b. two random homeless people who are both veterans?

13. Suppose that 20% of men ($P = .20$) and 15% of women ($P = .15$) are carriers of a particular genetic trait. This trait can only be inherited by a child if both parents are carriers. What is the probability that a child is born with this genetic trait?

14. A student taking a midterm exam in Ancient History comes to two questions pertaining to a class that he missed, and so he decides to take a random guess on both questions. One of the questions is true–false and the other is multiple choice with five possible answers. What is the probability of guessing
    a. the correct answer to the true–false question?
    b. the correct answer to the multiple-choice question?
    c. the correct answers to both the true–false and the multiple-choice questions?
    d. the incorrect answers to both the true–false and the multiple-choice questions?
    e. the correct answer to the true–false question and the incorrect answer to the multiple-choice question?
    f. the incorrect answer to the true–false question and the correct answer to the multiple-choice question?

15. Suppose that you purchase a lottery ticket that contains two numbers and a letter, such as

    3   7   P

    a. What is the probability that you match the first digit?
    b. What is the probability that you match the second digit?
    c. What is the probability that you do not match the first digit?
    d. What is the probability that you match both the first and second digits?
    e. What is the probability that you match the letter?
    f. What is the probability of a perfect match (both digits and the letter)?

16. A particular community wants to address the problem of racial conflict in the local high school. The mayor decides to set up a three-person commission of residents to advise him. To be completely fair in the selection process, the mayor decides to choose the commission members at random, designating the first selection as the chair. The racial–ethnic composition of the community is 50% white, 30% black, and 20% Latino. What is the probability that
    a. the chair is a white resident?
    b. the chair is a black resident?
    c. the chair is white or Latino?
    d. all three commission members are black?

17. Drawing one card at random from a standard deck of 52 cards, what is the probability of drawing
    a. the eight of diamonds?
    b. the eight of diamonds or the eight of hearts?
    c. an eight?
    d. a red card?
    e. a picture card (jack, queen, or king)?
    f. from six through nine, inclusive?
    g. an odd-numbered card?

18. Suppose that 31% of Democrats and 63% of Republicans support building more nuclear power plants in the United States. Now suppose that there are 40 Republicans and 35 Democrats in one room.
    a. What is the probability of randomly selecting a Republican from this room?
    b. What is the probability of randomly selecting a Democrat from this room?

c. What is the probability of randomly selecting a Republican who supports building more nuclear power plants?

d. What is the probability of randomly selecting a Democrat who supports building more nuclear power plants?

e. What is the probability of randomly selecting one Democrat and one Republican who both support building more nuclear power plants?

f. What is the probability of randomly selecting two Republicans who do not support building more nuclear power plants?

19. Research has shown that 6 out of 10 marriages end in divorce. What is the probability that
    a. a particular just-married couple stays married "until death do them part"?
    b. two couples married in a double ceremony both get divorced?

20. Suppose that 2% of convicted felons are in fact innocent.
    a. If a person is convicted of a felony, what is the probability that he is guilty?
    b. If two people are convicted of felonies, what is the probability that both are guilty?
    c. If three people are convicted of felonies, what is the probability that all three are guilty?
    d. If four people are convicted of felonies, what is the probability that all four are guilty?

21. Under any normal distribution of scores, what percentage of the total area falls
    a. between the mean (μ) and a score value that lies one standard deviation (1σ) above the mean?
    b. between a score value that lies one standard deviation below the mean and a score value that lies one standard deviation above the mean?
    c. between the mean and a score value that lies $+2\sigma$ above the mean?
    d. between a score value that lies -2σ below the mean and a score value that lies $+2\sigma$ above the mean?

22. The Scholastic Assessment Test (SAT) is standardized to be normally distributed with a mean $\mu = 500$ and a standard deviation $\sigma = 100$. What percentage of SAT scores falls
    a. between 500 and 600?
    b. between 400 and 600?
    c. between 500 and 700?
    d. between 300 and 700?
    e. above 600?
    f. below 300?

23. For the SAT, determine the z score (that is, the number of standard deviations and the direction) that each of the following scores falls from the mean:
    a. 500
    b. 400
    c. 650
    d. 570
    e. 750
    f. 380

24. Using the z scores calculated in Problem 23 and Table A, what is the percentage of SAT scores that falls
    a. 500 or above?
    b. 400 or below?
    c. between 500 and 650?
    d. 570 or above?

    e. between 250 and 750?

    f. 380 or above? (*Hint:* 50% of the area falls on either side of the curve.)

25. IQ scores are normally distributed with a mean $\mu = 100$ and a standard deviation $\sigma = 15$. Based on this distribution, determine

    a. the percentage of IQ scores between 100 and 120.

    b. the probability of selecting a person at random having an IQ between 100 and 120.

    c. the percentage of IQ scores between 88 and 100.

    d. the probability of selecting a person at random having an IQ between 88 and 100.

    e. the percentage of IQ scores that are 110 or above.

    f. the probability of selecting a person at random having an IQ of 110 or above.

    g. the probability of selecting two people at random both having an IQ of 110 or above.

    h. the percentile rank corresponding to an IQ score of 125.

26. Suppose that at a large state university graduate research assistants are paid by the hour. Data from the personnel office show the distribution of hourly wages paid to graduate students across the campus is approximately normal with a mean of $12.00 and a standard deviation of $2.50. Determine

    a. the percentage of graduate assistants who earn an hourly wage of $15 or more.

    b. the probability of selecting at random from personnel files a graduate assistant who earns an hourly wage of $15 or more.

    c. the percentage of graduate assistants who earn between $10 and $12 per hour.

    d. the probability of selecting at random from personnel files a graduate assistant who earns between $10 and $12 per hour.

    e. the percentage of graduate assistants who earn an hourly wage of $11 or less.

    f. the probability of selecting at random from personnel files a graduate assistant who earns an hourly wage of $11 or less.

    g. the probability of selecting at random from the personnel files a graduate assistant whose hourly wage is extreme in either direction—*either* $10 *or* below or $14 or above.

    h. the probability of selecting at random from the personnel files two graduate assistants whose hourly wages are less than average (mean).

    i. the probability of selecting at random from the personnel files two graduate assistants whose hourly wages are $13.50 or more.

27. Suppose probation officer caseloads have a mean of 115 offenders and a standard deviation of 10. Assuming caseload sizes are normally distributed, determine

    a. the probability that a particular probation officer has a caseload between 90 and 105.

    b. the probability that a particular probation officer has a caseload larger than 135.

    c. the probability that four probation officers have a caseload larger than 135.

28. Suppose the mean IQ for a group of psychopaths is 105 and the standard deviation is 7. Assuming a normal distribution, determine

    a. the percentage of IQ scores between 100 and 110.

    b. the probability of selecting a psychopath at random who has an IQ between 100 and 110.

    c. the percentage of IQ scores between 90 and 105.

    d. the probability of randomly selecting a psychopath who has an IQ between 90 and 105.

    e. the percentage of IQ scores that are 125 or above.

    f. the probability of randomly selecting a psychopath who has an IQ of 125 or above.

    g. the probability of randomly selecting two psychopaths who both have an IQ of 125 or above.

29. A major automobile company claims that its new model has an average rating of 25 mpg (miles per gallon). Company officials concede that some cars vary based on a variety of factors, and that the mpg performances have a standard deviation of 4 mpg. You are employed by a consumer protection group that routinely test drives cars. Taking five cars at random off the assembly line, your group finds them to have a poor mpg performance defined as 20 mpg or below.
   a. Assuming the company's claim to be true ($\mu = 25$ and $\sigma = 4$), what is the probability that a single car selected randomly performs poorly (mpg of 20 or below)?
   b. Assuming the company's claim to be true ($\mu = 25$ and $\sigma = 4$), what is the probability that five cars selected randomly all perform poorly (mpg of 20 or below)?
   c. Given the poor performance that your group observed with the five test cars, what conclusion can you draw about the company's mpg claim?

30. Assume that scores among Asian Americans on an alienation scale are normally distributed with a mean $\mu = 22$ and a standard deviation $\sigma = 2.5$ (higher scores reflect greater feelings of alienation). Based on this distribution, determine
   a. the probability of an Asian American having an alienation score between 22 and 25.
   b. the probability of an Asian American having an alienation score of 25 or more.

SPSS Exercises

1. School safety is an important issue. Using SPSS to analyze the Monitoring the Future Survey, find out how safe the typical student feels in American high schools (V1643). Determine the mean and standard deviation.
   a. Using this mean and standard deviation, what percentage of students are expected to feel satisfied or completely satisfied with their personal safety (6 or above)?
   b. What percentage of students are expected to feel neutral or below about how safe they feel in school?
   c. What is the probability of selecting at random a student who feels neutral or below about how safe he or she feels in school?

2. Using the frequencies procedure (ANALYZE, DESCRIPTIVE STATISTICS, FREQUENCIES), request a histogram of television viewing data from the General Social Survey (TVHOURS) with the normal curve overlay (from the Charts button), as well as the mean and standard deviation (from the Statistics button).
   a. Comment on the closeness of the actual distribution to the normal curve.
   b. Based on the normal distribution (using the z-score and Table A), what is the probability that a person watches three or more hours of television per day?
   c. Compare the probability from the normal distribution to the actual proportion of reports of three or more hours of television viewing per day.

3. Using the frequencies procedure (ANALYZE, DESCRIPTIVE STATISTICS, FREQUENCIES), request a histogram of the unemployment data from the Best Places Study with the normal curve overlay (from the Charts button), as well as the mean and standard deviation (from the Statistics button).
   a. Comment on the closeness of the actual distribution to the normal curve.
   b. Based on the normal distribution (using the z-score and Table A), what is the probability that a Metropolitan Statistical Area (MSA) has an unemployment rate of 6.5% or above?
   c. Compare the probability from the normal distribution to proportion of the actual scores of MSAs that are 6.5% or above.

6

Samples and Populations

The social researcher generally seeks to draw conclusions about large numbers of individuals. For instance, he or she might be interested in the 295 million residents of the United States, the 1,000 members of a particular labor union, the 10,000 African Americans who are living in a southern town, or the 25,000 students currently enrolled in an eastern university.

Until this point, we have been pretending that the social researcher investigates the entire group that he or she tries to understand. Known as a *population* or *universe,* this group consists of a set of individuals who share at least one characteristic, whether common citizenship, membership in a voluntary association, ethnicity, college enrollment, or the like. Thus, we might speak about the population of the United States, the population of labor union members, the population of African Americans residing in a southern town, or the population of university students.

Because social researchers operate with limited time, energy, and economic resources, they rarely study each and every member of a given population. Instead, researchers study only a *sample*—a smaller number of individuals from the population. Through the sampling process, social researchers seek to generalize from a sample (a small group) to the entire population from which it was taken (a larger group).

In recent years, for example, sampling techniques have allowed political pollsters to make fairly accurate predictions of election results based on samples of only a few hundred registered voters. For this reason, candidates for major political offices routinely monitor the effectiveness of their campaign strategy by examining sample surveys of voter opinion.

Sampling is an integral part of everyday life. How else would we gain much information about other people than by sampling those around us? For example, we might casually discuss political issues with other students to find out where students generally stand with respect to their political opinions; we might attempt to determine how our classmates are studying for a particular examination by contacting only a few members of the class beforehand; we might even invest in the stock market after finding that a small sample of our associates has made money through investments.

Sampling Methods

The social researcher's methods of sampling are usually more thoughtful and scientific than those of everyday life. He or she is primarily concerned with whether sample members are representative enough of the entire population to permit making accurate generalizations about that population. To make such inferences, the researcher selects an appropriate sampling method according to whether every member of the population has an equal chance of being drawn into the sample. If every population member is given an equal chance of sample selection, a *random* sampling method is being used; otherwise, a *nonrandom* type is employed.

Nonrandom Samples

The most popular nonrandom sampling method, *accidental* sampling, differs least from our everyday sampling procedures, because it is based exclusively on what is convenient for the researcher. That is, the researcher simply includes the most convenient cases in his or her sample and excludes the inconvenient cases. Most students can recall at least a few instances when an instructor, who was doing research, has asked all students in the class to take part in an experiment or to fill out a questionnaire. The popularity of this form of accidental sampling in psychology has provoked some observers to view psychology as "the science of the college sophomore," because so many college students are the subjects for research.

A new form of accidental sampling makes use of the telephone's 900 numbers to solicit public opinion on a wide variety of issues, both trivial and serious. Most of the time, these polls are presented with a disclaimer about their unscientific basis. Clearly, it is difficult to generalize from a sample of volunteer respondents who have paid to register their point of view (sometimes more than once).

Another nonrandom type is *quota* sampling. In this sampling procedure, diverse characteristics of a population, such as age, gender, social class, or ethnicity, are sampled in the proportions that they occupy in the population. Suppose, for instance, that we were asked to draw a quota sample from the students attending a university, where 42% were females and 58% were males. Using this method, interviewers are given a quota of students to locate, so that 42% of the sample consists of females and 58% of the sample consists of males. The same percentages are included in the sample as are represented in the larger population. If the total sample size is 200, then 84 female students and 116 male students are selected. Although gender may be properly represented in this sample, other characteristics such as age or race may not be. The inadequacy of quota sampling is precisely its lack of control over factors other than those set by quota.

A third variety of nonrandom sample is known as *judgment* or *purposive* sampling. In this type of sampling, logic, common sense, or sound judgment can be used to select a sample that is representative of a larger population. For instance, to draw a judgment sample of magazines that reflect middle-class values, we might, on an intuitive level, select *Reader's Digest, People,* or *Parade,* because articles from these titles *seem* to depict what most middle-class Americans desire (for example, the fulfillment of the American dream, economic success, and the like). In a similar way, those state districts that have traditionally voted for the winning candidates for state office might be polled in an effort to predict the outcome of a current state election.

Random Samples

As previously noted, random sampling gives every member of the population an equal chance of being selected for the sample.[1] This characteristic of random sampling indicates that every member of the population must be identified before the random sample is drawn, a requirement usually fulfilled by obtaining a list that includes every population member. A little thought will suggest that getting such a complete list of the population members is not always an easy task, especially if one is studying a large and diverse population. To take a relatively easy example, where could we get a *complete* list of students currently enrolled in a large university? Those social researchers who have tried will attest to its difficulty. For a more laborious assignment, try finding a list of every resident in a large city. How can we be certain of identifying everyone, even those residents who do not wish to be identified?

The basic type of random sample, *simple random* sampling, can be obtained by a process not unlike that of the now familiar technique of putting everyone's name on separate slips of paper and, while blindfolded, drawing only a few names from a hat. This procedure ideally gives every population member an equal chance for sample selection since one, and only one, slip per person is included. For several reasons (including the fact that the researcher would need an extremely large hat), the social researcher attempting to take a random sample

---

[1]Sometimes certain groups in a population are oversampled by giving their members a larger chance of selection. This inequality is compensated for in data analysis, however, and the sample is still considered random.

usually does not draw names from a hat. Instead, she or he uses a *table of random numbers*, such as Table B in Appendix C. A portion of a table of random numbers is shown:

Column Number

Row Number	**1**	**2**	**3**	**4**	**5**	**6**	**7**	**8**	**9**	**10**	**11**	**12**	**13**	**14**	**15**	**16**	**17**	**18**	**19**	**20**
1	2	3	1	5	7	5	4	8	5	9	0	1	8	3	7	2	5	9	9	3
2	6	2	4	9	7	0	8	8	6	9	5	2	3	0	3	6	7	4	4	0
3	0	*4*	*5*	5	5	0	*4*	*3*	*1*	0	5	3	7	4	3	5	0	8	9	0
4	1	1	8	3	7	4	4	1	0	9	6	2	2	1	3	4	3	1	4	8
5	1	6	0	3	5	0	3	2	4	0	4	3	6	2	2	2	3	5	0	0

A table of random numbers is constructed so as to generate a series of numbers having no particular pattern or order. As a result, the process of using a table of random numbers yields a representative sample similar to that produced by putting slips of paper in a hat and drawing names while blindfolded.

To draw a simple random sample by means of a table of random numbers, the social researcher first obtains a list of the population and assigns a unique identifying number to each member. For instance, if she is conducting research on the 500 students enrolled in Introduction to Sociology, she might secure a list of students from the instructor and give each student a number from 001 to 500. Having prepared the list, she proceeds to draw the members of her sample from a table of random numbers. Let us say the researcher seeks to draw a sample of 50 students to represent the 500 members of a class population. She might enter the random numbers table at any number (with eyes closed, for example) and move in any direction, taking appropriate numbers until she has selected the 50 sample members. Looking at the earlier portion of the random numbers table, we might arbitrarily start at the intersection of column 1 and row 3, moving from left to right to take every number that comes up between 001 and 500. The first numbers to appear at column 1 and row 3 are 0, 4, and 5. Therefore, student number 045 is the first population member chosen for the sample. Continuing from left to right, we see that 4, 3, and 1 come up next, so that student number 431 is selected. This process is continued until all 50 sample members have been taken. A note to the student: In using the table of random numbers, always disregard numbers that come up a second time or are higher than needed.

All random sample methods are actually variations of the simple random sampling procedure just illustrated. For instance, with *systematic* sampling, a table of random numbers is not required, because a list of population members is sampled by fixed intervals. By employing systematic sampling, then, every *n*th member of a population is included in a sample of that population. To illustrate, in drawing a sample from the population of 10,000 public housing tenants, we might arrange a list of tenants and then, beginning at a random place, take every tenth name on the list and come up with a sample of 1,000 tenants.

The advantage of systematic sampling is that a table of random numbers is not required. As a result, this method is less time consuming than the simple random procedure, especially for sampling from large populations. On the negative side, taking a systematic sample assumes that position on a list of population members does not influence randomness. If this assumption is not taken seriously, the result may be to overselect certain population members, while underselecting others. This can happen, for instance, when houses are systematically sampled from a list in which corner houses (which are generally more expensive than other houses on the block) occupy a fixed position, or when the names in a telephone directory are sampled by fixed intervals so that names associated with certain ethnic ties are underselected.

Another variation on simple random sampling, the *stratified* sample, involves dividing the population into more homogeneous subgroups or *strata* from which simple random samples are then taken. Suppose, for instance, we wish to study the acceptance of various birth control devices among the population of a certain city. Because attitudes toward birth control vary by religion and socioeconomic status, we might stratify our population on these variables, thereby forming more homogeneous subgroups with respect to the acceptance of birth control. More specifically, say, we could identify Catholics, Protestants, and Jews as well as upper-class, middle-class, and lower-class members of the population. Our stratification procedure might yield the following subgroups or strata:

> Upper-class Protestants
> Middle-class Protestants
> Lower-class Protestants
> Upper-class Catholics
> Middle-class Catholics
> Lower-class Catholics
> Upper-class Jews
> Middle-class Jews
> Lower-class Jews

Having identified our strata, we proceed to take a simple random sample from each subgroup or stratum (for example, from lower-class Protestants, from middle-class Catholics, and so on) until we have sampled the entire population. That is, each stratum is treated for sampling purposes as a complete population, and simple random sampling is applied. Specifically, each member of a stratum is given an identifying number, listed, and sampled by means of a table of random numbers. As a final step in the procedure, the selected members of each subgroup or stratum are combined to produce a sample of the entire population.

Stratification is based on the idea that a homogeneous group requires a smaller sample than does a heterogeneous group. For instance, studying individuals who are walking on a downtown street probably requires a larger sample than studying individuals living in a middle-class suburb. One can usually find individuals walking downtown who have any combination of characteristics. By contrast, persons living in a middle-class suburb are generally more alike with respect to education, income, political orientation, family size, and attitude toward work, to mention only a few characteristics.

On the surface, stratified random samples bear a striking resemblance to the nonrandom quota method previously discussed, because both procedures usually require the inclusion of sample characteristics in the exact proportions that they contribute to the population. Therefore, if 32% of our population is made up of middle-class Protestants, then exactly 32% of our sample must be drawn from middle-class Protestants; in the same way, if 11% of our population consists of lower-class Jews, then 11% of our sample must be similarly constituted, and so on. In the context of stratified sampling, an exception arises when a particular stratum is disproportionately well represented in the sample, making possible a more intensive subanalysis of that group. Such an occasion might arise, for example, when Asian Americans, who constitute a small proportion of a given population, are oversampled in an effort to examine their characteristics more closely.

Despite their surface similarities, quota and stratified samples are essentially different. Whereas members of quota samples are taken by whatever method is chosen by the investigator, members of stratified samples are always selected on a *random* basis, generally by means of a table of random numbers applied to a complete list of the population members.

Before leaving the topic of sampling methods, let us examine the nature of an especially popular form of random sampling known as the *cluster* or *multistage* sampling method. Such samplings are frequently used to minimize the costs of large surveys, which require interviewers to travel to many scattered localities. Employing the cluster method, at least two levels of sampling are put into operation:

1. The *primary sampling unit* or cluster, which is a well-delineated area that includes characteristics found in the entire population (for example, a city, census tract, city block, and so on)
2. The sample members within each cluster

For illustrative purposes, imagine that we wanted to interview a representative sample of individuals living in a large area of the city. Drawing a simple random, systematic, or stratified sample of respondents scattered over a wide area would entail a good deal of traveling, not to mention time and money. By means of multistage sampling, however, we would limit our interviewing to those individuals who are located within relatively few clusters. We might begin, for example, by treating the city block as our primary sampling unit or cluster. We might proceed, then, by obtaining a list of all city blocks within the area, from which we take a simple random sample of blocks. Having drawn our sample of city blocks, we might select the individual respondents (or households) on each block by the same simple random method. More specifically, all individuals (or households) on each of the selected blocks are listed and, with the help of a table of random numbers, a sample of respondents from each block is chosen. Using the cluster method, any given interviewer locates one of the selected city blocks and contacts more than one respondent who lives there.

On a much wider scale, the same cluster procedure can be applied to nationwide surveys by treating counties or cities as the primary sampling units initially selected and by interviewing a simple random sample of each of the chosen counties or cities. In this way, interviewers need not cover every county or city, but only a much smaller number of such areas that have been randomly selected for inclusion.

Sampling Error

Throughout the remainder of the text, we will be careful to distinguish between the characteristics of the samples we study and populations to which we hope to generalize. To make this distinction in our statistical procedures, we can therefore no longer use the same symbols to signify the mean and the standard deviation of both sample and population. Instead, we must employ different symbols, depending on whether we are referring to sample or population characteristics. We will always symbolize the mean of a *sample* as $\overline{X}$ and the mean of a *population* as μ. The standard deviation of a *sample* will be symbolized as s and the standard deviation of its *population* as σ. Because population distributions are rarely ever observed in full, just as with probability distributions, it is little wonder we use the symbols μ and σ for both the population and the probability distributions.

The social researcher typically tries to obtain a sample that is representative of the larger population in which he or she has an interest. Because random samples give every population member the same chance for sample selection, they are in the long run more representative of population characteristics than unscientific methods. As discussed briefly in Chapter 1, however, by chance alone, we can *always* expect some difference between a sample, random or otherwise, and the population from which it is drawn. A sample mean $(\overline{X})$ will almost never be exactly the same as the population mean (μ); a sample standard deviation *(s)* will hardly ever be exactly the same as the population standard deviation (σ). Known as *sampling error,* this difference results regardless of how well the sampling plan has been designed and carried out, under the researcher's best intentions, in which no cheating occurs and no mistakes have been made.

Even though the term *sampling error* may seem strange, you are probably more familiar with the concept than you realize. Recall that election polls, for example, typically generalize from a relatively small sample to the entire population of voters. When reporting the results, pollsters generally provide a margin of error. You might read that the Gallup or Roper organization predicts that candidate X will receive 56% of the vote, with a ±4% margin of error. In other words, the pollsters feel confident that somewhere between 52% (56% − 4%) and 60% (56% + 4%) of the vote will go candidate X's way. *Why can't they simply say that the vote will be precisely 56%?* The reason for the pollsters' uncertainty about the exact vote is because of the effect of sampling error. By the time you finish this chapter, not only will you understand this effect, but you will also be able to calculate the margin of error involved when you generalize from a sample to a population.

Table 6.1 illustrates the operation of sampling error. The table contains the population of 20 final examination grades and three samples A, B, and C drawn at random from

TABLE 6.1   *Population and Three Random Samples of Final Examination Grades*

Population			Sample A	Sample B	Sample C
70	80	93	96	40	72
86	85	90	99	86	96
56	52	67	56	56	49
40	78	57	<u>52</u>	<u>67</u>	<u>56</u>
89	49	48			
99	72	<u>30</u>			
96	94		$\overline{X} = 75.75$	$\overline{X} = 62.25$	$\overline{X} = 68.25$
	$\mu = 71.55$				

this population (each taken with the aid of a table of random numbers). As expected, there are differences among the sample means, and none of them equals the population mean ($\mu = 71.55$).

Sampling Distribution of Means

Given the presence of sampling error, the student may wonder how it is possible *ever* to generalize from a sample to a larger population. To come up with a reasonable answer, let us consider the work of a hypothetical sociologist interested in forces that counteract social isolation in modern America. He decides to focus on long-distance calling as a measure of the extent to which Americans "reach out and touch someone," specifically their friends and family in faraway places.

Rather than attempt to study the millions of Americans who own a phone, he monitors the long-distance calls of a sample of 200 households taken at random from the entire population. The sociologist observes each household to determine how many minutes are spent over a one-week period on long-distance phone calling. He finds in his sample of 200 households that weekly long-distance calling ranges from 0 to 240 minutes, with a mean of 101.55 minutes (see Figure 6.1).

It turns out that this hypothetical social researcher is mildly eccentric. He has a notable fondness—or, rather, compulsion—for drawing samples from populations. So intense is his enthusiasm for sampling that he continues to draw many additional 200-household

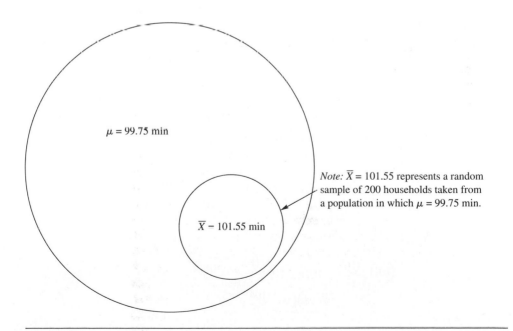

FIGURE 6.1   Mean long-distance phone time for a random sample taken from the hypothetical population

samples and to calculate the average (mean) time spent on long-distance calling for the households in each of these samples. Our eccentric researcher continues this procedure until he has drawn 100 samples containing 200 households *each*. In the process of drawing 100 random samples, he actually studies 20,000 households (200 × 100 = 20,000).

Let us assume, as shown in Figure 6.2, that among the entire population of households in America the average (mean) long-distance phone calling per week is 99.75 minutes. As also illustrated in Figure 6.2, let us suppose that the samples taken by our eccentric social researcher yield means that range between 92 and 109 minutes. In line with our previous discussion, this could easily happen simply on the basis of sampling error.

Frequency distributions of *raw scores* can be obtained from both samples and populations. In a similar way, we can construct a *sampling distribution of means,* a frequency distribution of a large number of random sample *means* that have been drawn from the same population. Table 6.2 presents the 100 sample means collected by our eccentric researcher in the form of a sampling distribution. As when working with a distribution of raw scores, the means in Table 6.2 have been arranged in consecutive order from high to low and their frequency of occurrence indicated in an adjacent column.

Characteristics of a Sampling Distribution of Means

Until this point, we have not directly come to grips with the problem of generalizing from samples to populations. The theoretical model known as the *sampling distribution of means*

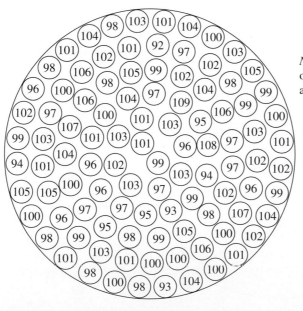

Note: Each $\overline{X}$ represents a sample of 200 households. The 100 $\overline{X}$'s average to 100.4.

FIGURE 6.2   Mean long-distance phone time in 100 random samples taken from a hypothetical population in which $\mu = 99.75$ min

TABLE 6.2 *Observed Sampling Distribution of Means (Long-Distance Phone Time) for 100 Samples*

Mean	f	
109	1	
108	1	
107	2	
106	4	
105	5	
104	7	
103	9	
102	9	
101	11	
100	11	Mean of 100 sample means = 100.4
99	9	
98	9	
97	8	
96	6	
95	3	
94	2	
93	2	
92	1	
	$N = 100$	

(approximated by the 100 sample means obtained by our eccentric social researcher) has certain properties, which give to it an important role in the sampling process. Before moving on to the procedure for making generalizations from samples to populations, we must first examine the characteristics of a sampling distribution of means:

1. *The sampling distribution of means approximates a normal curve.* This is true of all sampling distributions of means regardless of the shape of the distribution of raw scores in the population from which the means are drawn, as long as the sample size is reasonably large (over 30). If the raw data are normally distributed to begin with, then the distribution of sample means is normal regardless of sample size.

2. *The mean of a sampling distribution of means (the mean of means) is equal to the true population mean.* If we take a large number of random sample means from the same population and find the mean of all sample means, we will have the value of the true population mean. Therefore, the mean of the sampling distribution of means is the same as the mean of the population from which it was drawn. They can be regarded as interchangeable values.

3. *The standard deviation of a sampling distribution of means is smaller than the standard deviation of the population.* The sample mean is more stable than the scores that comprise it.

This last characteristic of a sampling distribution of means is at the core of our ability to make reliable inferences from samples to populations. As a concrete example from everyday life, consider how you might compensate for a digital bathroom scale that tends to give you different readings of your weight, even when you immediately step back on it. Obviously, your actual weight doesn't change, but the scale says otherwise. More likely, the scale is very sensitive to where your feet are placed or how your body is postured. The best approach to determining your weight, therefore, might be to weigh yourself four times and take the mean. Remember, the mean weight will be more reliable than any of the individual readings that go into it (see Box 6.1).

Let's now return to the eccentric researcher and his interest in long-distance phone calling. As illustrated in Figure 6.3, the variability of a sampling distribution is always smaller than the variability in either the entire population or any one sample. Figure 6.3(a) shows the population distribution of long-distance phone time with a mean (μ) of 99.75 (ordinarily, we would not have this information). The distribution is skewed to the right: More families spend less than the mean of 99.75 minutes calling long distance, but a few at the right tail seem never to get off the phone. Figure 6.3(b) shows the distribution of long-distance telephone time within *one* particular sample of 200 households. Note that it

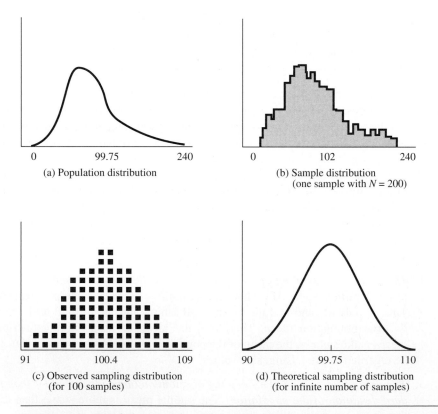

FIGURE 6.3   Population, sample, and sampling distributions

is similar in shape and somewhat close in mean ($\overline{X} = 102$) to the population distribution. Figure 6.3(c) shows the sampling distribution of means (the means from our eccentric researcher's 100 samples). It appears fairly normal rather than skewed, has a mean (100.4) almost equal to the population mean, and has far less variability than either the population distribution in (a) or the sample distribution in (b), which can be seen by comparing the base-line values. Had the eccentric researcher continued forever to take samples of 200 households, a graph of the means of these samples would look like a normal curve, as in Figure 6.3(d). This is the true sampling distribution.

Let's think about the diminished variability of a sampling distribution in another way. In the population, there are many households in which long-distance calling is used sparingly, for less than 30 minutes a month, for example. How likely would it be to get a sample of 200 households with a mean of under 30 minutes? Given that $\mu = 99.75$, it would be virtually impossible. We would have to obtain by random draw a huge number of phonephobics and very few phonaholics. The laws of chance make it highly unlikely that this would occur.

The Sampling Distribution of Means as a Normal Curve

As indicated in Chapter 5, if we define probability in terms of the likelihood of occurrence, then the normal curve can be regarded as a probability distribution (we can say that probability decreases as we travel along the base line away from the mean in either direction).

With this notion, we can find the probability of obtaining various raw scores in a distribution, given a certain mean and standard deviation. For instance, to find the probability associated with obtaining someone with an hourly wage between $10 and $12 in a population having a mean of $10 and a standard deviation of $1.5, we translate the raw score $12 into a z score ($+1.33$) and go to Table A in Appendix C to get the percent of the distribution falling between the z score 1.33 and the mean. This area contains 40.82% of the raw scores. Thus, $P = .41$ rounded off that we will find an individual whose hourly wage lies between $10 and $12. If we want the probability of finding someone whose income is $12/hr or more, we must go to column c of Table A which subtracts the percent obtained in column b of Table A from 50%—that percentage of the area that lies on either side of the mean. From column c of Table A, we learn that 9.18% falls at or beyond $12. Therefore, when the decimal is moved two places to the left, we can say $P = .09$ (9 chances in 100) that we would find an individual whose hourly wage is $12 or more.

In the present context, we are no longer interested in obtaining probabilities associated with a distribution of *raw scores*. Instead, we find ourselves working with a distribution of *sample means,* which have been drawn from the total population of scores, and we wish to make probability statements about these sample means.

As illustrated in Figure 6.4, because the sampling distribution of means takes the form of the normal curve, we can say that probability decreases as we move farther away from the mean of means (true population mean). This makes sense because, as the student may recall, the sampling distribution is a product of chance differences among sample means (sampling error). For this reason, we would expect by chance and chance alone that most sample means will fall close to the value of the true population mean, and relatively few sample means will fall far from it.

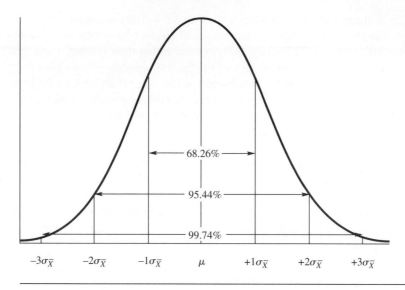

FIGURE 6.4   The sampling distribution of means as a probability distribution

It is critical that we distinguish clearly between the standard deviation of raw scores in the population (σ) and the standard deviation of the sampling distribution of sample means. For this reason, we denote the standard deviation of the sampling distribution by $\sigma_{\overline{X}}$. The use of the Greek letter σ reminds us that the sampling distribution is an unobserved or theoretical probability distribution, and the subscript $\overline{X}$ signifies that this is the standard deviation among all possible sample means.

Figure 6.4 indicates that about 68% of the sample means in a sampling distribution fall between $-1\sigma_{\overline{X}}$ and $+1\sigma_{\overline{X}}$ from the mean of means (true population mean). In probability terms, we can say that $P = .68$ of any given sample mean falling within this interval. In the same way, we can say the probability is about .95 (95 chances out of 100) that any sample mean falls between $-2\sigma_{\overline{X}}$ and $+2\sigma_{\overline{X}}$ from the mean of means, and so on.

Because the sampling distribution takes the form of the normal curve, we are also able to use z scores and Table A to get the probability of obtaining any sample mean, not just those that are exact multiples of the standard deviation. Given a mean of means (μ) and standard deviation of the sampling distribution ($\sigma_{\overline{X}}$), the process is identical to that used in the previous chapter for a distribution of raw scores. Only the names have been changed.

Imagine, for example, that a certain university claims its recent graduates earn an average (μ) annual income of $25,000. We have reason to question the legitimacy of this claim and decide to test it out on a random sample of 100 alumni who graduated within the last two years. In the process, we get a sample mean of only $23,500. We now ask this question: How probable is it that we would get a sample mean of $23,500 or less if the true population mean is actually $25,000? Has the university told the truth? Or is this only an attempt to propagandize to the public to increase enrollments or endowments? Figure 6.5 illustrates the area for which we seek a solution. Because the sample size is fairly large

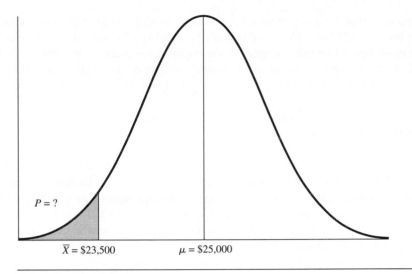

$P = ?$

$\overline{X} = \$23,500$ $\mu = \$25,000$

FIGURE 6.5 The probability associated with obtaining a sample mean of $23,500 or less if the true population mean is $25,000 and the standard deviation is $700

($N = 100$), the sampling distribution of means is approximately normal, even if the distribution of incomes of the individual graduates is not.

To locate a sample mean in the sampling distribution in terms of the number of standard deviations it falls from the center, we obtain the z score:

$$z = \frac{\overline{X} - \mu}{\sigma_{\overline{X}}}$$

where $\overline{X}$ = sample mean in the distribution

μ = mean of means (equal to the university's claim as to the true population mean)

$\sigma_{\overline{X}}$ = standard deviation of the sampling distribution of means

Suppose we know hypothetically that the standard deviation of the sampling distribution is $700. Following the standard procedure, we translate the sample mean $23,500 into a z score as follows:

$$z = \frac{23,500 - 25,000}{700} = -2.14$$

The result of the previous procedure is to tell us that a sample mean of $23,500 lies exactly 2.14 standard deviations below the claimed true population mean of $25,000. Going to column b of Table A in Appendix C we see that 48.38% of the sample means fall between $23,500 and $25,000. Column c of Table A gives us the percent of the distribution that represents sample means of $23,500 or less, if the true population mean is $25,000. This figure is 1.62%. Therefore, the probability is .02 rounded off

(2 chances out of 100) of getting a sample mean of $23,500 or less, when the true population mean is $25,000. With such a small probability of being wrong, we can say with some confidence that the true population mean is *not* actually $25,000. It is doubtful whether the university's report of the annual income of its recent graduates represents anything but bad propaganda.

Standard Error of the Mean

Up until now, we have pretended that the social researcher actually has firsthand information about the sampling distribution of means. We have acted as though he or she, like the eccentric researcher, really has collected data on a large number of sample means, which were randomly drawn from some population. If so, it would be a simple enough task to make generalizations about the population, because the mean of means takes on a value that is equal to the true population mean.

In actual practice, the social researcher rarely collects data on more than one or two samples, from which he or she still expects to generalize to an entire population. Drawing a sampling distribution of means requires the same effort as it might take to study every population member. As a result, the social researcher does not have actual knowledge as to the mean of means or the standard deviation of the sampling distribution. However, the standard deviation of a theoretical sampling distribution (the distribution that would exist in theory if the means of all possible samples were obtained) can be derived. This quantity—known as the *standard error of the mean* $(\sigma_{\bar{X}})$—is obtained by dividing the population standard deviation by the square root of the sample size. That is,

$$\sigma_{\bar{X}} = \frac{\sigma}{\sqrt{N}}$$

To illustrate, the IQ test is standardized to have a population mean (μ) of 100 and a population standard deviation (σ) of 15. If one were to take a sample size of 10, the sample mean would be subject to a standard error of

$$\sigma_{\bar{X}} = \frac{15}{\sqrt{10}}$$

$$= \frac{15}{3.1623}$$

$$= 4.74$$

Thus, whereas the population of IQ scores has a standard deviation $\sigma = 15$, the sampling distribution of the sample mean for $N = 10$ has a standard error (theoretical standard deviation) $\sigma_{\bar{X}} = 4.74$.

As previously noted, the social researcher who investigates only one or two samples cannot know the mean of means, the value of which equals the true population mean. He

BOX 6.1 • *Practical and Statistical: Weigh It Again, Sam*

I just threw out my digital bathroom scale, which I have owned for almost 20 years (it was the first model on the market). When you stepped on the scale, the digital reading would start at "00" and start counting quickly until it locked on your weight. There was only one problem with the scale: it wasn't very reliable. If I stepped on it twice, it wouldn't give me the same weight twice. Apparently, it was very sensitive to how close my feet were to its center and to how I was postured.

That didn't discourage me, however. I simply weighed myself four times each morning and took the mean as my weight. I counted on the fact that the standard deviation among the means

would be less than the standard deviation of the individual readings. In other words, the mean is more reliable than the values that go into it. Had I been more compulsive, I could have weighed myself 25 times each morning, which would have given me a very reliable mean.

What made me throw away the scale, however, was that I went for a checkup last week and on the doctor's scale I was five pounds heavier! The theory of sampling distributions can ensure a precise or reliable mean, but it cannot guarantee a valid or accurate one. From the look of things, when it came to eating, the samples I took were too large!

or she only has the obtained sample mean, which differs from the true population mean as the result of sampling error. But have we not come full circle to our original position? How is it possible to estimate the true population mean from a single sample mean, especially in light of such inevitable differences between samples and populations?

We have, in fact, traveled quite some distance from our original position. Having discussed the nature of the sampling distribution of means, we are now prepared to esti- mate the value of a population mean. With the aid of the standard error of the mean, we can find the range of mean values within which our true population mean is likely to fall. We can also estimate the probability that our population mean actually falls within that range of mean values. This is the concept of the *confidence interval*.

Confidence Intervals

To explore the procedure for finding a *confidence interval,* let us continue with the case of IQ scores. Suppose that the dean of a certain private school wants to estimate the mean IQ of her student body without having to go through the time and expense of administering tests to all 1,000 students. Instead, she selects 25 students at random and gives them the test. She finds that the mean for her sample is 105. She also realizes that because this value of $\overline{X}$ comes from a sample rather than the entire population of students, she cannot be sure that $\overline{X}$ is actually reflective of the student population. As we have already seen, after all, sampling error is the inevitable product of only taking a portion of the population.

We do know, however, that 68.26% of all random sample means in the sampling distribution of means will fall between −1 standard error and +1 standard error from the true population mean. In our case (with IQ scores for which $\sigma = 15$), we have a standard error of

$$\sigma_{\overline{X}} = \frac{\sigma}{\sqrt{N}}$$

$$= \frac{15}{\sqrt{25}}$$

$$= \frac{15}{5}$$

$$= 3$$

Therefore, using 105 as an *estimate* of the mean for all students (an estimate of the true population mean), we can establish a range within which there are 68 chances out of 100 (rounded off) that the true population mean will fall. Known as the *68% confidence interval* (CI), this range of mean IQs is graphically illustrated in Figure 6.6.

The 68% confidence interval can be obtained in the following manner:

$$68\% \text{ CI} = \overline{X} \pm \sigma_{\overline{X}}$$

where $\overline{X}$ = sample mean
       $\sigma_{\overline{X}}$ = standard error of the sample mean

By applying this formula to the problem at hand,

$$68\% \text{ CI} = 105 \pm 3$$
$$= 102 \text{ to } 108$$

The dean can therefore conclude with 68% confidence that the mean IQ for the entire school (μ) is 105, give or take 3. In other words, there are 68 chances out of 100 ($P = .68$)

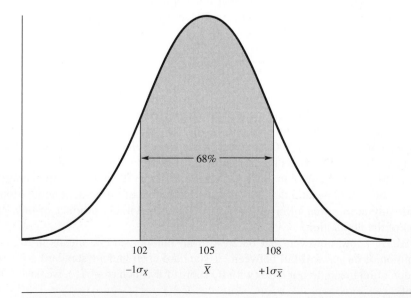

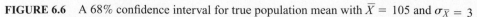

FIGURE 6.6   A 68% confidence interval for true population mean with $\overline{X} = 105$ and $\sigma_{\overline{X}} = 3$

that the true population mean lies within the range 102 to 108. This estimate is made despite sampling error, but with a ±3 margin of error and at a specified probability level (68%), known as the *level of confidence.*

Confidence intervals can technically be constructed for any level of probability. Social researchers are not confident enough to estimate a population mean knowing there are only 68 chances out of 100 of being correct (68 out of every 100 sample means fall within the interval between 102 and 108). As a result, it has become a matter of convention to use a *wider,* less precise confidence interval having a *better probability* of making an accurate or true estimate of the population mean. Such a standard is found in the *95% confidence interval* (CI), whereby the population mean is estimated, knowing there are 95 chances out of 100 of being right; there are 5 chances out of 100 of being wrong (95 out of every 100 sample means fall within the interval). Even when using the 95% confidence interval, however, it must always be kept firmly in mind that the researcher's sample mean could be one of those five sample means that fall outside the established interval. In statistical decision making, one never knows for certain.

How do we go about finding the 95% confidence interval? We already know that 95.44% of the sample means in a sampling distribution lie between $-2\sigma_{\bar{X}}$ and $+2\sigma_{\bar{X}}$ from the mean of means. Going to Table A, we can make the statement that 1.96 standard errors in both directions cover exactly 95% of the sample means (47.50% on either side of the mean of means). To find the 95% confidence interval, we must first multiply the standard error of the mean by 1.96 (the interval is 1.96 units of $\sigma_{\bar{X}}$ in either direction from the mean). Therefore,

$$95\% \text{ CI} = \bar{X} \pm 1.96\sigma_{\bar{X}}$$

where $\bar{X}$ = sample mean
   $\sigma_{\bar{X}}$ = standard error of the sample mean

If we apply the 95% confidence interval to our estimate of the mean IQ of a student body, we see that the

$$
\begin{aligned}
95\% \text{ CI} &= 105 \pm (1.96)(3) \\
&= 105 \pm 5.88 \\
&= 99.12 \text{ to } 110.88
\end{aligned}
$$

Therefore, the dean can be 95% confident that the population mean lies in the interval 99.12 to 110.88. Note that, if asked whether her students are above the norm in IQ (the norm is 100), she could not quite conclude that to be the case with 95% confidence. This is because the true population mean of 100 is within the 95% realm of possibilities based on these results. However, given the 68% confidence interval (102 to 108), the dean could assert with 68% confidence that students at her school average above the norm in IQ.

An even more stringent confidence interval is the *99% confidence interval.* From Table A in Appendix C, we see that the *z* score 2.58 represents 49.50% of the area on either side of the curve. Doubling this amount yields 99% of the area under the curve; 99% of the sample means fall into that interval. In probability terms, 99 out of every 100 sample means fall between $-2.58\sigma_{\bar{X}}$ and $+2.58\sigma_{\bar{X}}$ from the mean. Conversely, only 1 out of every 100 means falls outside of the interval. By formula,

$$99\% \text{ CI} = \bar{X} \pm 2.58\sigma_{\bar{X}}$$

where $\bar{X}$ = sample mean
$\qquad$ $\sigma_{\bar{X}}$ = standard error of the sample mean

With regard to estimating the mean IQ for the population of students,

$$\begin{aligned}
99\% \text{ CI} &= 105 \pm (2.58)(3) \\
&= 105 \pm 7.74 \\
&= 97.26 \text{ to } 112.74
\end{aligned}$$

Consequently, based on the sample of 25 students, the dean can infer with 99% confidence that the mean IQ for the entire school is between 97.26 and 112.74.

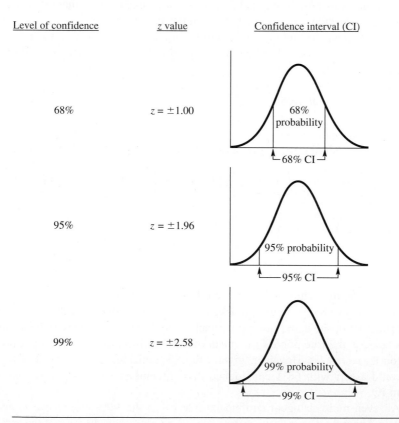

FIGURE 6.7   Levels of confidence

Note that the 99% confidence interval consists of a wider band (97.26 to 112.74) than does the 95% confidence interval (99.12 to 110.88). The 99% interval encompasses more of the total area under the normal curve and therefore a larger number of sample means. This wider band of mean scores gives us greater confidence that we have accurately estimated the true population mean. Only a single sample mean in every 100 lies outside the interval. On the other hand, by increasing our level of confidence from 95% to 99%, we have also sacrificed a degree of precision in pinpointing the population mean. Holding sample size constant, the social researcher must choose between greater precision or greater confidence that he or she is correct.

The precision of an estimate is determined by the *margin of error*, obtained by multiplying the standard error by the *z* score representing a desired level of confidence. This is the extent to which the sample mean is expected to vary from the population mean due to sampling error alone.

Figure 6.7 compares confidence intervals for the 68%, the 95%, and the 99% levels of confidence. The greater the level of confidence that the interval includes the true population mean, the larger the *z* score, the larger the margin of error, and the wider the confidence interval.

BOX 6.2 • *Step-by-Step Illustration: 95% Confidence Interval Using z*

Let's summarize the method for finding the 95% confidence interval using a step-by-step illustration. Suppose that a certain automobile company wishes to determine the expected miles per gallon for one of its new models. Based on years of experience with cars, the company statistician realizes that not all cars are equal and that a standard deviation of 4 miles per gallon ($\sigma = 4$) is to be expected due to variations in parts and workmanship. To estimate the mean miles per gallon for the new model, he test runs a random sample of 100 cars off the assembly line and obtains a sample mean of 26 miles per gallon.

We follow these steps for obtaining a 95% confidence interval for the mean miles per gallon for all cars of this model.

Step 1   Obtain the mean for a random sample (in this problem, it is given).

$$N = 100 \quad \bar{X} = 26$$

Step 2   Calculate the standard error of the mean (knowing that $\sigma = 4$).

$$\sigma_{\bar{X}} = \frac{\sigma}{\sqrt{N}}$$

$$= \frac{4}{\sqrt{100}}$$

$$= \frac{4}{10}$$

$$= .4$$

BOX 6.2   Continued

Step 3   Compute the margin of error by multiplying the standard error of the mean by 1.96, the value of z for a 95% confidence interval.

$$\text{Margin of error} = 1.96\sigma_{\bar{X}}$$
$$= (1.96)(.4)$$
$$= .78$$

Step 4   Add and subtract the margin of error from the sample mean to find the range of mean scores within which the population mean is expected, with 95% confidence, to fall.

$$99\% \text{ CI} = \bar{X} \pm 1.96\sigma_{\bar{X}}$$
$$= 26 \pm .78$$
$$= 25.22 \text{ to } 26.78$$

Thus, the statistician can be 95% confident that the true mean miles per gallon for this model (μ) is between 25.22 and 26.78.

BOX 6.3 • *Step-by-Step Illustration: 99% Confidence Interval Using z*

Reporting these data to his superiors, the statistician is informed that 95% confidence is not confident enough for their needs. To be 99% confident, the statistician need not collect more data but only perform some additional calculations for a 99% confidence interval using a different value of z.

Step 1   Obtain the mean for a random sample (this is the same as with the 95% confidence interval).

$$N = 100 \qquad \bar{X} = 26$$

Step 2   Calculate the standard error of the mean (this is the same as with the 95% confidence interval).

$$\sigma_{\bar{X}} = \frac{\sigma}{\sqrt{N}}$$
$$= \frac{4}{\sqrt{100}}$$
$$= \frac{4}{10}$$
$$= .4$$

Step 3   Compute the margin of error by multiplying the standard error of the mean by 2.58, the value of z for a 99% confidence interval (we begin to see a change from the 95% confidence interval).

$$\text{Margin of error} = 2.58\sigma_{\overline{X}}$$
$$= (2.58)(.4)$$
$$= 1.03$$

Step 4 Add and subtract the margin of error from the sample mean to find the range of mean scores within which the population mean is expected, with 99% confidence, to fall.

$$99\% \text{ CI} = \overline{X} \pm 2.58\sigma_{\overline{X}}$$
$$= 26 \pm 1.03$$
$$= 24.97 \text{ to } 27.03$$

Thus, the statistician is 99% confident that the true mean miles per gallon for this model (μ) is between 24.97 and 27.03. Reporting this with 99% certainty to his superiors, they complain that the interval is now wider than it was before and that the estimate is less precise. He explains to them that the greater the level of confidence, the larger the interval, so that 99% of the possible sample means are encompassed, rather than just 95%. They are still not pleased, so the statistician decides to go back and increase the sample size, which will decrease the standard error and, as a result, will narrow the confidence intervals.

The *t* Distribution

Thus far, we have only dealt with situations in which the standard error of the mean was known or could be calculated from the population standard deviation by the formula,

$$\sigma_{\overline{X}} = \frac{\sigma}{\sqrt{N}}$$

If you think about it realistically, it makes little sense that we would know the standard deviation of our variable in the population (σ), but not know and need to estimate the population mean (μ). Indeed, there are very few cases when the population standard deviation (and thus the standard error of the mean $\sigma_{\overline{X}}$) is known. In certain areas of education and psychology, the standard deviations for standardized scales such as the SAT and IQ scores are determined by design of the test. Usually, however, we need to estimate not only the population mean from a sample, but also the standard error from the same sample.

To obtain an *estimate* of the standard error of the mean, one might be tempted simply to substitute the sample standard deviation *(s)* for the population standard deviation in the previous standard error formula. This, however, would have the tendency to underestimate the size of the true standard error ever so slightly. This problem arises because the sample standard deviation tends to be a bit smaller than the population standard deviation.

Recall from Chapter 3 that the mean is the point of balance within a distribution of scores; the mean is the point in a distribution around which the scores above it perfectly balance with those below it, as in the lever and fulcrum analogy in Figure 3.2. As a result, the sum of squared deviations (and, therefore, the variance and standard deviation) computed around the mean is smaller than from any other point of comparison.

Thus, for a given sample drawn from a population, the sample variance and standard deviation (s^2 and s) are smaller when computed from the sample mean than they would be if one actually knew and used the population mean (μ) in place of the sample mean. In a sense, the sample mean is custom tailored to the sample, whereas the population mean is off the rack; it fits the sample data fairly well but not perfectly like the sample mean does. Thus, the sample variance and the standard deviation are slightly biased estimates (tend to be too small) of the population variance and standard deviation.

It is necessary, therefore, to let out the seam a bit, that is, to inflate the sample variance and standard deviation slightly to produce more accurate estimates of the population variance and population standard deviation. To do so, we divide by $N - 1$ rather than N. That is, unbiased estimates of the population variance and the population standard deviation are given by

$$\hat{\sigma}^2 = \frac{\Sigma(X - \overline{X})^2}{N - 1}$$

and

$$\hat{\sigma} = \sqrt{\frac{\Sigma(X - \overline{X})^2}{N - 1}}$$

The caret over the Greek letter σ indicates that it is an unbiased sample estimate of this population value.[2] Note that in large samples this correction is trivial (s^2 and s are almost equivalent to $\hat{\sigma}^2$ and $\hat{\sigma}$). This should be the case because in large samples the sample mean tends to be a very reliable (close) estimate of the population mean.

The distinction between the sample variance and standard deviation using the sample size N as the denominator versus the sample estimate of the population variance and standard deviation using $N - 1$ as the denominator may be small computationally but is important theoretically. That is, it makes little difference in terms of the final numerical result whether we divide by N or $N - 1$, especially if the sample size N is fairly large. Still, there are two very different purposes for calculating the variance and standard deviation: (1) to describe the extent of variability within a sample of cases or respondents and (2) to make an inference or generalize about the extent of variability within the larger population of cases from which a sample was drawn. It is likely that an example would be helpful right about now.

Suppose that an elementary school teacher is piloting a new language-based math curriculum that teaches math skills through word problems and logical reasoning, rather than through rote memorization of math facts. Just before the end of the school year, she administers a math test to her class of 25 pupils to determine the extent to which they have learned the material using the novel teaching strategy. Her interest lies not only in the average performance of the class (mean score), but also in whether the new approach tends to

---

[2]Alternatively, $\hat{\sigma}^2$ and $\hat{\sigma}$ can by calculated from s^2 and s by multiplying by a bias correction factor, $N/(N - 1)$. Specifically,

$$\hat{\sigma}^2 = s^2\frac{N}{N - 1} \quad \text{and} \quad \hat{\sigma} = s\sqrt{\frac{N}{N - 1}}$$

be easy for some pupils but difficult for others (standard deviation). In fact, she suspects that the curriculum may be a good one, but not for all kinds of learners. She calculates the sample variance (and standard deviation) using the N denominator because her sole interest is in her particular class of pupils. She has no desire to generalize to pupils elsewhere.

As it turns out, this same class of students had been identified by the curriculum design company as a "test case." Because it would not be feasible to assemble a truly random selection of fourth-graders from around the country into the same classroom, this particular class was viewed as "fairly representative" of fourth-graders. The designers' interest extends well beyond the walls of this particular classroom, of course. Their interest is in using this sample of 25 fourth-graders to estimate the central tendency and variability in the overall population (that is, to generalize to all fourth-graders were they to have had this curriculum). The sample mean test score for the class could be used to generalize to the population mean, but the sample variance and standard deviation would have to be adjusted slightly. Specifically, using $N-1$ in the denominator provides an unbiased or fair estimate of the variability that would exist in the entire population of fourth-graders.

At this point, we have only passing interest in estimating the population standard deviation. Our primary interest here is in estimating the standard error of the mean based on a sample of N scores. The same correction procedure applies, nevertheless. That is, an unbiased estimate of the standard error of the mean is given by replacing σ by s and N by $N-1$,

$$s_{\overline{X}} = \frac{s}{\sqrt{N-1}}$$

where s is the sample standard deviation, as obtained in Chapter 4, from a distribution of raw scores or from a frequency distribution. Technically, the unbiased estimate of the standard error should be symbolized by $\hat{\sigma}_{\overline{X}}$ rather than $s_{\overline{X}}$. However, for the sake of simplicity, $s_{\overline{X}}$ can be used without any confusion as the unbiased estimate of the standard error.

One more problem arises when we estimate the standard error of the mean. The sampling distribution of means is no longer quite normal if we do not know the population standard deviation. That is, the ratio

$$\frac{X - \mu}{s_{\overline{X}}}$$

with an estimated standard error in the denominator, does not quite follow the z or normal distribution. The fact that we estimate the standard error from sample data adds an extra amount of uncertainty in the distribution of sample means beyond that which arises due to sampling variability alone. In particular, the sampling distribution of means when we estimate the standard error is a bit wider (more dispersed) than a normal distribution, because of this added source of variability (that is, the uncertainty in estimating the standard error of the mean). The ratio

$$t = \frac{\overline{X} - \mu}{s_X}$$

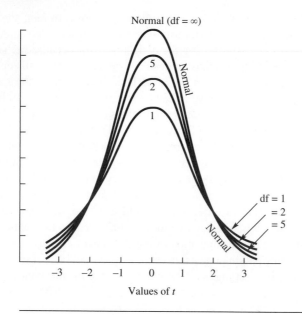

FIGURE 6.8   Family of *t* distributions

follows what is known as the *t* distribution, and thus it is called the *t ratio*. There is actually a whole family of *t* distributions (see Figure 6.8). A concept known as *degrees of freedom* (which we will encounter often in later chapters) is used to determine which of the *t* distributions applies in a particular instance. The degrees of freedom indicate how close the *t* distribution comes to approximating the normal curve.[3] When estimating a population mean, the degrees of freedom are one less than the sample size; that is,

$$\text{df} = N - 1$$

The greater the degrees of freedom, the larger the sample size and the closer the *t* distribution gets to the normal distribution. This makes good sense, because the extent of uncertainty that causes us to use a *t* ratio rather than a *z* score diminishes as the sample size gets larger. In other words, the quality or reliability of our estimate of the standard error of the mean increases as our sample size increases, and so the *t* ratio approaches a *z* score. Recall that the only difference between the *t* ratio and the *z* score is that the former uses an estimate of the standard error based on sample data. We repeat for the sake of emphasis

---

[3]Another way to look at the concept of degrees of freedom is a bit more subtle. Degrees of freedom are the number of observations that are free rather than fixed. When calculating the sample variance for use in determining the estimate of the standard error $(s_{\bar{X}})$, we really do not have N free observations. Because the sample mean is an element in calculating the sample standard deviation, this must be considered a fixed quantity. Then, once we have all but the last observation ($N - 1$ of them), the last observation is predetermined. For example, for the set of data 2, 3, and 7, we take as given that the mean is 4 when we calculate the sample standard deviation. If someone told you that the mean of three cases was 4 and that two of the cases were 2 and 3, you would then know that the last case had to be 7. This is because, for the mean of three observations to be 4, $\Sigma X = 12$.

that, as the sample size and thus the degrees of freedom increase, the *t* distribution becomes a better approximation to the normal or *z* distribution.

When dealing with the *t* distribution, we use Table C rather than Table A. Unlike Table A, for which we had to search out values of *z* corresponding to 95% and 99% areas under the curve, Table C is calibrated for special areas. More precisely, Table C is calibrated for various levels of α (the Greek letter alpha). The α value represents the area in the tails of the *t* distribution. Thus, the α value is *1 minus the level of confidence*. That is,

$$\alpha = 1 - \text{level of confidence}$$

For example, for a 95% level of confidence, $\alpha = .05$. For a 99% level of confidence, $\alpha = .01$.

We enter Table C (the first page of the table that indicates two-tailed test in the header—more on this later) with two pieces of information: (1) the degrees of freedom (which, for estimating a sample mean, is $N - 1$) and (2) the alpha value, the area in the tails of the distribution. For example, if we wanted to construct a 95% confidence interval with a sample size of 20, we would have 19 degrees of freedom (df = $20 - 1 = 19$), $\alpha = .05$ area combined in the two tails, and, as a result, a *t* value from Table C of 2.093.

What would one do, however, for larger samples for which the degrees of freedom may not appear in Table C? For instance, a sample size of 50 produces 49 degrees of freedom. The *t* value for 49 degrees of freedom and $\alpha = .05$ is somewhere between 2.021 (for 40 df) and 2.000 (for 60 df). Given that these two values of *t* are so close, it makes little practical difference what we decide on for a compromise value. However, to be on the safe side, it is recommended that one go with the more modest degrees of freedom (40) and thus the larger value of *t* (2.021).

The reason *t* is not tabulated for all degrees of freedom over 30 is that they become so close that it would be like splitting hairs. Note that the values of *t* get smaller and tend to converge as the degrees of freedom increase. For example, the *t* values for $\alpha = .05$ begin at 12.706 for 1 df, decrease quickly to just under 3.0 for 4 df, gradually approach a value of 2.000 for 60 df, and finally approach a limit of 1.960 for infinity degrees of freedom (that is, an infinitely large sample). This limit of 1.960 is also the .05 value for *z* we found earlier from Table A. Again, we see that the *t* distribution approaches the *z* or normal distribution as the sample size increases.

Thus, for cases in which the standard error of the mean is estimated, we can construct confidence intervals using an appropriate table value of *t* as follows:

$$\boxed{\text{Confidence interval} = \overline{X} \pm t s_{\overline{X}}}$$

BOX 6.4 • *Step-by-Step Illustration: Confidence Interval Using t*

With a step-by-step example, let's see how the use of the *t* distribution translates into constructing confidence intervals. Suppose that a researcher wanted to examine the extent of cooperation among kindergarten children. To do so, she unobtrusively observes a group of

(continued)

BOX 6.4   Continued

children at play for 30 minutes and notes the number of cooperative acts engaged in by each child. Here are the number of cooperative acts exhibited by each child:

X
1
5
2
3
4
1
2
2
4
3

Step 1   Find the mean of the sample.

X	
1	
5	$\bar{X} = \dfrac{\Sigma X}{N}$
2	
3	$= \dfrac{27}{10}$
4	
1	$= 2.7$
2	
2	
4	
3	
$\Sigma X = 27$	

Step 2   Obtain the standard deviation of the sample (we will use the formula for raw scores).

X	X^2	
1	1	$s = \sqrt{\dfrac{\Sigma X^2}{N} - \bar{X}^2}$
5	25	
2	4	
3	9	$= \sqrt{\dfrac{89}{10} - (2.7)^2}$
4	16	
1	1	$= \sqrt{8.9 - 7.29}$
2	4	
2	4	$= \sqrt{1.61}$
4	16	
3	9	$= 1.2689$
	$\Sigma X^2 = 89$	

Step 3   Obtain the estimated standard error of the mean.

$$s_{\bar{X}} = \frac{s}{\sqrt{N-1}}$$

$$= \frac{1.2689}{\sqrt{10-1}}$$

$$= \frac{1.2689}{3}$$

$$= .423$$

Step 4   Determine the value of t from Table C.

$$\text{df} = N - 1 = 10 - 1 = 9$$
$$\alpha = .05$$

Thus,

$$t = 2.262$$

Step 5   Obtain the margin of error by multiplying the standard error of the mean by 2.262.

$$\text{Margin of error} = ts_{X}$$
$$= (2.262)(.423)$$
$$= .96$$

Step 6   Add and subtract this product from the sample mean to find the interval within which we are 95% confident the population mean falls:

$$95\% \text{ CI} = \bar{X} \pm ts_{\bar{X}}$$
$$= 2.7 \pm .96$$
$$= 1.74 \text{ to } 3.66$$

Thus, we can be 95% certain that the mean number of cooperative acts for all kindergartners is between 1.74 and 3.66.

To construct a 99% confidence interval, Steps 1 through 3 would remain the same. Next, with df = 9 and $\alpha = .01$ (that is, $1 - .99 = .01$), from Table B, we find $t = 3.250$. The 99% confidence interval is then

$$99\% \text{ CI} = \bar{X} \pm ts_{\bar{X}}$$
$$= 2.7 \pm (3.250)(.423)$$
$$= 2.7 \pm 1.37$$
$$= 1.33 \text{ to } 4.07$$

Thus, we can be 99% confident that the population mean (mean number of cooperative acts) is between 1.33 and 4.07. This interval is somewhat wider than the 95% interval (1.74 to 3.66), but for this trade-off we gain greater confidence in our estimate.

Estimating Proportions

Thus far, we have focused on procedures for estimating population means. The social researcher often seeks to come up with an estimate of a population *proportion* strictly on the basis of a proportion obtained in a random sample. A familiar circumstance is the pollster whose data suggest that a certain proportion of the vote will go to a particular political issue or candidate for office. When a pollster reports that 45% of the vote will be in favor of a certain candidate, he does so with the realization that he is less than 100% certain. In general, he is 95% or 99% confident that his estimated proportion falls within the range of proportions (for example, between 40% and 50%).

We estimate proportions by the procedure that we have just used to estimate means. All statistics—including means and proportions—have their sampling distributions, and the sampling distribution of a proportion is normal. Just as we earlier found the standard error of the mean, we can now find the *standard error of the proportion*. By formula,

$$s_p = \sqrt{\frac{P(1-P)}{N}}$$

where s_P = standard error of the proportion (an estimate of the standard deviation of the sampling distribution of proportions)

   P = sample proportion

   N = total number in the sample

For illustrative purposes, let us say 45% of a random sample of 100 college students report they are in favor of the legalization of all drugs. The standard error of the proportion would be

$$s_P = \sqrt{\frac{(.45)(.55)}{100}}$$

$$= \sqrt{\frac{.2475}{100}}$$

$$= \sqrt{.0025}$$

$$= .05$$

The *t* distribution was used previously for constructing confidence intervals for the population mean when *both* the population mean (μ) and the population standard deviation (σ) were unknown and had to be estimated. When dealing with proportions, however, only *one* quantity is unknown: We estimate the population proportion π (the Greek letter *pi*) by the sample proportion *P*. Consequently, we use the *z* distribution for constructing confidence intervals for the population proportion (π) (with $z = 1.96$ for a 95% confidence interval and $z = 2.58$ for a 99% confidence interval), rather than the *t* distribution.

To find the 95% confidence interval for the population proportion, we multiply the standard error of the proportion by 1.96 and add and subtract this product to and from the sample proportion:

$$95\% \text{ CI} = P \pm 1.96 s_P$$

where P = sample proportion
  s_P = standard error of the proportion

If we seek to estimate the proportion of college students in favor of the legalization of drugs,

$$
\begin{aligned}
95\% \text{CI} &= .45 \pm (1.96)(.05) \\
&= .45 \pm .098 \\
&= .352 \text{ to } .548
\end{aligned}
$$

We are 95% confident that the true population proportion is neither smaller than .352 nor larger than .548. More specifically, somewhere between 35% and 55% of this population of college students are in favor of the legalization of all drugs. There is a 5% chance we are wrong; 5 times out of 100 such confidence intervals will not contain the true population proportion.

BOX 6.5 • *Step-by-Step Illustration: Confidence Interval for Proportions*

One of the most common applications of confidence intervals for proportions arises in election polling. Polling organizations routinely report not only the proportion (or percentage) of a sample of respondents planning to vote for a particular candidate, but also the margin of error—that is, z multiplied by the standard error.

Suppose that a local polling organization contacted 400 registered voters by telephone and asked the respondents whether they intended to vote for candidate A or B. Suppose that 60% reported their intention to vote for candidate A. Let us now derive the standard error, margin of error, and 95% confidence interval for the proportion indicating preference for candidate A.

Step 1   Obtain the standard error of the proportion.

$$
\begin{aligned}
s_P &= \sqrt{\frac{P(1-P)}{N}} \\
&= \sqrt{\frac{(.60)(1-.60)}{400}} \\
&= \sqrt{\frac{.24}{400}}
\end{aligned}
$$

(continued)

BOX 6.5   Continued

$$= \sqrt{.0006}$$
$$= .0245$$

Step 2   Multiply the standard error of the proportion by $z = 1.96$ to obtain the margin of error.

$$\text{Margin of error} = (1.96)s_P$$
$$= (1.96)(.0245)$$
$$= .048$$

Step 3   Add and subtract the margin of error to find the confidence interval.

$$95\%\text{CI} = P \pm (1.96)s_P$$
$$= .60 \pm .048$$
$$= .552 \text{ to } .648$$

Thus, with a sample size of 400, the poll has a margin of error of ±4.8%, or about 5%. Given the resulting confidence interval (roughly, 55% to 65%), candidate A can feel fairly secure about her prospects for receiving more than 50% of the vote.

Summary

Researchers rarely work directly with every member of a population. Instead, they study a smaller number of cases known as a sample. In this chapter we explored the key concepts and procedures related to generalizing from a sample to the entire group (that is, population) that a social researcher attempts to understand. If every population member is given an equal chance of being drawn into the sample, a random sampling method is being applied; otherwise, a nonrandom type is employed. It was suggested that sampling error—the inevitable difference between a sample and a population based on chance—can be expected to occur despite a well-designed and well-executed random sampling plan. As a result of sampling error, it is possible to characterize a sampling distribution of means, a theoretical distribution in the shape of a normal curve whose mean (the mean of means) equals the true population mean. With the aid of the standard error of the mean, it is possible to estimate the standard deviation of the sampling distribution of means. Armed with this information, we can construct confidence intervals for means (or proportions) within which we have confidence (95% or 99%) that the true population mean (or proportion) actually falls. In this way, it is possible to generalize from a sample to a population. The confidence interval establishes a margin of error when estimating population parameters (such as μ and π) based on sample statistics (such as $\overline{X}$ and P). The size of the margin of error depends on three factors: the level of confidence (95% or 99%) used, the magnitude of the standard deviation (s) of the variable being analyzed, and the number of cases in the sample (N). In this chapter we also introduced the t distribution for the many circumstances in which the population standard deviation is unknown and must be estimated from sample data. The t distribution will play a major role in the hypothesis tests presented in the following chapter.

Terms to Remember

Population (universe)
Sample
Nonrandom sample
  Accidental
  Quota
  Judgment or purposive
Random sample
  Simple random sample
  Systematic sample
  Stratified sample
  Cluster or multistage sample
Sampling error
Sampling distribution of means

Standard error of the mean
Confidence interval
Level of confidence
95% confidence interval
99% confidence interval
Margin of error
t distribution
Degrees of freedom
Alpha (α)
Standard error of the proportion

Questions and Problems

1. In _____ sampling, every member of a population is given an equal chance of being selected for the sample.
   a. accidental
   b. quota
   c. judgment
   d. random

2. The inevitable difference between the mean of a sample and the mean of a population based on chance alone is a
   a. random sample.
   b. confidence interval.
   c. sampling error.
   d. probability.

3. A frequency distribution of a random sample of means is a
   a. random sample.
   b. confidence interval.
   c. sampling distribution of means.
   d. standard error of the mean.

4. Why does a sampling distribution of means take the shape of the normal curve?
   a. Random sample
   b. Probability
   c. Confidence interval
   d. Sampling error

5. Alpha represents the area
   a. in the tails of a distribution.
   b. toward the center of the distribution.
   c. higher than the mean of the distribution.
   d. lower than the mean of the distribution.

6. Suppose that the population standard deviation (σ) for a normally distributed standardized test of achievement is known to be 7.20. What would the standard error of the sample mean ($\sigma_{\bar{X}}$) be if we were to draw a random sample of 16 test scores?

7. Suppose that the random sample in problem 6 yielded these observed scores:

6	5	6	12	5	10	11	13
12	10	9	20	23	20	28	18

   a. Find the 95% confidence interval for the mean.
   b. Find the 99% confidence interval for the mean.

8. Now suppose that we did not feel comfortable assuming that $\sigma = 7.20$. Use the scores in problem 7 to
   a. estimate the standard error of the sample mean ($s_{\bar{X}}$).
   b. find the 95% confidence interval for the mean.
   c. find the 99% confidence interval for the mean.

9. Estimate the standard error of the mean with the following sample of 30 responses on a 7-point scale, measuring whether an extremist hate group should be given a permit to demonstrate (1 = strongly oppose through 7 = strongly favor):

3	5	1	4
3	3	6	6
2	3	3	1
1	2	2	1
5	2	1	3
4	3	1	4
5	2	2	3
3	4		

10. With the sample mean in problem 9, find (a) the 95% confidence interval and (b) the 99% confidence interval.

11. Estimate the standard error of the mean with the following sample of 34 scores on a 10-item objective test of political name recognition:

10	1	4	8
10	7	5	5
5	6	6	10
7	6	3	8
5	7	4	7
4	6	5	5
6	5	6	4
7	3	5	4
8	5		

12. With the sample mean in problem 11, find (a) the 95% confidence interval and (b) the 99% confidence interval.

13. Estimate the standard error of the mean with the following sample of 37 scores on a 7-point scale of views on affirmative action (with 1 being the most in favor of affirmative action and 7 being the most opposed to affirmative action):

7	1	4	6
5	1	5	2
4	2	5	3
6	5	2	1
6	7	1	5
2	4	1	7
1	3	3	2
1	5	2	
3	2	4	
6	4	7	

14. Estimate the standard error of the mean with the following sample of 36 responses on a 7-point scale measuring whether sexual orientation should be protected by civil rights laws (with 1 being strongly opposed and 7 being strongly in favor):

1	3	7	4	5	2
2	7	7	4	6	5
1	3	6	5	2	4
3	5	6	1	7	2
7	5	3	1	7	6
3	5	2	2	5	4

15. With the sample mean in problem 14, find (a) the 95% confidence interval and (b) the 99% confidence interval.

16. To determine the views of students on a particular campus about fraternities, an 11-point attitude scale was administered to a random sample of 40 students. This survey yielded a sample mean of 6 (the higher the score, the more favorable the view of the fraternities) and a standard deviation of 1.5.
    a. Estimate the standard error of the mean.
    b. Find the 95% confidence interval for the population mean.
    c. Find the 99% confidence interval for the population mean.

17. A smoking researcher is interested in estimating the average age when cigarette smokers first began to smoke. Taking a random sample of 25 smokers, she determines a sample mean of 16.8 years and a sample standard deviation of 1.5 years. Construct a 95% confidence interval to estimate the population mean age of the onset of smoking.

18. A medical researcher wants to determine how long patients survive once diagnosed with a particular form of cancer. Using data collected on a group of 40 patients with the disease, she observes an average survival time (time until death) of 38 months with a standard deviation of 9 months. Using a 95% level of confidence, estimate the population mean survival time.

19. A medical researcher is interested in the prenatal care received by pregnant women in inner cities. She interviews 35 randomly selected women with children on the streets of Baltimore and finds that the average number of gynecological checkups per pregnancy was 3, with a standard deviation of 1. Using a 95% level of confidence, estimate the population mean number of gynecological visits per pregnancy.

20. An educational researcher sought to estimate the average number of close friends that students on a particular campus made during their first year of school. Questioning a random sample of 50 students completing their freshman year, he finds a sample mean of 3 close friends and a sample standard deviation of 1. Construct a 95% confidence interval to estimate the mean number of close friends made by the population of college students during their first year on campus.

21. An administrator in charge of undergraduate education on a large campus wanted to estimate the average number of books required by instructors. Using bookstore data, she drew a random sample of 25 courses for which she obtained a sample mean of 2.8 books and a sample standard deviation of .4. Construct a 99% confidence interval to estimate the mean number of books assigned by instructors on campus.

22. A local police department attempted to estimate the average rate of speed (μ) of vehicles along a strip of Main Street. With hidden radar, the speed of a random selection of 25 vehicles was measured, which yielded a sample mean of 42 mph and a standard deviation of 6 mph.
    a. Estimate the standard error of the mean.
    b. Find the 95% confidence interval for the population mean.
    c. Find the 99% confidence interval for the population mean.

23. To estimate the proportion of students on a particular campus who favor a campuswide ban on alcohol, a social researcher interviewed a random sample of 200 students from the college population. She found that 36% of the sample favored banning alcohol (sample proportion = .36). With this information, (a) find the standard error of the proportion and (b) find a 95% confidence interval for the population proportion.

24. A polling organization interviewed by phone 400 randomly selected fans about their opinion on random drug testing of professional baseball players and found that 68% favored such a regulation.
    a. Find the standard error of the proportion.
    b. Find the 95% confidence interval for the population proportion.
    c. Find the 99% confidence interval for the population proportion.

25. A major research organization conducted a national survey to determine what percent of Americans feel that things are getting better for them economically. Asking 1,200 respondents called at random if their own economic situation was better today than last year, 45% reported that they were in fact better off than before.
    a. Find the standard error of the proportion.
    b. Find the 95% confidence interval for the population proportion.
    c. Find the 99% confidence interval for the population proportion.

26. A local school district wants to survey parental attitudes toward the proposed elimination of after-school sports as a cost-cutting move. Rather than send a questionnaire home with the students, the school committee decides to conduct a phone survey. Out of 120 parents questioned, 74 supported the plan to cut the sports program.
    a. Find the standard error of the proportion.
    b. Find the 95% confidence interval for the population proportion.
    c. Find the 99% confidence interval for the population proportion.

27. A political pollster surveyed a random sample of 500 registered voters, asking whether they intended to vote for candidate A or candidate B. She found that 54% preferred candidate A. Using a 95% confidence interval, determine whether the pollster is justified in predicting that candidate A will win the election.

28. These days many parents are having children far later in life than the generation of parents before them. A researcher interested in whether teenagers can relate well to older parents interviewed 120 high school students with at least one parent 40 or more years their senior and found that 35% felt that they could not relate well to them.
    a. Find the standard error of the proportion.
    b. Find the 95% confidence interval for the population proportion.
    c. Find the 99% confidence interval for the population proportion.

29. To estimate the proportion of troops who support the U.S. military's "Don't Ask, Don't Tell" policy of discharging people who are openly homosexual, a political researcher interviewed a random sample of 50 troops from the population at a particular army base. She found that 69% of the sample favored this policy. With this information, (a) find the standard error of the proportion and (b) find the 95% confidence interval for the population proportion.

SPSS Exercises

1. Using the Monitoring the Future Study, calculate 95% and 99% confidence intervals for the population mean for American high school students' perception of the risk of five or more drinks once or twice each weekend (V1779). Hint: ANALYZE, COMPARE MEANS, ONE-SAMPLE T-TEST, and set options to desired percentages.

2. Use SPSS to recode lifetime use of marijuana or hashish (V115) by high school seniors as a yes or no variable. Using this recoded variable, calculate a 95% and 99% confidence intervals for the proportion of American high school seniors using marijuana or hashish. Hint: TRANSFORM, RECODE INTO DIFFERENT VARIABLE.

3. Using the General Social Survey, calculate a 95% confidence interval for American public opinion on assisted suicide if a person is tired of living (SUICIDE4).

4. Using the Best Places Study, calculate a 95% confidence interval for the average commute time to work (in minutes) for people living in Metropolitan Statistical Areas (COMMUTE).

5. Choose a variable from either the Monitoring the Future Survey or the General Social Survey to calculate a 95% confidence interval. Remember it has to be an interval level variable or a dichotomous (two category) variable.

Looking at the Larger Picture: Generalizing from Samples to Populations

At the end of Part I, we described characteristics of the survey of high school students regarding cigarette and alcohol use. We determined that 62% of the respondents smoked and that, among smokers, the average daily consumption was 16.9 cigarettes. In terms of alcohol, the mean number of occasions on which respondents had had a drink in the past months averaged 1.58.

Recognizing that these survey respondents are but a sample drawn from a larger population, we can now estimate the population proportions and means along with confidence intervals for this larger population. But what exactly is the population or universe that this sample can represent? In the strictest sense, the population is technically the entire high school student body. But since it may be safe to assume that this high school is fairly typical of urban public high schools around the country, we might also be able to generalize the findings to urban public high school students in general. By contrast, it would be hard to assume that the students selected at this typical urban high school could be representative of all high school students, even those in suburban and rural areas or private and sectarian schools. Thus, all we can reasonably hope for is that the sample drawn here (250 students from a typical urban high school) can be used to make inferences about urban public high school students in general.

As shown, 62% of the students within the entire Sample smoked. From this we can calculate the standard error of the proportion as 3.1% and then generate a 95% confidence interval for the population proportion who smoke. We find that we can be 95% certain that the population proportion (π) is between 55.9% and 68.1%, indicating that somewhere between 56% and 68% of all urban public high school students smoke. Moving next to daily smoking habits for the smokers alone, we can use

Confidence Intervals

Variable	Statistic	
If a smoker		
	N	250
	%	62.0%
	SE	3.1%
	95% CI	55.9% to 68.1%
Daily cigarettes		
	N	155
	Mean	16.9
	SE	0.84
	95% CI	15.3 to 18.6
Occasions drinking		
	N	250
	Mean	1.58
	SE	0.07
	95% CI	1.44 to 1.73

the sample mean (16.9) to estimate with 95% confidence the population mean (μ). The standard error of the mean is .84, producing a 95% confidence interval between 15.3 and 18.6, indicating that the average smoker in urban public high schools consumes between 15 and almost 19 cigarettes daily. Finally, for drinking, the mean is 1.58 and the standard error is .07, yielding a 95% confidence interval which extends from 1.44 to 1.73 occasions, a fairly narrow band. Thus, the average student drinks on less than two occasions per month. The following table summarizes these confidence intervals.

At the end of Part II, we also looked at differences in the distribution of daily smoking for male and female students. Just as we did overall, we can construct confidence intervals separately for each gender. As shown in the following table, the confidence interval for percentage of males

who smoke (44.1% to 60.1%) is entirely below the corresponding confidence interval for the percentage of females who smoke (63.5% to 79.5%). Although we will encounter a formal way to test these differences in Part III, it does seem that we can identify with some confidence a large gender difference in the percentage who smoke. In terms of the extent of smoking for male and female smokers, we are 95% confident that the population mean for males is between 13.5 and 18.3 cigarettes daily, and we are also 95% confident that the population mean for females is between 15.4 and 20.0 cigarettes daily. Since these two confidence intervals overlap (that is, for example, the population mean for both males and for females could quite conceivably be about 17), we cannot feel so sure about a real gender difference in the populations. We will also take on this question again in Part III.

Confidence Intervals by Gender

Variable	Statistic	Group	
		Males	*Females*
If a smoker			
	N	127	123
	%	52.8%	71.5%
	SE	4.4%	4.1%
	95% CI	44.1% to 61.5%	63.5% to 79.5%
Daily smoking			
	N	67	88
	Mean	15.9	17.7
	SE	1.24	1.14
	95% CI	13.5 to 18.3	15.4 to 20.0

Part III

Decision Making

7

Testing Differences between Means

In Chapter 6, we saw that a population mean or proportion can be estimated from the information we gain in a single sample. For instance, we might estimate the level of anomie in a certain city, the proportion of aged persons who are economically disadvantaged, or the mean attitude toward racial separation among a population of Americans.

Although the descriptive, fact-gathering approach of estimating means and proportions has obvious importance, it *does not* constitute the primary decision-making goal or activity of social research. Quite to the contrary, most social researchers are preoccupied with the task of *testing hypotheses.*

Suppose, for example, a local restaurant announces a contest designed to enhance its lunchtime business. According to the promotion, 20% of the computer-generated meal checks, selected at random, will have a red star printed on them, signifying that lunch is "on the house." You have eaten at the restaurant four times since the promotion was started, and still no free lunch. Should you begin to question the authenticity of the restaurant's promotion? How about if you go 0 for 8, or 0 for 16; is it time to complain, or should you pass it off as your bad luck?

Based on what we know from the rules of probability presented in Chapter 5, your chance of not getting a free meal is 0.8 each time you dine. Four winless lunches has a $(.8)^4 = .410$ probability, based on the multiplication rule—not at all unlikely. The chance of losing eight times in a row is $(.8)^8 = .168$—less likely, but still hardly justification for a complaint to the Better Business Bureau. But 16 losses without a free meal is very improbable, $(.8)^{16} = .028$, assuming that the restaurant's claim that a random 20% of the meals are free is indeed valid.

Indeed, 16 losses is so improbable that you should question whether the assumption that 20% of checks have a lucky red star is true. Of course, there is that small probability (.028) that you are "cursed," but the social scientist (not believing in curses) would conclude instead that the assumption that 20 percent of the checks have red stars should be rejected.

This is the logic (though not quite the mechanics) of hypothesis testing. We establish a hypothesis about populations, collect sample data, and see how likely the sample results are, given our hypothesis about the populations. If the sample results are reasonably plausible under the hypothesis about the populations, we retain the hypothesis and attribute any departure from our expected results to pure chance based on sampling error. On the other hand, if the sample results are so unlikely (for example, less than 5 chances in 100), we then reject the hypothesis.

In the social sciences, hypotheses typically pertain to differences between groups. When testing these differences, social researchers ask such questions as these: Do Germans differ from Americans with respect to obedience to authority? Do Protestants or Catholics have a higher rate of suicide? Do political conservatives discipline their children more severely than political liberals? Note that each research question involves making a *comparison* between two groups: conservatives versus liberals, Protestants versus Catholics, Germans versus Americans.

Take a more concrete example. Suppose that a gerontologist is interested in comparing two methods for enhancing the memory of nursing home residents. She selects 10 residents and then randomly divides them into two groups. One group is assigned to Method A and the other to Method B.

Suppose further that following the memory-enhancement training, all 10 subjects are administered the same test of recall. The sample mean score for 5 subjects under Method A is 82, and the sample mean for the Method B group is 77. Is Method A better at enhancing recall? Perhaps. Perhaps not. It is impossible to draw any conclusion until we know more about the data.

Let us say just for a moment that the sets of recall scores for the two groups of nursing home residents were as follows:

Method A	Method B
82	78
83	77
82	76
80	78
83	76
Mean = 82	Mean = 77

In the Method A group, the recall scores are consistently in the low 80s, so we would have a good deal of confidence that the population mean would be close to the sample mean of 82. Similarly, we would say with confidence that the population mean for the Method B group is likely near the sample mean of 77. Given the homogeneity of the scores and thus the sample means, we could probably conclude that the difference between sample means is more than just a result of pure chance or sampling error. In fact, Method A appears to be more effective than Method B.

Now suppose instead that the following sets of scores produced the two sample means 82 and 77. It is clear that in both groups there is wide variability or spread among the recall scores. As a result, both sample means are relatively unstable estimates of their respective population means. Given the heterogeneity of the sample scores and the unreliability of the sample means, we therefore could not conclude that the difference between sample means is anything more than a result of pure chance or sampling error. In fact, there is not enough evidence to conclude that Method A is more effective than Method B.

Method A	Method B
90	70
98	90
63	91
74	56
85	78
Mean = 82	Mean = 77

The Null Hypothesis: No Difference between Means

It has become conventional in statistical analysis to set out testing the *null hypothesis,* which says that two samples have been drawn from equivalent populations. According to the null hypothesis, any observed difference between samples is regarded as a chance occurrence resulting from sampling error alone. Therefore, an obtained difference between two sample means does not represent a true difference between their population means.

In the present context, the null hypothesis can be symbolized as

$$\mu_1 = \mu_2$$

where μ_1 = mean of the first population
    μ_2 = mean of the second population

Let us examine null hypotheses for the research questions posed earlier:

1. Germans are no more or less obedient to authority than Americans.
2. Protestants have the same suicide rate as Catholics.
3. Political conservatives and liberals discipline their children to the same extent.

It should be noted that the null hypothesis does not deny the possibility of obtaining differences between *sample* means. On the contrary, it seeks to explain such differences between sample means by attributing them to the operation of sampling error. In accordance with the null hypothesis, for example, if we find that a *sample* of female teachers earns less money ($\overline{X}$ = $32,000) than a *sample* of male teachers ($\overline{X}$ = $33,000), we do not, on that basis, conclude that the *population* of female teachers earns less money than the *population* of male teachers. Instead, we treat our obtained sample difference ($33,000 − $32,000 = $1,000) as a product of sampling error—the difference that inevitably results from the process of sampling from a given population. As we shall see later, this aspect of the null hypothesis provides an important link with sampling theory.

To conclude that sampling error is responsible for our obtained difference between sample means is to retain the null hypothesis. The use of the term *retain* does not imply that we have proved the population means are equal ($\mu_1 = \mu_2$) or even that we believe it. Technically, we are merely unable to reject the null hypothesis due to lack of contradictory evidence. Therefore, the phrase *retain the null hypothesis* will be used throughout the text when we are unable to reject it.

The Research Hypothesis: A Difference between Means

The null hypothesis is generally (although not necessarily) set up with the hope of nullifying it. This makes sense, for most social researchers seek to establish relationships between variables. That is, they are often more interested in finding differences than in determining that differences do not exist. Differences between groups—whether expected on theoretical or empirical grounds—often provide the rationale for research.

If we reject the null hypothesis, if we find our hypothesis of no difference between means probably does not hold, we automatically accept the *research hypothesis* that a true population difference does exist. This is often the hoped for result in social research. The research hypothesis says that the two samples have been taken from populations having different means. It says that the obtained difference between sample means is too large to be accounted for by sampling error.

The research hypothesis for mean differences is symbolized by

$$\mu_1 \neq \mu_2$$

where μ_1 = mean of the first population
         μ_2 = mean of the second population

Note: $\neq$ is read *does not equal.*

The following research hypotheses can be specified for the research questions posed earlier:

1. Germans differ from Americans with respect to obedience to authority.
2. Protestants do not have the same suicide rate as Catholics.
3. Political liberals differ from political conservatives with respect to permissive child-rearing methods.

Sampling Distribution of Differences between Means

In the preceding chapter, we saw that the 100 means from the 100 samples drawn by our eccentric social researcher could be plotted in the form of a sampling distribution of means. In a similar way, let us now imagine that the same eccentric social researcher studies not one but two samples at a time to make comparisons between them.

Suppose, for example, that our eccentric researcher is teaching a course on the sociology of the family. He is curious about whether there are gender differences in attitudes toward child rearing. Specifically, he is interested in determining whether males and females differ in terms of their child-rearing permissiveness.

To test for differences, he first constructs a multi-item scale, which includes several questions about the appropriateness of spanking, chores for young children, and obedience to parental demands. His permissiveness scale ranges from a minimum of 1 (not at all permissive) to 100 (extremely permissive). Next, he selects a random sample of 30 females and a random sample of 30 males from the student directory and administers his questionnaire to all 60 students. As graphically illustrated in Figure 7.1, our eccentric researcher finds his sample of females is more permissive ($\overline{X} = 58.0$) than his sample of males ($\overline{X} = 54.0$).

Before our eccentric researcher concludes that women are actually more permissive than men, he might ask this question: In light of sampling error, can we expect a difference between 58.0 and 54.0 ($58.0 - 54.0 = +4.0$) strictly on the basis of chance and chance alone? Based solely on the luck of the draw, could the female sample have been comprised of more permissive people than the male sample? Must we retain the null hypothesis of no population difference or is the obtained sample difference $+4.0$ large enough to indicate a true population difference between females and males with respect to their child-rearing attitudes?

In Chapter 2 we were introduced to frequency distributions of raw scores from a given population. In Chapter 6 we saw it was possible to construct a sampling distribution of mean scores, a frequency distribution of sample means. In addressing ourselves to the question at hand, we must now take the notion of frequency distribution a step further and

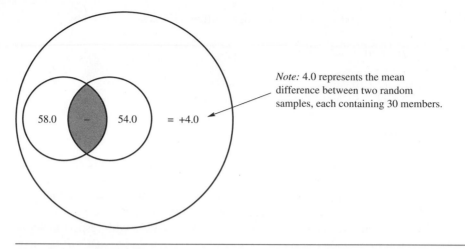

Note: 4.0 represents the mean difference between two random samples, each containing 30 members.

FIGURE 7.1   Mean difference in permissiveness between samples of females and males taken from a hypothetical population

examine the nature of a *sampling distribution of differences between means*—that is, a frequency distribution of a large number of *differences* between sample means that have been randomly drawn from a given population.

To illustrate the sampling distribution of differences between means, let us return to the compulsive activities of our eccentric researcher whose passion for drawing random samples has once again led him to continue the sampling process beyond its ordinary limits. Rather than draw a single sample of 30 females and a single sample of 30 males, he studies 70 *pairs* of such samples (70 pairs of samples, *each* containing 30 females and 30 males), feeling fortunate that he teaches at a large school.

For each pair of samples, the eccentric researcher administers the same scale of child-rearing permissiveness. He then calculates a sample mean for each female sample and a sample mean for each male sample. Thus, he has a female mean and a male mean for each of his 70 pairs of samples.

Next, he derives a difference between means score by subtracting the mean score for males from the mean score for females for each pair of samples. For example, his first comparison produced a difference between means of +4.0. His second pair of means might be 57.0 for the female sample and 56.0 for the male sample, yielding a difference between means score of +1.0. Likewise, the third pair of samples may have produced a mean of 60.0 for the females and a mean of 64.0 for the males, and the difference between means would be −4.0. Obviously, the larger the difference score, the more the two samples of respondents differ with respect to permissiveness. Note that we always subtract the second sample mean from the first sample mean (in the present case, we subtract the mean score for the male sample from the mean score for the female sample). The 70 difference between means scores derived by our eccentric social researcher have been illustrated in Figure 7.2.

Let us suppose that we know that the populations of females and males do not differ at all with respect to permissiveness in child-rearing attitudes. Let us say for the sake of argument that $\mu = 57.0$ in both the female and male populations. If we assume the null

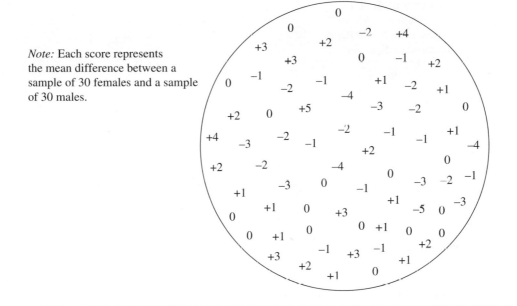

Note: Each score represents the mean difference between a sample of 30 females and a sample of 30 males.

FIGURE 7.2   Seventy mean difference scores representing differences in permissiveness between samples of females and males taken at random from a hypothetical population

hypothesis is correct and that females and males are identical in this respect, we can use the 70 mean differences obtained by our eccentric researcher to illustrate the sampling distribution of differences between means. This is true because the sampling distribution of differences between means makes the assumption that all sample pairs differ only by virtue of sampling error and not as a function of true population differences.

The 70 scores representing differences between means shown in Figure 7.2 have been rearranged in Table 7.1 as a sampling distribution of differences between means. Like the scores in other types of frequency distributions, these have been arranged in consecutive order from high to low, and frequency of occurrence is indicated in an adjacent column.

To depict the key properties of a sampling distribution of differences between means, the frequency distribution from Table 7.1 is graphically presented in Figure 7.3. As illustrated therein, we see that the *sampling distribution of differences between means approximates a normal curve whose mean (mean of differences between means) is zero.*[1] This makes sense because the positive and negative differences between means in the distribution tend to cancel out one another (for every negative value, there tends to be a positive value of equal distance from the mean).

As a normal curve, most of the differences between sample means in this distribution fall close to zero—its middlemost point; there are relatively few differences between means having extreme values in either direction from the mean of these differences. This is to be expected, because the entire distribution of differences between means is a product of sampling

[1]This assumes we have drawn large random samples from a given population of raw scores.

TABLE 7.1 *Sampling Distribution of Mean Differences for 70 Pairs of Random Samples*

Mean Difference[a]	f
+5	1
+4	2
+3	5
+2	7
+1	10
0	18
−1	10
−2	8
−3	5
−4	3
−5	1
	$N = 70$

[a]These difference scores include fractional values (for example, −5 includes the values −5.0 through −5.9).

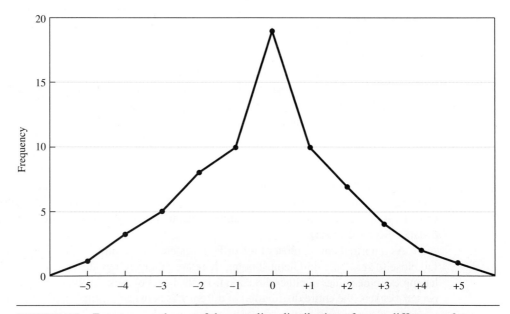

FIGURE 7.3   Frequency polygon of the sampling distribution of mean differences from Table 7.1

error, rather than actual population differences between females and males. In other words, if the actual mean difference between the populations of females and males is zero, we also expect the mean of the sampling distribution of differences between sample means to be zero.

Testing Hypotheses with the Distribution of Differences between Means

In earlier chapters we learned to make probability statements regarding the occurrence of both raw scores and sample means. In the present case we seek to make statements of probability about the difference scores in the sampling distribution of differences between means. As pointed out earlier, this sampling distribution takes the form of the normal curve and, therefore, can be regarded as a probability distribution. We can say that probability decreases as we move farther and farther from the mean of differences (zero). More specifically, as illustrated in Figure 7.4, we see that 68.26% of the mean differences fall between $+1\sigma_{\bar{X}_1-\bar{X}_2}$ and $-1\sigma_{\bar{X}_1-\bar{X}_2}$ from zero. (The notation $\sigma_{\bar{X}_1-\bar{X}_2}$ represents the standard deviation of the differences between $\bar{X}_1$ and $\bar{X}_2$.) In probability terms, this indicates $P = .68$ that any difference between sample means falls within this interval. Similarly, we can say the probability is roughly .95 (95 chances in 100) that any sample mean difference falls between $-2\sigma_{\bar{X}_1-\bar{X}_2}$ and $+2\sigma_{\bar{X}_1-\bar{X}_2}$ from a mean difference of zero, and so on.

The sampling distribution of differences provides a sound basis for testing hypotheses about the difference between two sample means. Unlike the eccentric researcher who compulsively takes pairs of samples, one after another, most normal researchers study only one pair of samples to make inferences on the basis of just one difference between means.

Suppose, for instance, that a normal social researcher wanted to test the eccentric researcher's hypothesis, at least a variation of it, in a realistic way. She agrees that there

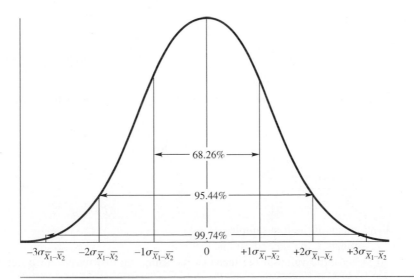

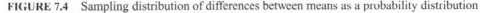

FIGURE 7.4 Sampling distribution of differences between means as a probability distribution

may be gender differences in child-rearing attitudes, but among real parents rather than would-be parents.

She selects at random 30 mothers and 30 fathers (not necessarily married couples) and administers the eccentric researcher's permissiveness scale to all 60 parents. She obtains mean permissiveness scores of 45 for the mothers and 40 for the fathers. The researcher's reasoning then goes like this: If the obtained difference between means of 5 ($45 - 40 = 5$) lies so far from a difference of zero that it has only a small probability of occurrence in the sampling distribution of differences between means, we reject the null hypothesis, which says that the obtained difference between means is a result of sampling error. If, on the other hand, our sample mean difference falls so close to zero that its *probability* of occurrence is large, we must retain the null hypothesis and treat our obtained difference between means merely as a result of sampling error.

Therefore, we seek to determine how far our obtained difference between means (in this case, 5) lies from a mean difference of zero. In so doing, we must first translate our obtained difference into units of standard deviation.

Recall that we translate *raw scores* into units of standard deviation by the formula

$$z = \frac{X - \mu}{\sigma}$$

where X = raw score
μ = mean of the distribution of raw scores
σ = standard deviation of the distribution of raw scores

Likewise, we translate the *mean scores* in a distribution of sample means into units of standard deviation by the formula

$$z = \frac{\overline{X} - \mu}{\sigma_{\overline{X}}}$$

where $\overline{X}$ = sample mean
μ = population mean (mean of means)
$\sigma_{\overline{X}}$ = standard error of the mean (standard deviation of the distribution of means)

In the present context, we similarly seek to translate our sample mean difference $(\overline{X}_1 - \overline{X}_2)$ into units of standard deviation by the formula

$$z = \frac{(\overline{X}_1 - \overline{X}_2) - 0}{\sigma_{\overline{X}_1 - \overline{X}_2}}$$

where $\overline{X}_1$ = mean of the first sample
$\overline{X}_2$ = mean of the second sample
0 = zero, the value of the mean of the sampling distribution of differences between means (we assume that $\mu_1 - \mu_2 = 0$)
$\sigma_{\overline{X}_1 - \overline{X}_2}$ = standard deviation of the sampling distribution of differences between means

Because the value of the mean of the distribution of differences between means is assumed to be zero, we can drop it from the z-score formula without altering our result. Therefore,

$$z = \frac{\overline{X}_1 - \overline{X}_2}{\sigma_{\overline{X}_1 - \overline{X}_2}}$$

With regard to permissiveness between the mother and father samples, we must translate our obtained difference between means into its z-score equivalent. If the standard deviation of the sampling distribution of differences between means is $\sigma_{\overline{X}_1 - \overline{X}_2}$ (more on how to get this number later), we obtain the following z score:

$$z = \frac{45 - 40}{2}$$

$$= \frac{5}{2}$$

$$= +2.5$$

Thus, a difference of 5 between the means for the two samples falls 2.5 standard deviations from a mean difference of zero in the distribution of differences between means.

What is the probability that a difference of 5 or more between sample means can happen strictly on the basis of sampling error? Going to column c of Table A in Appendix C, we learn that $z = 2.5$ cuts off .62% of the area in each tail, or 1.24% in total (see Figure 7.5). Rounding off, $P = .01$ that the mean difference of 5 (or greater than 5) between samples can happen strictly on the basis of sampling error. That is, a mean difference of 5 or more

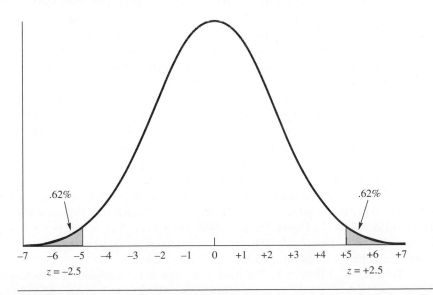

FIGURE 7.5 Graphic representation of the percent of total area in the distribution of differences beyond $z = 2.5$ or $z = -2.5$

occurs by sampling error (and therefore appears in the sampling distribution) *only once* in every 100 mean differences. Knowing this, would we not consider rejecting the null hypothesis and accepting the research hypothesis that a population difference actually exists between mothers and fathers with respect to child-rearing permissiveness? One chance out of 100 represents pretty good odds.

Given that situation, most of us would choose to reject the null hypothesis, even though we might be wrong in doing so (don't forget that 1 chance out of 100 still remains). However, the decision is not always so clear-cut. Suppose, for example, we learn our difference between means happens by sampling error 10 ($P = .10$), 15 ($P = .15$), or 20 ($P = .20$) times out of 100. Do we still reject the null hypothesis? Or do we play it safe and attribute our obtained difference to sampling error?

We need a consistent cutoff point for deciding whether a difference between two sample means is so large that it can no longer be attributed to sampling error. We need a method for determining when our results show a *statistically significant difference.* When a difference is found to be significant, it is then regarded as real. That is, the difference is large enough to be generalized to the populations from which the samples were derived. Thus, statistically significant does not necessarily mean substantively important; nor does it indicate anything about the size of the difference in the population. In large samples, a small difference can be statistically significant; in small samples, a large difference can be a result of sampling error.

Levels of Significance

To establish whether our obtained sample difference is statistically significant—the result of a real population difference and not just sampling error—it is customary to set up a *level of significance,* which we denote by the Greek letter α (alpha). The alpha value is the level of probability at which the null hypothesis can be rejected with confidence and the research hypothesis can be accepted with confidence. Accordingly, we decide to reject the null hypothesis if the probability is very small (for example, less than 5 chances out of 100) that the sample difference is a product of sampling error. Conventionally, we symbolize this small probability by $P < .05$.

It is a matter of convention to use the $\alpha = .05$ *level of significance.* That is, we are willing to reject the null hypothesis if an obtained sample difference occurs by chance less then 5 times out of 100. The .05 significance level is graphically depicted in Figure 7.6. As shown, the .05 level of significance is found in the small areas of the tails of the distribution of mean differences. These are the areas under the curve that represent a distance of plus or minus 1.96 standard deviations from a mean difference of zero. In this case (with an $\alpha = .05$ level of significance), the z scores 1.96 are called *critical values*; if we obtain a z score that exceeds 1.96 (that is, $z > 1.96$ or $z < -1.96$), it is called statistically significant. The shaded regions in Figure 7.6 are called the *critical* or *rejection regions*, because a z score within these areas leads us to reject the null hypothesis (the top portion of the figure shows the critical regions for a .05 level of significance).

To understand better why this particular point in the sampling distribution represents the .05 level of significance, we might turn to column c of Table A in Appendix C to determine the percent of total frequency associated with 1.96 standard deviations from the mean. We see that 1.96 standard deviations in *either* direction represent 2.5% of the differences in sample

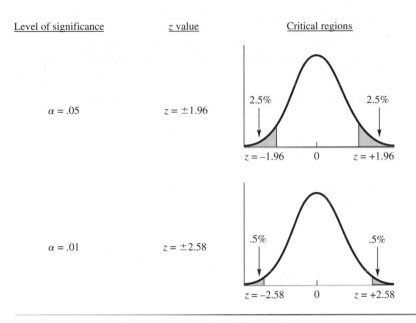

Level of significance	*z* value	Critical regions

FIGURE 7.6 The .05 and .01 levels of significance

means. In other words, 95% of these differences fall between $-1.96\sigma_{\bar{X}_1-\bar{X}_2}$ and $+1.96\sigma_{\bar{X}_1-\bar{X}_2}$ from a mean difference of zero; only 5% fall at or beyond this point $(2.5\% + 2.5\% = 5\%)$.

Significance levels can be set up for any degree of probability. For instance, a more stringent level is the *.01 level of significance*, whereby the null hypothesis is rejected if there is less than 1 chance out of 100 that the obtained sample difference could occur by sampling error. The .01 level of significance is represented by the area that lies 2.58 standard deviations in both directions from a mean difference of zero (see Figure 7.6).

Levels of significance do not give us an *absolute* statement as to the correctness of the null hypothesis. Whenever we decide to reject the null hypothesis at a certain level of significance, we open ourselves to the chance of making the wrong decision. Rejecting the null hypothesis when we should have retained it is known as *Type I error* (see Figure 7.7). A Type I error can only arise when we reject the null hypothesis, and its probability varies according to the level of significance we choose. For example, if we reject the null hypothesis at the .05 level of significance and conclude that there are gender differences in child-rearing attitudes, then there are 5 chances out of 100 we are wrong. In other words, $P = .05$ that we have committed Type I error and that gender actually has no effect at all. Likewise, if we choose the $\alpha = .01$ level of significance, there is only 1 chance out of 100 $(P = .01)$ of making the wrong decision regarding the difference between genders. Obviously, the more stringent our level of significance (the farther out in the tail it lies), the less likely we are to make Type I error. To take an extreme example, setting up a .001 significance level means that Type I error occurs only 1 time in every 1,000. The probability of Type I error is symbolized by α.

The farther out in the tail of the curve our critical value falls, however, the greater the risk of making another kind of error known as *Type II error*. This is the error of retaining the null hypothesis when we should have rejected it. Type II error indicates that our research hypothesis may still be correct, despite the decision to reject it and retain the null hypothesis.

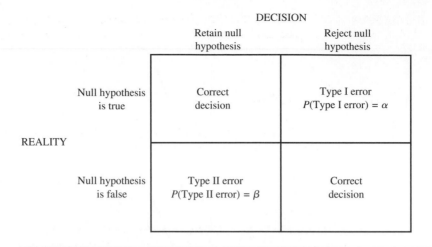

FIGURE 7.7   Type I and type II errors

One method for reducing the risk of committing Type II error is to increase the size of the samples so that a true population difference is more likely to be represented. The probability of Type II error is denoted by β (beta).

We can never be certain that we have not made a wrong decision with respect to the null hypothesis, for we examine only differences between sample means, not between means of the complete population. As long as we do not have knowledge of true population means, we take the risk of making either a Type I or a Type II error, depending on our decision. This is the risk of statistical decision making that the social researcher must be willing to take.

Choosing a Level of Significance

We have seen that the probabilities of Type I error and Type II error are inversely related: The larger one error is, the smaller the other. In practice, a researcher does not have actual control of the likelihood of Type II error (β) directly. That is, she or he cannot set the probability of a Type II error to whatever level is desired. On the other hand, the chance of a Type I error is a quantity directly controlled by the researcher, because it is precisely the level of significance (α) he or she chooses for the hypothesis test. Of course, the larger the chosen level of significance (say, .05 or even .10), the larger the chance of Type I error and the smaller the chance of Type II error. The smaller the chosen significance level (say, .01 or even .001), the smaller the chance of Type I error, but the greater the likelihood of Type II error.

We predetermine our level of significance for a hypothesis test on the basis of which type of error (Type I or Type II) is more costly or damaging and therefore riskier. If in a particular instance it would be far worse to reject a true null hypothesis (Type I error) than to retain a false null hypothesis (Type II error), we should opt for a small level of significance (for example, $\alpha = .01$ or .001) to minimize the risk of Type I error, even at the expense of increasing the chance of Type II error. If, however, it is believed that Type II error would be worse, we would set a large significance level ($\alpha = .05$ or $.10$) to produce a lower chance of Type II error, that is, a lower beta value.

Suppose, for example, a researcher was doing research on gender differences in SAT performance for which she administered an SAT to a sample of males and a sample of females. Before deciding upon a level of significance, she should pause and ask herself, which is worse—claiming that there is a true gender difference on the basis of results distorted by excessive sampling error (Type I error) or not claiming a difference when there is in fact one between the population of males and the population of females? In this instance, a Type I error would probably be far more damaging—could even be used as a basis for discriminating unfairly against women—and so she should elect a small alpha value (say, .01).

Let's consider a reverse situation—one in which Type II error is far more worrisome. Suppose a researcher is testing the effects of marijuana smoking on SAT performance, and he compares a sample of smokers with a sample of nonsmokers. If there was even a modest indication that marijuana smoking affected one's performance, this information should be disseminated. We would not want a researcher to retain the null hypothesis of no population difference between smokers and nonsmokers in spite of an observed difference in sample means just because the difference was not quite significant. This Type II error could have a serious impact on public health. A Type I error, by which the researcher was misled to believe marijuana smoking altered performance when the sample difference was only due to chance, would certainly not be as problematic. Given this situation, the researcher would be advised to select a large alpha level (like .10) to avoid the risk of a serious Type II error. That is, he should be less stringent in rejecting the null hypothesis.

Although some social scientists might debate the point, a majority of social science research involves minimal cost associated with either Type I or Type II error. With the exception of certain policy-sensitive areas, much, but by no means all, social research revolves around relatively nonsensitive issues so that errors in hypothesis testing do not cause any great problems for individuals or society as a whole and may only affect the professional reputation of the researcher. Indeed, one occasionally sees in journals a series of studies that are contradictory; some of this may be due to testing error. In any event, because of the innocuous nature of much research, it has become customary to use a modest level of significance, usually $\alpha = .05$.

What Is the Difference between P and α?

The difference between P and α can be a bit confusing. To avoid confusion, let's compare the two quantities directly. Put simply, P is the exact probability of obtaining the sample data when the null hypothesis is true; the alpha value is the threshold below which the probability is considered so small that we decide to reject the null hypothesis. That is, we reject the null hypothesis if the P value is less than the alpha value and otherwise retain it.

In testing hypotheses, a researcher decides ahead of time on the alpha value. This choice is made on the basis of weighing the implications of Type I and Type II errors or is simply made by custom—that is, $\alpha = .05$. The alpha value refers to the size of the tail regions under the curve (of z, for example) that will lead us to reject the null hypothesis. That is, alpha is the area to the right of the tabled critical value of $+z$ and to the left of the tabled value of $-z$. With $\alpha = .05$, these regions are to the right of $z = +1.96$ and to the

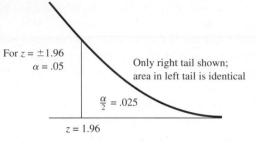

For $z = \pm 1.96$
$\alpha = .05$

Only right tail shown;
area in left tail is identical

$\frac{\alpha}{2} = .025$

$z = 1.96$

Critical value of $z = \pm 1.96$ cuts off an $\alpha = .05$ rejection area

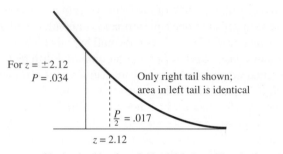

For $z = \pm 2.12$
$P = .034$

Only right tail shown;
area in left tail is identical

$\frac{P}{2} = .017$

$z = 2.12$

Obtained value of $z = 2.12$ yields $P < .05$ and rejects null hypothesis

FIGURE 7.8   Differences between P and α

left of -1.96 (see Figure 7.8). (With the t distribution, the critical values depend also on the degrees of freedom.) Thus, alpha represents the chance of Type I error that we are willing to tolerate.

In contrast, P is the actual probability of obtaining the sample data if the null hypothesis is true. If this probability is small enough (that is, if the null hypothesis is very unlikely), we tend to reject the null hypothesis. Unlike the alpha value, which is determined by the researcher in advance, the P value is determined by the data themselves—specifically by the computed value of the test statistic, such as the z score—and is not set by the researcher. The P value is the area to the right of the calculated $+z$ on the positive side plus the area to the left of the calculated $-z$ on the negative side. Thus, if after data collection, a z score of 2.12 is obtained, we learn from column c of Table A that $P = .034$ (.017 in each tail). If this were the result, we would conclude that $P < .05$, and so we would reject the null hypothesis (see Figure 7.8).

In practice, one does not regularly look up the actual value of P, as we just did from Table A. One only needs to look at whether the calculated z value exceeds the critical value for the chosen α level. Note that the quickest way to determine the critical value for z is to look at the bottom row of the t table. That is, with infinity degrees of freedom, t is equal to z.

If the calculated z exceeds the critical value, we simply say $P < .05$ (if .05 were the significance level preselected) and that the results are statistically significant at the .05 level. If the calculated z does not exceed the critical value, we would say the result (or difference) was not significant. In other words, one does not need to determine the actual value of P to make a decision on the hypothesis. With most statistical software currently available, the exact P values are automatically calculated from elaborate formulas. Therefore, in the future we may see more people giving the actual P value, rather than just saying $P < .05$.

Standard Error of the Difference between Means

We rarely have firsthand knowledge of the standard deviation of the distribution of mean differences. And just as in the case of the sampling distribution of means (Chapter 6), it would be a major effort if we were actually to draw a large number of sample pairs to calculate it. Yet this standard deviation plays an important role in the method for testing hypotheses about mean differences and therefore cannot be ignored.[2]

Fortunately, we do have a simple method whereby the standard deviation of the distribution of differences can be estimated on the basis of just two samples that we actually draw. The sample estimate of the standard deviation of the sampling distribution of differences between means, referred to as the *standard error of the difference between means* and symbolized by $s_{\overline{X}_1 - \overline{X}_2}$, is

$$s_{\overline{X}_1 - \overline{X}_2} = \sqrt{\left(\frac{N_1 s_1^{\,2} + N_2 s_2^{\,2}}{N_1 + N_2 - 2}\right)\left(\frac{N_1 + N_2}{N_1 N_2}\right)}$$

where $s_1^{\,2}$ and $s_2^{\,2}$ are the variances of the two samples first introduced in Chapter 4:

$$s_1^{\,2} = \frac{\Sigma X_1^{\,2}}{N_1} - \overline{X}_1^{\,2}$$

$$s_2^{\,2} = \frac{\Sigma X_2^{\,2}}{N_2} - \overline{X}_2^{\,2}$$

The formula for $s_{\overline{X}_1 - \overline{X}_2}$ combines information from the two samples. Thus, the variance for each sample in addition to the respective sample sizes goes into our estimate of how different $\overline{X}_1$ and $\overline{X}_2$ can be due to sampling error alone. A large difference between $\overline{X}_1$ and $\overline{X}_2$ can result if (1) one mean is very small, (2) one mean is very large, or (3) one mean is moderately small while the other is moderately large. The likelihood of any of these conditions occurring is dictated by the variances and sample sizes present in the respective samples.

[2]In the last chapter, it was pointed out that the true population standard deviation (σ) and standard error ($\sigma_{\overline{X}}$) are rarely known. Also in the two sample cases, the true standard error of the difference is generally unknown. However, for the rare situation in which the standard errors of both sample means are known, the true standard error of the difference between means is

$$\sigma_{\overline{X}_1 - \overline{X}_2} = \sqrt{\sigma_{\overline{X}_1}^{\,2} + \sigma_{\overline{X}_2}^{\,2}} = \sqrt{\frac{\sigma_1^{\,2}}{N_1} + \frac{\sigma_1^{\,2}}{N_1}}$$

Testing the Difference between Means

Suppose a medical sociologist has devised a multi-item scale to measure support for health care reform. She obtained the following data for a sample of 25 liberals and 35 conservatives on the health care reform scale:

Liberals	Conservatives
$N_1 = 25$	$N_2 = 35$
$\overline{X}_1 = 60$	$\overline{X}_2 = 49$
$s_1 = 12$	$s_2 = 14$

From this information, we can calculate the estimate of the standard error of the difference between means:

$$
\begin{aligned}
s_{\overline{X}_1 - \overline{X}_2} &= \sqrt{\left(\frac{N_1 s_1^2 + N_2 s_2^2}{N_1 + N_2 - 2}\right)\left(\frac{N_1 + N_2}{N_1 N_2}\right)} \\
&= \sqrt{\left[\frac{(25)(12)^2 + (35)(14)^2}{25 + 35 - 2}\right]\left(\frac{25 + 35}{(25)(35)}\right)} \\
&= \sqrt{\left(\frac{3{,}600 + 6{,}860}{58}\right)\left(\frac{60}{875}\right)} \\
&= \sqrt{(180.3448)(.0686)} \\
&= \sqrt{12.3717} \\
&= 3.52
\end{aligned}
$$

The standard error of the difference between means (our estimate of the standard deviation of the theoretical sampling distribution of differences between means) turns out to be 3.52. If we were testing the difference in support for health care reform between liberals (mean of 60) and conservatives (mean of 49), we could use our standard error result to translate the difference between sample means into a t ratio:

$$
\boxed{t = \frac{\overline{X}_1 - \overline{X}_2}{s_{\overline{X}_1 - \overline{X}_2}}}
$$

Here

$$
\begin{aligned}
t &= \frac{60 - 49}{3.52} \\
&= \frac{11}{3.52} \\
&= 3.13
\end{aligned}
$$

We use t rather than z because we do not know the true population standard deviations for liberals and conservatives. Because we are estimating both σ_1 and σ_2 from s_1 and s_2, respectively, we compensate by using the wider t distribution, with degrees of freedom $N_1 + N_2 - 2$. For each population standard deviation that we estimate, we lose 1 degree of freedom from the total number of cases. Here, we have 60 cases from which we subtract 2 to obtain the 58 degrees of freedom.

Turning to Table C in Appendix C, we use the critical value for 40 degrees of freedom, the next lowest to 58, which is not given explicitly. Our calculated t value of 3.13 exceeds all the standard critical points, except that for the .001 level. Therefore, we could reject the null hypothesis at the .10, .05, or .01 level, whichever we had established for the alpha value at the start of our study. Had a .001 alpha value been established for whatever reason (there would seem little justification for choosing such a stringent test in this instance), we would have to retain the null hypothesis despite the large t value. Our chance of a Type II error would run quite high as a consequence.

BOX 7.1 • *Step-by-Step Illustration: Test of Difference between Means*

To provide a step-by-step illustration of the foregoing procedure for testing a difference between sample means, let us say that we wanted to test the null hypothesis at the $\alpha = .05$ significance level that male and female college students are equally concerned about employment prospects after graduation (that is, $\mu_1 = \mu_2$). To test this hypothesis, suppose that we surveyed random samples of 35 male and 35 female college students using a scale of employment concern ranging from 1 for "not at all concerned" to 10 for "extremely concerned."

The resulting employment concern scores and, for the sake of our calculations, their squared values are as follows:

Males ($N_1 = 35$)		Females ($N_2 = 35$)	
X_1	X_1^2	X_2	X_2^2
8	64	4	16
7	49	3	9
6	36	1	1
10	100	7	49
2	4	6	36
1	1	9	81
4	16	10	100
3	9	4	16
5	25	3	9
6	36	6	36
7	49	4	16
5	25	3	9
9	81	6	36
8	64	8	64
10	100	6	36
6	36	5	25

(continued)

BOX 7.1   Continued

Males ($N_1 = 35$)		Females ($N_2 = 35$)	
X_1	$X_1{}^2$	X_2	$X_2{}^2$
6	36	7	49
7	49	3	9
4	16	4	16
5	25	6	36
9	81	9	81
10	100	8	64
2	4	4	16
4	16	5	25
3	9	8	64
5	25	2	4
4	16	7	49
8	64	1	1
7	49	5	25
4	16	6	36
9	81	4	16
8	64	8	64
9	81	7	49
10	100	6	36
6	36	4	16
$\Sigma X_1 = 217$	$\Sigma X_1{}^2 = 1{,}563$	$\Sigma X_2 = 189$	$\Sigma X_2{}^2 = 1{,}195$

Step 1   Find the mean for each sample.

$$\overline{X}_1 = \frac{\Sigma X_1}{N_1} \quad \overline{X}_2 = \frac{\Sigma X_2}{N_2}$$

$$= \frac{217}{35} \qquad\quad = \frac{189}{35}$$

$$= 6.2 \qquad\qquad = 5.4$$

Step 2   Find the variance for each sample.

$$s_1{}^2 = \frac{\Sigma X_1{}^2}{N_1} - \overline{X}_1{}^2 \qquad s_2{}^2 = \frac{\Sigma X_2{}^2}{N_2} - \overline{X}_2{}^2$$

$$= \frac{1{,}563}{35} - (6.2)^2 \qquad = \frac{1{,}195}{35} - (5.4)^2$$

$$= 44.66 - 38.44 \qquad = 34.14 - 29.16$$

$$= 6.22 \qquad\qquad\qquad = 4.98$$

Step 3   Find the standard error of the difference between means.

$$s_{\overline{X}_1 - \overline{X}_2} = \sqrt{\left(\frac{N_1 s_1{}^2 + N_2 s_2{}^2}{N_1 + N_2 - 2}\right)\left(\frac{N_1 + N_2}{N_1 N_2}\right)}$$

$$= \sqrt{\left[\frac{35(6.22) + 35(4.98)}{35 + 35 - 2}\right]\left(\frac{35 + 35}{35 \times 35}\right)}$$

$$= \sqrt{\left(\frac{217.7 + 174.3}{68}\right)\left(\frac{70}{1,225}\right)}$$

$$= \sqrt{\left(\frac{392}{68}\right)\left(\frac{70}{1,225}\right)}$$

$$= \sqrt{(5.7647)(.0571)}$$

$$= \sqrt{.3294}$$

$$= .574$$

Step 4   Compute the t ratio by dividing the difference between means by the standard error of the difference between means.

$$t = \frac{\overline{X}_1 - \overline{X}_2}{s_{\overline{X}_1 - \overline{X}_2}}$$

$$= \frac{6.2 - 5.4}{.574}$$

$$= 1.394$$

Step 5   Determine the critical value for t.

$$df = N_1 + N_2 - 2 = 68$$

Because the exact value for degrees of freedom ($df = 68$) is not provided in Table C, we use the next lowest value ($df = 60$) to find

$$\alpha = .05$$
$$\text{table } t = 2.000$$

Step 6   Compare the calculated and table t values.

    The calculated t (1.394) does not exceed the table t (2.000) in either a positive or negative direction, and so we retain the null hypothesis of no difference in population means. That is, the observed difference in mean level of employment concern between samples of male and female college students could easily have occurred as a result of sampling error.

    Thus, even though the sample means are indeed different (6.2 for the males and 5.4 for the females), they were not sufficiently different to conclude that the

(continued)

BOX 7.1 Continued

populations of male and female college students differ in mean concern about employment after graduation. Of course, we could be committing a Type II error (retaining a false null hypothesis), in that there could in fact be a difference between population means. But these sample results are not disparate enough ($\bar{X}_1 - \bar{X}_2$ is not large enough) nor are the sample sizes large enough to allow us to infer that the sample difference would hold in the population.

Adjustment for Unequal Variances

The formula for estimating the standard error of the difference between means that we presented earlier pools or combines together variance information from both samples. In doing so, it is assumed that the population variances are the same for the two groups, that is, $\sigma_1^2 = \sigma_2^2$. Of course, we don't really know the magnitude of the two population variances; how then can we determine if it is reasonable to assume them to be equal?

The answer to this dilemma comes from the sample variances, s_1^2 and s_2^2, and whether they are reasonably similar or wildly dissimilar. After all, these sample variances do give us some indication as to the size of the population variances. If the sample variances are quite different in magnitude, then pooling them together in order to obtain an estimate of the overall standard error of the difference between means would be like combining apples and oranges to make anything other than mixed fruit.

Like other comparisons that we make, the determination of whether s_1^2 and s_2^2 are so different in size depends on sample size (or degrees of freedom). SPSS and other software products utilize the *Levene Test* to assess whether the sample variances are so dissimilar that we must reject the notion that the population variances are the same. Without delving into these details, we can use as a fairly useful rule of thumb that if either sample variance is more than twice as large as the other, we may be wise to seek an alternative to pooling. Specifically, in those instances when either of the sample variances is more than double the other, the standard error is calculated without pooling by the following formula:

$$s_{\bar{X}_1 - \bar{X}_2} = \sqrt{\frac{s_1^2}{N_1 - 1} + \frac{s_2^2}{N_2 - 1}}$$

Notice that the two sample variances are separate under the square root radical, rather than blended as we do in the usual formula for the standard error of the difference between means.

The formula for the degrees of freedom when not combining variances is rather complex, and generally produces fractional values. However, a simple and safe substitute for degrees of freedom is the smaller of N_1 and N_2.

As an example, suppose that a professor asks his class how many hours it takes to do a particularly difficult homework assignment requiring the use of a computer and whether they used a PC or Mac. The results are as follows:

	PC	Mac
N	36	23
Mean	6.5	5.6
Variance	7.8	3.6
Standard deviation	2.8	1.8

We see from the sample results that the PC users took, on average, 0.9 hours longer. More relevant to this illustration is the wider variability among the PC users, with the sample variance more than twice that for Mac users. Using the alternative, non-pooled standard error formula,

$$s_{\overline{X}_1 \; \overline{X}_2} = \sqrt{\frac{s_1^{\,2}}{N_1 - 1} + \frac{s_2^{\,2}}{N_2 - 1}}$$

$$= \sqrt{\frac{7.8}{36 - 1} + \frac{3.6}{23 - 1}}$$

$$= \sqrt{.228 + .165}$$

$$= \sqrt{.393}$$

$$= .626$$

The t ratio is calculated the same as before, but with the non-pooled estimate of the standard error of the difference between means in the denominator:

$$t = \frac{\overline{X}_1 - \overline{X}_2}{s_{\overline{X}_1 - \overline{X}_2}}$$

$$= \frac{6.5 - 5.6}{.626}$$

$$= 1.457$$

For degrees of freedom, we use the smaller of the two sample sizes: df = 23. Consulting Table C in Appendix C for df = 3 and a .05 level of significance, we determine that the t ratio must exceed 2.069 to reject the null hypothesis of equal population means. Therefore, although PC users took nearly one hour longer on average to complete the assignment, we cannot rule out the possibility that this difference occurred as a result of sampling error.

Comparing Dependent Samples

So far, we have discussed making comparisons between two *independently* drawn samples (for example, males versus females, blacks versus whites, or liberals versus conservatives).

Before leaving this topic, we must now introduce a variation of the two mean comparison referred to as a *before–after* or *panel* design: the case of a *single* sample measured at two different points in time (time 1 versus time 2). For example, a researcher may seek to measure hostility in a single sample of children both before and after they watch a certain television program. In the same way, we might want to measure differences in attitudes toward capital punishment before and after a highly publicized murder trial.

Keep in mind the important distinction between studying the same sample on two different occasions versus sampling from the same population on two different occasions. The *t* test of difference between means for the same sample measured twice generally assumes that the same people are examined repeatedly—in other words, each respondent is compared to himself or herself at another point in time.

For example, a polling organization might interview the same 1,000 Americans both in 1995 and 2000 to measure their change in attitude over time. Because the same sample is measured twice, the *t* test of difference between means for the same sample measured twice is appropriate.

Suppose, instead, that this polling organization administered the same survey instrument to one sample of 1,000 Americans in 1995 and a different sample of 1,000 Americans in 2000. Even though the research looks at change in attitude over time, the two samples would have been chosen independently—that is, the selection of respondents in 2000 would not have depended in any way on who was selected in 1995. Although the same population (all Americans) would have been sampled twice, the particular people interviewed would be different, and thus the *t* test of difference between means for independent groups would apply.

BOX 7.2 • *Step-by-Step Illustration: Test of Difference between Means for Same Sample Measured Twice*

To provide a step-by-step illustration of a before–after comparison, let us suppose that several individuals have been forced by a city government to relocate their homes to make way for highway construction. As social researchers, we are interested in determining the impact of forced residential mobility on feelings of neighborliness (that is, positive feelings about neighbors in the *pre*relocation neighborhood versus feelings about neighbors in the *post*relocation neighborhood). In this case, then, μ_1 is the mean score of neighborliness at time 1 (*before* relocating), and μ_2 is the mean score of neighborliness at time 2 (*after* relocating). Therefore,

Null hypothesis: $(\mu_1 = \mu_2)$	*The degree of neighborliness does not differ before and after the relocation.*
Research hypothesis: $(\mu_1 \neq \mu_2)$	*The degree of neighborliness differs before and after the relocation.*

To test the impact of forced relocation on neighborliness, we interview a random sample of six individuals about their neighbors both before and after they are forced to move. Our interviews yield the following scores of neighborliness (higher scores from 1 to 4 indicate greater neighborliness):

Respondent	Before Move (X_1)	After Move (X_2)	Difference $(D = X_1 - X_2)$	(Difference)^2 (D^2)
Stephanie	2	1	1	1
Myron	1	2	−1	1
Carol	3	1	2	4
Inez	3	1	2	4
Leon	1	2	−1	1
David	4	1	3	9
	$\Sigma X_1 = 14$	$\Sigma X_2 = 8$		$\Sigma D^2 = 20$

As the table shows, making a before–after comparison focuses our attention on the *difference* between time 1 and time 2, as reflected in the formula to obtain the standard deviation (for the distribution of before–after difference scores):

$$s_D = \sqrt{\frac{\Sigma D^2}{N} - (\bar{X}_1 - \bar{X}_2)^2}$$

where s_D = standard deviation of the distribution of before–after difference scores
D = after-move raw score subtracted from before-move raw score
N = number of cases or respondents in sample

From this we obtain the standard error of the difference between means:

$$s_{\bar{D}} = \frac{s_D}{\sqrt{N - 1}}$$

Step 1   Find the mean for each point in time.

$$\bar{X}_1 = \frac{\Sigma X_1}{N} \quad \bar{X}_2 = \frac{\Sigma X_2}{N}$$

$$= \frac{14}{6} \qquad = \frac{8}{6}$$

$$= 2.33 \qquad = 1.33$$

Step 2   Find the standard deviation for the difference between time 1 and time 2.

$$s_D = \sqrt{\frac{\Sigma D^2}{N} - (\bar{X}_1 - \bar{X}_2)^2}$$

(continued)

BOX 7.2   Continued

$$= \sqrt{\frac{20}{6} - (2.33 - 1.33)^2}$$

$$= \sqrt{\frac{20}{6} - (1.00)^2}$$

$$= \sqrt{3.33 - 1.00}$$

$$= \sqrt{2.33}$$

$$= 1.53$$

Step 3   Find the standard error of the difference between means.

$$s_{\bar{D}} = \frac{s_D}{\sqrt{N - 1}}$$

$$= \frac{1.53}{\sqrt{6 - 1}}$$

$$= \frac{1.53}{2.24}$$

$$= .68$$

Step 4   Translate the sample mean difference into units of standard error of the difference between means.

$$t = \frac{\bar{X}_1 - \bar{X}_2}{s_{\bar{D}}}$$

$$= \frac{2.33 - 1.33}{.68}$$

$$= \frac{1.00}{.68}$$

$$= 1.47$$

Step 5   Find the number of degrees of freedom.

$$df = N - 1$$

$$= 6 - 1$$

$$= 5$$

Note: N refers to the total number of *cases,* not the number of scores, for which there are two per case or respondent.

Step 6   Compare the obtained t ratio with the appropriate t ratio in Table C.

$$\text{obtained } t = 1.47$$

table $t = 2.571$

df $= 5$

$\alpha = .05$

To reject the null hypothesis at the .05 significance level with 5 degrees of freedom, we must obtain a calculated t ratio of 2.571. Because our t ratio is only 1.47—less than the required table value—we retain the null hypothesis. The obtained sample difference in neighborliness before and after the relocation was probably a result of sampling error.

Testing differences in means for the same sample measured twice is just one of several applications of the t test for dependent samples. Any time that one group is sampled based on the cases in another group, this special t test should be used. For example, comparing the development of identical twins raised apart; comparing the attitudes of women and their husbands to public spending on after-school programs; and comparing the impact of custodial and noncustodial treatment programs for delinquents in groups matched by age, race, gender, and crime type all offer the opportunity to measure and test the average difference between scores among pairs of subjects. In fact, using groups that are matched or paired in some way can offer increased control and power over drawing two samples completely independent of each other (as in the t test for independent samples).

BOX 7.3 • *Step-By-Step Illustration: Test of Difference between Means for Matched Samples*

Criminologists often compare the homicide rates in states with capital punishment and the states without it. Of course, states differ from each other in countless ways that may impact the homicide rate other than whether or not a death penalty statute is on the books. One approach that has been used to overcome the extraneous impact of the other variables is to match death penalty and non-death penalty states in terms of geographic similarity.

Shown below are 2005 homicide rates for seven pairs of states, differing in whether or not capital punishment is on the books but similar in terms of geography and other critical demographic and socioeconomic variables. In addition, the differences and squared differences in homicide rates between the seven pairs are calculated.

Death Penalty		**No Death Penalty**		*D*	D^2
Nebraska	2.5	Iowa	1.3	1.2	1.44
Indiana	5.7	Michigan	6.1	−0.4	0.16
South Dakota	2.3	North Dakota	1.1	1.2	1.44
Connecticut	2.9	Rhode Island	3.2	−0.3	0.09
New Hampshire	1.4	Vermont	1.3	0.1	0.01
Kentucky	4.6	West Virginia	4.4	0.2	0.04
Minnesota	2.2	Wisconsin	3.5	−1.3	1.69
$\Sigma X_1 = 21.6$		$\Sigma X_2 = 20.9$		$\Sigma D = 0.7$	$\Sigma D^2 = 4.87$

(continued)

BOX 7.3   Continued

Using these data, we can test the equality of mean homicide rates for the two groups of paired states, using the following hypotheses:

Null hypothesis: $(\mu_1 = \mu_2)$ — *There is no difference in mean homicide rate between death penalty and non-death penalty states.*

Research hypothesis: $(\mu_1 \neq \mu_2)$ — *There is a difference in mean homicide rate between death penalty and non-death penalty states.*

Step 1   Find the mean for both groups. Note that N refers to the number of matched pairs of scores.

$$\overline{X}_1 = \frac{\Sigma X_1}{N} \qquad \overline{X}_2 = \frac{\Sigma X_2}{N}$$

$$= \frac{21.6}{7} \qquad\quad = \frac{20.9}{7}$$

$$= 3.09 \qquad\qquad = 2.99$$

Step 2   Find the standard deviation of the differences between each pair of cases.

$$s_D = \sqrt{\frac{\Sigma D^2}{N} - (\overline{X}_1 - \overline{X}_2)^2}$$

$$= \sqrt{\frac{4.87}{7} - (3.09 - 2.99)^2}$$

$$= \sqrt{\frac{4.87}{7} - (.10)^2}$$

$$= \sqrt{.70 - .01}$$

$$= \sqrt{.69}$$

$$= .83$$

Step 3   Find the standard error of the mean difference.

$$s_{\overline{D}} = \frac{s_D}{\sqrt{N - 1}}$$

$$= \frac{.83}{\sqrt{7 - 1}}$$

$$= \frac{.83}{2.45}$$

$$= .34$$

Step 4   Calculate the t ratio.

$$t = \frac{\overline{X}_1 - \overline{X}_2}{s_{\overline{D}}}$$

$$= \frac{3.09 - 2.99}{.34}$$

$$= \frac{.10}{.34}$$

$$= .29$$

Step 5   Find the degrees of freedom.

$$df = N - 1$$
$$= 7 - 1$$
$$= 6$$

Step 6   Compare the obtained t ratio with the appropriate value from Table C.

$$\text{Obtained } t = .29$$
$$\text{Table } t = 2.447$$
$$df = 6$$
$$\alpha = .05$$

To reject the null hypothesis of no difference in mean homicide rate between death penalty and non-death penalty states, the t ratio would need to be at least 2.447. Because our obtained t ratio of .29 is less than that, we must retain the hypothesis of no difference.

Two Sample Test of Proportions

In the previous chapter we learned how to construct confidence intervals for means and for proportions using the notion of the standard error of the mean and of the proportion. In this chapter we shifted attention to the difference between samples and employed a standard error of the difference between means. It would seem logical also to consider the sampling distribution of the difference between proportions.

The important role of testing the difference between proportions goes far beyond simply a desire to have our presentation symmetrical and complete. So many important measures used in the social sciences are cast in terms of proportions. We are often interested in knowing if two groups (for example, males–females, whites–blacks, northerners–southerners, etc.) differ in the percentage who favor some political issue, who have some characteristic or attribute, or who succeed on some test.

Fortunately, the logic of testing the difference between two proportions is the same as for testing the difference between means. The only change is in the symbols and formulas used for calculating sample proportions and the standard error of the difference. As before, our statistic (z) is the difference between sample statistics divided by the standard error of the difference.

Rather than the z formula we used for testing the difference between means (with known σ_1 and σ_2), we use

$$z = \frac{P_1 - P_2}{s_{P_1 - P_2}}$$

where P_1 and P_2 are the respective sample proportions. The standard error of the difference in proportions is given by

$$s_{P_1 - P_2} = \sqrt{P^*(1 - P^*)\left(\frac{N_1 + N_2}{N_1 N_2}\right)}$$

where P^* is the combined sample proportion:

$$P^* = \frac{N_1 P_1 + N_2 P_2}{N_1 + N_2}$$

BOX 7.4 • *Step-by-Step Illustration: Test of Difference between Proportions*

We will describe the necessary steps for this test by illustration. A social psychologist is interested in how personality characteristics are expressed in the car someone drives. In particular, he wonders whether men express a greater need for control than women by driving big cars. He takes a sample of 200 males and 200 females over age 18 and determines whether they drive a full-sized car. Any respondent who does not own a car is dropped from the samples but not replaced by someone else. Consequently, the final sample sizes for analysis are not quite 200 for each sex. The following hypotheses concern the population proportion of men with big cars (π_1) and the population proportion of women with big cars (π_2):

Null hypothesis:
($\pi_1 = \pi_2$)

The proportions of men and women who drive big cars are equal.

Research hypothesis:
($\pi_1 \neq \pi_2$)

The proportions of men and women who drive big cars are not equal.

The social psychologist obtains the following data:

	Male	Female	Overall
Sample size (N)	180	150	330
Drive big car (f)	81	48	129
Proportion with big car (P)	0.45	0.32	0.39

Step 1   Compute the two sample proportions and the combined sample proportion.

$$P_1 = \frac{f_1}{N_1} = \frac{81}{180} = .45$$

$$P_2 = \frac{f_2}{N_2} = \frac{48}{150} = .32$$

$$P^* = \frac{N_1P_1 + N_2P_2}{N_1 + N_2}$$

$$= \frac{(180)(.45) + (150)(.32)}{180 + 150}$$

$$= \frac{81 + 48}{180 + 150}$$

$$= \frac{129}{330}$$

$$= .39$$

Step 2   Compute the standard error of the difference.

$$s_{P_1-P_2} = \sqrt{P^*(1 - P^*)\left(\frac{N_1 + N_2}{N_1N_2}\right)}$$

$$= \sqrt{(.39)(1 - .39)\left[\frac{180 + 150}{(180)(150)}\right]}$$

$$= \sqrt{(.39)(.61)\left(\frac{330}{27,000}\right)}$$

$$= \sqrt{\frac{78.507}{27,000}}$$

$$= \sqrt{.002908}$$

$$= .0539$$

Step 3   Translate the difference between proportions into units of the standard error of the difference.

$$z = \frac{P_1 - P_2}{s_{P_1-P_2}}$$

$$= \frac{.45 - .32}{.0539}$$

$$= .241$$

(continued)

BOX 7.4   Continued

Step 4	Compare the obtained z value with the critical value in Table A (or from the bottom row of Table C).

For $\alpha = .05$, the critical value of $z = 1.96$. Because the obtained $z = 2.41$, we reject the null hypothesis. Because the difference between sample proportions was statistically significant, the social psychologist was able to conclude that men and women generally tend to drive different-sized cars. |

One-Tailed Tests

The tests of significance covered thus far are known as *two-tailed tests*: We can reject the null hypothesis at both tails of the sampling distribution. A very large *t* ratio in *either* the positive *or* negative direction leads to rejecting the null hypothesis. A *one-tailed test*, in contrast, rejects the null hypothesis at only one tail of the sampling distribution.

Many statistics texts, particularly older editions, stress both one- and two-tailed tests. The one-tailed test has been deemphasized in recent years because statistical software packages (for example, SPSS) routinely produce two-tailed significance tests as part of their standard output.

Still, on some occasions a one-tailed test is appropriate. We will briefly discuss the differences between one- and two-tailed tests for those instances. It should be emphasized, however, that the only changes are in the way the hypotheses are stated and which *t* table is used; fortunately, none of the formulas presented thus far changes in any way.

Suppose that an educational researcher wishes to test whether a particular remedial math program significantly improves math skills. She decides to use a before–after design in which nine youngsters targeted for remediation are administered a math test and are then given a similar test six months later following participation in the remedial math program. The approach discussed earlier (two-tailed test) would have us set up our hypotheses as follows:

Null hypothesis: *Math ability does not differ before and after remediation.*
$(\mu_1 = \mu_2)$

Research hypothesis: *Math ability differs before and after remediation.*
$(\mu_1 \neq \mu_2)$

The research hypothesis covers two possible results—that the students do far better (the mean score on the posttest is higher than on the pretest) and that they do far worse (the posttest mean is lower than the pretest mean). However, in evaluating whether the remedial program is worthwhile (that is, produces a significant improvement in math skills), the researcher would not be excited by a significant reduction in performance.

A one-tailed test is appropriate when a researcher is only concerned with a change (for a sample tested twice) or difference (between two independent samples) in one prespecified direction or when a researcher anticipates the *direction* of the change or difference. For example, an attempt to show that black defendants receive harsher sentences (mean sentence) than whites indicates the need for a one-tailed test. If, however, a researcher is just looking for differences in sentencing by race, whether it is blacks or whites who get harsher sentences, he or she would instead use a two-tailed test.

In our example of remedial education, the following hypotheses for a one-tailed test are appropriate:

Null hypothesis: *Math ability does not improve after remediation.*
$(\mu_1 \geq \mu_2)$

Research hypothesis: *Math ability improves after remediation.*
$(\mu_1 < \mu_2)$

Note carefully the difference in these hypotheses from those stated earlier. The null hypothesis includes the entire sampling distribution to the right of the critical value; the research hypothesis only includes a change in one direction. This translates into a critical region on the distribution of t with only one tail. Moreover, the tail is larger than with a two-tailed test, because the entire area (for example, $\alpha = .05$) is loaded into one tail rather than divided into the two sides. As a consequence, the table value of t that permits the rejection of the null hypothesis is somewhat lower and easier to achieve. However, any t value in the opposite direction—no matter how extreme—would not allow rejection of the null hypothesis.

The differences between two-tailed and one-tailed tests are summarized in Figure 7.9. As a general rule, to construct a one-tailed test from two-tailed probabilities, simply use a table value with twice the area of the two-tailed test. Thus, for example, the critical value for a two-tailed test with $\alpha = .10$ is identical to that for a one-tailed test with $\alpha = .05$. For convenience, however, a separate table of one-tailed critical values for t is given in Table C.

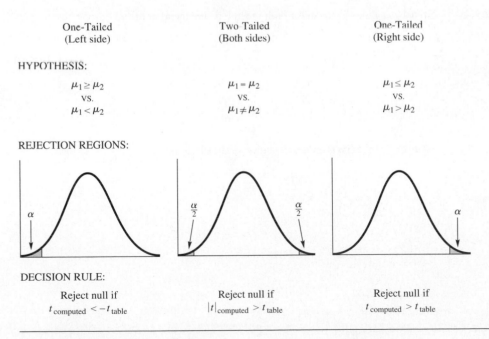

FIGURE 7.9   One-and two-tailed tests

BOX 7.5 • *Step-by-Step Illustration: One-Tailed Test of Means for Same Sample Measured Twice*

As we carry out our example, note that the mechanics for calculating t are unchanged when we employ a one-tailed rejection region. Suppose the before and after math scores for a sample of nine remedial students are as follows:

Student	Before (X_1)	After (X_2)	Difference $(D = X_1 - X_2)$	(Difference)^2 (D^2)
1	58	66	−8	64
2	63	68	−5	25
3	66	72	−6	36
4	70	76	−6	36
5	63	78	−15	225
6	51	56	−5	25
7	44	69	−25	625
8	58	55	3	9
9	50	55	−5	25
	$\Sigma X_1 = 523$	$\Sigma X_2 = 595$		$\Sigma D^2 = 1{,}070$

We can compute the t value exactly as we did previously. The change in procedure for a one-tailed test is just the use of the second page of Table C rather than the first.

Step 1   Find the mean for both the before and the after tests.

$$\overline{X}_1 = \frac{\Sigma X_1}{N_1} \qquad \overline{X}_2 = \frac{\Sigma X_2}{N_2}$$

$$= \frac{523}{9} \qquad\qquad = \frac{595}{9}$$

$$= 58.11 \qquad\qquad = 66.11$$

Step 2   Find the standard deviation of the differences.

$$s_D = \sqrt{\frac{\Sigma D^2}{N} - (\overline{X}_1 - \overline{X}_2)^2}$$

$$= \sqrt{\frac{1{,}070}{9} - (58.11 - 66.11)^2}$$

$$= \sqrt{118.89 - 64}$$

$$= \sqrt{54.89}$$

$$= 7.41$$

Step 3   Find the standard error of the difference between means.

$$s_{\overline{D}} = \frac{s_D}{\sqrt{N-1}}$$

$$= \frac{7.41}{\sqrt{9-1}}$$

$$= \frac{7.41}{2.83}$$

$$= 2.62$$

Step 4   Translate the sample mean difference into units of the standard error of the difference.

$$t = \frac{\overline{X}_1 - \overline{X}_2}{s_{\overline{D}}}$$

$$= \frac{58.11 - 66.11}{2.62}$$

$$= \frac{-8.00}{2.62}$$

$$= -3.05$$

Step 5   Find the degrees of freedom.

$$df = N - 1$$

$$= 9 - 1$$

$$= 8$$

Step 6   Compare the obtained t ratio with the critical value from the second page of Table C.

$$\text{obtained } t = -3.05$$
$$\text{table } t = -1.86$$
$$\alpha = .05$$

Because we hypothesize that the posttest mean should be higher than the pretest mean, t should be negative. Therefore, we use the negative critical value. The obtained t (-3.05) is more extreme in the negative direction than the critical value (-1.86), so we reject the null hypothesis. Thus, the remedial math program has produced a statistically significant improvement in math ability.

All the tests of sample differences that we have presented in this chapter can be modified into one-tailed tests. If the researcher who tested for differences in the proportions of males and females that drive large cars had anticipated that the proportion for males would be larger, a one-tailed test could have been employed.

By the same token, the *t* test of difference in mean employment concern between males and females discussed earlier could have been structured as a one-tailed test, if the researcher's theory suggested, for example, that men would be more concerned on the average. Changing to a one-tailed test only affects the hypotheses and the critical value of *t*, but not any of the calculations. The null hypothesis would instead be that men (group 1) are no more concerned than females (group 2) or, symbolically, $\mu_1 \leq \mu_2$, and the research hypothesis would be that men are more concerned ($\mu_1 > \mu_2$). With a .05 level of significance and again using 60 degrees of freedom in lieu of the actual 68, we obtain now from Table C a critical value $t = 1.67$. Locating the entire .05 rejection region into one tail lessens somewhat the size of *t* needed for significance (from 2.00 to 1.67), but still the calculated *t* ratio ($t = +1.394$, the same as before) is not significant.

BOX 7.6 • *Step-by-Step Illustration: Independent Groups, One-Tailed Test*

A professor who teaches a section of an English course required of all first-year students has a hunch that students who attended private or sectarian high schools are better prepared in English than students who attended public high schools. The professor decides to test her hypothesis concerning English preparedness using her class of 72 students.

During the first week of class, the professor gives a test covering grammar and vocabulary; she also asks on the test which type of high school the student had attended, private–sectarian or public. The professor learns that there are 22 private–sectarian school graduates and 50 public school graduates in her class. She then calculates descriptive statistics separately for the two groups:

Private–Sectarian School	Public School
$N_1 = 22$	$N_2 = 50$
$\overline{X}_1 = 85$	$\overline{X}_2 = 82$
$s_1 = 6$	$s_2 = 8$

Because she anticipated that the graduates of private–sectarian high schools would score better on the test, the professor set up her hypotheses as follows:

Null hypothesis:
($\mu_1 \leq \mu_2$)

English preparedness is not greater among private–sectarian high school graduates than among public high school graduates.

Research hypothesis:
($\mu_1 > \mu_2$)

English preparedness is greater among private–sectarian high school graduates than among public high school graduates.

Step 1   Obtain the sample means (these are given as 85 and 82, respectively).

$$\overline{X}_1 = \frac{\Sigma X_1}{N_1} = 85$$

$$\overline{X}_2 = \frac{\Sigma X_2}{N_2} = 82$$

Step 2   Obtain sample standard deviations (the sample standard deviations are given as 6 and 8, respectively).

$$s_1 = \sqrt{\frac{\Sigma X_1^2}{N_1} - \overline{X}_1^2} = 6$$

$$s_2 = \sqrt{\frac{\Sigma X_2^2}{N_2} - \overline{X}_2^2} = 8$$

Step 3   Calculate the standard error of the difference between means.

$$s_{\overline{X}_1 - \overline{X}_2} = \sqrt{\left(\frac{N_1 s_1^2 + N_2 s_2^2}{N_1 + N_2 - 2}\right)\left(\frac{N_1 + N_2}{N_1 N_2}\right)}$$

$$= \sqrt{\left[\frac{(22)(6)^2 + (50)(8)^2}{22 + 50 - 2}\right]\left[\frac{22 + 50}{(22)(50)}\right]}$$

$$= \sqrt{\left(\frac{(22)(36) + (50)(64)}{70}\right)\left(\frac{72}{1,100}\right)}$$

$$= \sqrt{\left(\frac{3,992}{70}\right)\left(\frac{72}{1,100}\right)}$$

$$= \sqrt{3.7328}$$

$$= 1.93$$

Step 4   Translate the sample mean difference into units of the standard error of the difference.

$$t = \frac{\overline{X}_1 - \overline{X}_2}{s_{\overline{X}_1 - \overline{X}_2}}$$

$$= \frac{85 - 82}{1.93}$$

$$= \frac{3}{1.93}$$

$$= 1.55$$

Step 5   Determine the degrees of freedom.

$$df = N_1 + N_2 - 2$$

(continued)

BOX 7.6   Continued

$$= 22 + 50 - 2$$
$$= 70$$

Step 6   Compare the obtained t ratio with the appropriate t ratio from the second page of Table C.

$$\text{obtained } t = 1.55$$
$$\text{table } t = 1.671$$
$$df = 70$$
$$\alpha = 0.5$$

Because the calculated t (1.55) does not exceed the table value (1.671), the professor cannot reject the null hypothesis. Therefore, although the difference between sample means was consistent with the professor's expectations (85 versus 82), the difference was not larger than she could have observed by chance alone. Furthermore, using a one-tailed test with the entire 5% critical region on one side of the sampling distribution made it easier to find a significant difference by lowering the critical value. Still, the results obtained by the professor were not quite statistically significant.

Requirements for Testing the Difference between Means

As we shall see throughout the remainder of this text, every statistical test should be used only if the social researcher has at least considered certain requirements, conditions, or assumptions. Employing a test inappropriately may confuse an issue and mislead the investigator. As a result, the following requirements should be kept firmly in mind when considering the appropriateness of the z score or t ratio as a test of significance:

1. *A comparison between two means.* The z score and t ratio are employed to make comparisons between two means from independent samples or from a single sample measured twice (repeated measures).
2. *Interval data.* The assumption is that we have scores at the interval level of measurement. Therefore, we cannot use the z score or t ratio for ranked data or data that can only be categorized at the nominal level of measurement (see Chapter 1).
3. *Random sampling.* We should have drawn our samples on a random basis from a population of scores.
4. *A normal distribution.* The t ratio for small samples requires that the sample characteristic we have measured be normally distributed in the underlying population (the t ratio for large samples is not much affected by failure to meet this assumption). Often, we cannot be certain that normality exists. Having no reason to believe otherwise, many researchers pragmatically assume their sample characteristic to be normally distributed. However, if the researcher has reason to suspect that normality cannot be assumed and the sample size is not large, he or she is best advised that the t ratio may be an inappropriate test.

5. *Equal variances.* The *t* ratio for independent samples assumes that the population variances are equal. The sample variances, of course, may differ as a result of sampling. A moderate difference between the sample variances does not invalidate the results of the *t* ratio. But when this difference in sample variances is extreme (for example, when one sample variance is twice as large as the other), the *t* ratio presented here may not be completely appropriate. An adjusted standard error formula is available for situations when the sample variances are very different.

Summary

In Chapter 6 we saw how a population mean or proportion can be estimated from the information we obtain from a single sample. In this chapter we turned our attention to the task of testing hypotheses about differences *between* sample means or proportions (see Figure 7.10). Studying one sample at a time, we previously focused on characteristics of the sampling distribution of means. As we saw in this chapter, there is also a probability distribution for comparing mean differences. As a result of sampling error, the sampling distribution of differences between means consists of a large number of differences between means, randomly selected, that approximates a normal curve whose mean (mean of differences between means) is zero. With the aid of this sampling distribution and the standard error of the difference between means (our estimate of the standard deviation of the sampling distribution based on the two samples we actually draw), we were able to make a probability statement about the occurrence of a difference between means. We asked, what is the probability that the sample mean difference we obtain in our study could have happened strictly on the basis of sampling error? If the difference falls close to the center of the sampling distribution (that is, close to a mean difference of zero), we retain the null hypothesis. Our result is treated as merely a product of sampling error. If, however, our mean difference is so large that it falls a considerable distance from the sampling distribution's mean of zero, we instead reject the null hypothesis and accept the idea that we have

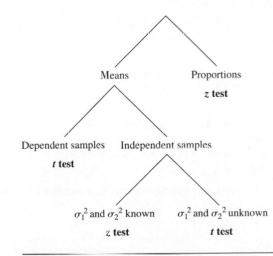

FIGURE 7.10 Tests of differences

found a true population difference. Our mean difference is too large to be explained by sampling error. But at what point do we retain or reject the null hypothesis? To establish whether our obtained sample difference reflects a real population difference (that is, constitutes a statistically significant difference), we need a consistent cutoff point for deciding when to retain or reject the null hypothesis. For this purpose, it is conventional to use the .05 level of significance (also the stricter .01 level of significance). That is, we are willing to reject the null hypothesis if our obtained sample difference occurs by chance less than 5 times out of 100 (1 time per 100 at the .01 level). We determine whether or not the null hypothesis is plausible by calculating degrees of freedom and a significance test known as the *t* ratio. We then compare our obtained *t* ratio against the *t* ratios at varying degrees of freedom given in Table C in Appendix C. Using the same logic, there is also a test of significance for differences between proportions. In either case, the researcher must choose between a one- or two-tailed test of significance, depending on whether he or she can anticipate in advance the direction of the difference.

Terms to Remember

Null hypothesis
Research hypothesis
Sampling distribution of differences between means
Level of significance
.05 level of significance
.01 level of significance
t ratio

Type I error
Type II error
Standard error of the difference between means
Statistically significant difference
Two-tailed test
One-tailed test

Questions and Problems

1. Retaining the null hypothesis assumes which of the following?
   a. There is no difference between sample means.
   b. There is no difference between population means.
   c. The difference between means is significant.
   d. The difference between means is too large to be a sampling error.

2. The larger the *z* score,
   a. the greater the distance of our mean difference from zero.
   b. the more likely we are to reject the null hypothesis.
   c. the more likely that our result is significant.
   d. All of the above

3. At $P = .01$, a mean difference in a sampling distribution occurs
   a. by sampling error once in every 100 mean differences.
   b. by sampling error more than 99 times in every 100 mean differences.
   c. very frequently by sampling error.

4. The level of probability at which the null hypothesis can be rejected with confidence is known as the
   a. level of significance.
   b. distribution.
   c. degrees of freedom.
   d. All of the above

5. The farther out in the tail of a distribution our critical value falls, the greater the risk of making a
   a. Type 1 error.
   b. Type 2 error.
   c. Type 1 and Type 2 errors.
   d. Type 3 error.

6. The size of the tail regions under the curve that will lead us to reject the null hypothesis is (are)
   a. the alpha value.
   b. *P*.
   c. the standard error of the difference.
   d. the degrees of freedom.

7. The standard error of the difference between means is defined as
   a. the standard deviations of the two samples we have drawn.
   b. the standard deviations in the two populations from which are two samples were drawn.
   c. our estimate of the standard deviation of the theoretical sampling distribution of differences between means.

8. The larger the *t* value, the more likely we are to
   a. retain the null hypothesis.
   b. reject the null hypothesis.
   c. conclude that our result is not statistically significant.

9. In a statistical sense, "significant" means
   a. important.
   b. large.
   c. real.
   d. unlikely to be found in the population.

10. When you predict that a result will be in one or another direction (for example, that a particular treatment will reduce recidivism), you would use a
    a. one-tailed test.
    b. two tailed test.
    c. three-tailed test.
    d. no-tailed test.

11. Which of the following is *not* a requirement for employing a *t* test?
    a. A comparison between two means
    b. Random sampling
    c. A normal distribution
    d. Nominal data

12. The Scholastic Assessment Test (SAT) is standardized to have a population mean $\mu = 500$ and a population standard deviation $\sigma = 100$. Suppose that a researcher gives the SAT to a random sample of 50 males and 50 females, yielding sample means of 511 and 541, respectively. Based on these samples' sizes, the researcher has already calculated the true standard deviation of the sampling distribution of the difference between means $\sigma_{\bar{X}_1 - \bar{X}_2}$ to be 20. Based on the areas under the normal curve given in Table A, find the probability of obtaining a sample mean for females that is at least 30 points higher than the sample mean for males.

13. Two groups of subjects participated in an experiment designed to test the effect of frustration on aggression. The experimental group of 40 subjects received a frustrating puzzle to solve, whereas the control group of 40 subjects received an easy, nonfrustrating version of the

same puzzle. Level of aggression was then measured for both groups. Whereas the experimental group (frustration) had a mean aggression score $\bar{X}_1 = 4.0$ and a standard deviation of $s_1 = 2.0$, the control group (no frustration) had a mean aggression score $\bar{X}_2 = 3.0$ and a standard deviation of $s_2 = 1.5$ (higher mean scores indicate greater aggression). Using these results, test the null hypothesis of no difference with respect to aggression between the frustration and no-frustration conditions. What does the outcome of this test indicate?

14. A researcher is interested in finding out if college-bound high school students tend to smoke cigarettes less than high school students who are not college-bound. He distributes a questionnaire and finds that for a group of 57 college-bound students the mean number of cigarettes smoked per week is $\bar{X}_1 = 4.0$, with a standard deviation of $s_1 = 2.0$. For 36 non-college-bound students, he finds that the mean number of cigarettes smoked per week is $\bar{X}_2 = 9.0$, with a standard deviation of $s_2 = 3.0$. Using these results, test the null hypothesis of no difference between college-bound and noncollege-bound high school students with respect to smoking. What does the outcome of this test indicate?

15. A criminologist was interested in whether there was any disparity in sentencing based on the race of the defendant. She selected at random 18 burglary convictions and compared the prison terms given to the 10 whites and 8 blacks sampled. The sentence lengths (in years) are shown for the white and black offenders. Using these data, test the null hypothesis that whites and blacks convicted of burglary in this jurisdiction do not differ with respect to prison sentence length.

Whites	Blacks
3	4
5	8
4	7
7	3
4	5
5	4
6	5
4	4
3	
2	

16. In a field experiment on the effects of perceived control, residents on one floor of a nursing home were given opportunities for increased control in their lives (for example, arrange their own furniture, decide how to spend free time, choose and take care of a plant), whereas the residents on another floor were treated as usual. That is, the staff took care of these details. The feelings of well-being (on a 21-point scale) follow for the conditions of increased and no increased control. Using these data, test the null hypothesis that this minimal manipulation of perception of control had no effect on the residents' feelings of well-being.

Increased Control	No Increased Control
15	8
17	7
14	9
11	12
13	14

17. In a test of the hypothesis that females smile at others more than males do, females and males were videotaped while interacting and the number of smiles emitted by each sex was noted. Using the following number of smiles in the 5-minute interaction, test the null hypothesis that there are no sex differences in the number of smiles.

Males	Females
8	15
11	19
13	13
4	11
2	18

18. A social psychologist was interested in sex differences in the sociability of teenagers. Using the number of good friends as a measure, he compared the sociability of eight female and seven male teenagers. Test the null hypothesis of no difference with respect to sociability between females and males. What do your results indicate?

Females	Males
8	1
3	5
1	8
7	3
7	2
6	1
8	2
5	

19. A personnel consultant was hired to study the influence of sick-pay benefits on absenteeism. She randomly selected samples of hourly employees who do not get paid when out sick and salaried employees who receive sick pay. Using the following data on the number of days absent during a 1-year period, test the null hypothesis that hourly and salaried employees do not differ with respect to absenteeism. What do your results indicate?

Hourly	Salaried
1	2
1	0
2	4
3	2
3	10
6	12
7	5
4	4
0	5
4	8

20. A professor is conducting research on memory. She shows her subjects (volunteers from her Intro Psych course) a list of words on the computer, with some of the words being food items and the rest of the words being completely unrelated. She then gives each subject 30 seconds to recall as many words from the list as possible, with her theory being that the subjects will be able to remember the food-related words better than the unrelated words. Listed below are the scores representing the number of food-related and nonfood-related words that each subject was able to recall:

Food Related	Nonfood Related
6	0
5	1
4	4
7	3
9	5
3	2
5	1
3	4
6	3
7	2
5	7

Test the null hypothesis of no significant difference between the number of food-related and nonfood-related words that the subjects could remember. What do your results indicate?

21. A psychologist was interested in whether men tend to show signs of Alzheimer's disease at an earlier age than women. Following are the ages of onset of Alzheimer's symptoms for a sample of men and women. Test the null hypothesis that there is no difference between the age of onset of Alzheimer's symptoms for men and women.

Men	Women
67	70
73	68
70	57
62	66
65	74
59	67
80	61
66	72
	64

22. Samples of Republicans and Democrats were asked to rate on a scale from 1 to 7 their opinion of prayer in public schools, with 1 being the most opposed to school prayer and 7 being

the most in favor of school prayer. Given the scores below, test the null hypothesis of no difference between Republican and Democratic views of prayer in public schools.

Republicans	Democrats
5	1
7	2
4	1
3	4
6	3
5	1
7	4
3	5
6	1
5	2
4	2
6	

23. Using Durkheim's theory as a basis, a sociologist calculated the following suicide rates (number of suicides per 100,000 population, rounded to the nearest whole number) for 10 "high-anomie" and 10 "low-anomie" metropolitan areas. Anomie (normlessness) was indicated by the presence of a large number of newcomers or transients in an area.

High Anomie	Low Anomie
19	15
17	20
22	11
18	13
25	14
29	16
20	14
18	9
19	11
23	14

Test the null hypothesis that high-anomie metropolitan areas do not have higher suicide rates than low-anomie areas. What do your results indicate?

24. An educator was interested in cooperative versus competitive classroom activities as they might influence the ability of students to make friends among their classmates. On a random basis, he divided his 20 students into two different styles of teaching: a cooperative approach in which students relied on one another in order to get a grade and a competitive approach in which students worked individually to outperform their classmates.

Use the following data indicating the number of classmates chosen by students as their friends to test the null hypothesis that cooperation has no effect on students' friendship choices. What do your results indicate?

Competition	Cooperation
3	7
4	4
1	6
1	3
5	10
8	7
2	6
7	9
5	8
4	10

25. A computer company conducted a "new and improved" course designed to train its service representatives in learning to repair personal computers. Twenty trainees were split into two groups on a random basis: 10 took the customary course and 10 took the new course. At the end of 6 weeks, all 20 trainees were given the same final examination.

    For the following results, test the null hypothesis that the new course was no better than the customary course in terms of teaching service representatives to repair personal computers (higher scores indicate greater repair skills). What do your results indicate?

Customary	"New and Improved"
3	8
5	5
7	9
9	3
8	2
9	6
7	4
4	5
9	2
9	5

26. An employee in a large office wants to find out if her employer discriminates against women in terms of salary raises. Listed below are the numbers of years that a sample of men and women worked in this office before receiving their first raise. Using these data, test the null hypothesis that men and women working in this office do not differ with respect to salary raises.

Men	Women
1.00	3.25
2.50	2.00
1.50	3.50
3.25	1.00
2.00	2.25
2.50	3.50
1.00	3.25
1.50	2.00

3.50	2.25
4.00	1.00
2.25	

27. A researcher is interested in whether union workers earned significantly higher wages than nonunion workers last year. She interviewed 12 union workers and 15 nonunion workers, and their wages are shown below. Test the null hypothesis of no difference between the wages of union workers and nonunion workers.

Union	Nonunion
$16.25	$11.20
15.90	10.95
14.00	13.50
17.21	15.00
16.00	12.75
15.50	11.80
18.24	13.00
16.35	14.50
13.90	10.75
15.45	15.80
17.80	13.00
19.00	11.65
	10.50
	14.00
	15.40

28. How influential can a movie be? A curious researcher asked a random group of people aged 18 and over to rate the Bush administration on a scale from 1 to 10 (with 1 being the least supportive and 10 being the most supportive). She then had them view the anti-Bush movie *Fahrenheit 9/11* by Michael Moore after it had just been released and asked them the same question again. Given the results below, test the null hypothesis that there is no difference in the before and after support for the Bush administration.

Before *Fahrenheit 9/11*	After *Fahrenheit 9/11*
4	2
7	5
1	1
10	6
9	9
2	1
3	1
6	2
8	5
9	5
7	6
5	2

29. The following table shows the mean number of 911 calls to the police per month for six city blocks that set up a neighborhood watch program.

Block	Before Neighborhood Watch	After Neighborhood Watch
A	11	6
B	15	11
C	17	13
D	10	8
E	16	12
F	21	17

Test the significance of the difference in the mean number of 911 calls to the police before and after the neighborhood watch program was established.

30. The racism of eight white young adults, all convicted of committing an anti-Asian hate crime, was measured both before and after they had seen a film designed to reduce their racist attitudes. Using the following scores on an anti-Asian prejudice scale obtained by the eight subjects, test the null hypothesis that the film did not result in a reduction of racist attitudes. What do your results indicate?

	Before	After
A	36	24
B	25	20
C	26	26
D	30	27
E	31	18
F	27	19
G	29	27
H	31	30

31. The Center for the Study of Violence wants to determine whether a conflict-resolution program in a particular high school alters aggressive behavior among its students. For 10 students, aggression was measured both before and after they participated in the conflict-resolution course. Their scores were the following (higher scores indicate greater aggressiveness):

Before Participating	After Participating
10	8
3	3
4	1
8	5
8	7

9	8
5	1
7	5
1	1
7	6

Test the null hypothesis that aggression does not differ as a result of participation in the conflict-resolution program. What do your results indicate?

32. The short-term effect of a lecture on attitudes toward illicit drug use was studied by measuring 10 students' attitudes about drug abuse both before and after they attended a persuasive antidrug lecture given by a former addict. Using the following attitude scores (higher scores indicate more favorable attitudes toward drug use), test the null hypothesis that the antidrug lecture makes no difference in students' attitudes. What do your results indicate?

Student	Before	After
A	5	1
B	9	7
C	6	5
D	7	7
E	3	1
F	9	6
G	9	5
H	8	7
I	4	4
J	5	5

33. Even in the darkest areas of human behavior, we tend to imitate our heroes. A suicidologist studied the incidence of suicide in five randomly selected communities of moderate size both before and after publicity was given to the suicide of a famous singer. Using the following data on the number of suicides, test the null hypothesis that publicity about suicide had no effect on suicide.

Community	Before	After
A	3	6
B	4	7
C	9	10
D	7	9
E	5	8

34. A researcher believes that alcohol intoxication at even half the legal limit, that is, .05 blood alcohol instead of .10, might severely impair driving ability. To test this, he subjects 10 volunteers to a driving simulation test first while sober and then after drinking sufficient to

raise the blood alcohol level to .05. The researcher measures performance as the number of simulated obstacles with which the driver collides. Thus, the higher the number, the poorer the driving. The obtained results are as follows:

Before Drinking	After Drinking
1	4
2	2
0	1
0	2
2	5
1	3
4	3
0	2
1	4
2	3

Test the null hypothesis that there is no difference in driving ability before and after alcohol consumption to the .05 blood alcohol level (use the .05 significance level).

35. A national polling organization conducts a telephone survey of American adults concerning whether they believe the president is doing a good job. A total of 990 persons surveyed had voted in the previous presidential election. Of the 630 who had voted for the president, 72% said they thought he was doing a good job. Of the 360 who did not vote for the president, 60% reported that they thought he was doing a good job. Test the null hypothesis of no difference in population proportions who believe the president is doing a good job (use a .05 level of significance).

36. A pollster interviews by telephone a statewide random sample of persons aged 18 and over about their attitudes toward stricter gun control. Using the following set of results, test the significance of the difference between the proportions of men and women who support stricter controls.

	Males	Females
Favor	92	120
Oppose	74	85
N	166	205

37. A researcher is interested in gender differences in attitudes toward flying. Polling a sample of 100 men and 80 women, he finds that 36% of the men and 40% of the women are fearful of flying. Test the significance of the difference in sample proportions.

SPSS Exercises

1. Using SPSS to analyze the Best Places Study, determine whether there is a difference in property crime rates (CRIMEP) comparing southern metropolitan statistical areas (SOUTH) to other MSAs. Hint: ANALYZE, COMPARE MEANS, INDEPENDENT SAMPLES *T* TEST, and set SOUTH as the grouping variable with "South" category as "1" and the "Non-south" category as "2." Which *t* test is used? Why? What is the result of the comparison?

2. The Best Places Study has information about the average number of alcoholic drinks per month that an adult consumes. Use SPSS to test if people in southern MSAs drink more or less than people in other areas. Which *t* test is used? Why? What do the results of the independent samples *t* test mean?

3. Is there a gender difference in how safe boys and girls feel while at their high school? Use SPSS to analyze the Monitoring the Future Study to find out whether boys and girls (V150) differ in satisfaction with personal safety (V1643).

4. Using the Monitoring the Future Study, find out whether there is a statistically significant difference in marijuana use comparing males to females (V115 and V150).

5. In problem 31, the Center for the Study of Violence wanted to determine whether a conflict-resolution program in a particular high school alters aggressive behavior among its students. For 10 students, aggression was measured both before and after they participated in the conflict-resolution course. The scores were the following (higher scores indicating greater aggressiveness):

Before Participating	After Participating
10	8
3	3
4	1
8	5
8	7
9	8
5	1
7	5
1	1
7	6

   a. Use SPSS to test the null hypothesis that aggression does not differ as a result of the participation in the conflict resolution program. Hint: First, enter the data into a new data set. Second, select ANALYZE, COMPARE MEANS, PAIRED-SAMPLES *T* TEST, and then chose variables.
   b. What do your results indicate?

6. Use SPSS to reanalyze problem 17 which you had calculated by hand. Code men as 1 and women as 2 and enter into SPSS the following data on number of smiles emitted by each

sex in the 5-minute interaction. Test the null hypothesis that there are no gender differences in the number of smiles emitted:

Smiles	Sex
8	1
11	1
13	1
4	1
2	1
15	2
19	2
13	2
11	2
18	2

7. Use ABCalc to reanalyze the means and standard deviations obtained in the question above to test the null hypothesis that there are no gender differences in time spent interviewing witnesses.

8. Compare your hand calculations in problem 17 to results from SPSS in question 6 and to those using ABCalc in question 7.
   a. Find the *t* value from your hand calculation. Compare it to the result from SPSS and the result from ABCalc. Are the results within rounding error?
   b. What are the advantages and disadvantages of working with raw data?
   c. What are the advantages and disadvantages of using SPSS?

8

Analysis of Variance

Blacks versus whites, males versus females, and liberals versus conservatives represent the kind of two-sample comparisons that occupied our attention in the previous chapter. Yet social reality cannot always be conveniently sliced into two groups; respondents do not always divide themselves in so simple a fashion.

As a result, the social researcher often seeks to make comparisons among three, four, five, or more samples or groups. To illustrate, he or she may study the influence of racial/ethnic identity (black, Latino, white, or Asian) on job discrimination, degree of economic deprivation (severe, moderate, or mild) on juvenile delinquency, or subjective social class (upper, middle, working, or lower) on achievement motivation.

You may wonder whether we can use a *series* of t ratios to make comparisons among three or more sample means. Suppose, for example, we want to test the influence of subjective social class on achievement motivation. Can we simply compare all possible pairs of social class in terms of our respondents' levels of achievement motivation and obtain a *t* ratio for each comparison? Using this method, four samples of respondents would generate six paired combinations for which six *t* ratios must be calculated:

1. Upper class versus middle class
2. Upper class versus working class
3. Upper class versus lower class
4. Middle class versus working class
5. Middle class versus lower class
6. Working class versus lower class

Not only would the procedure of calculating a series of *t* ratios involve a good deal of work, but it has a major statistical limitation as well. This is because it increases the probability of making a Type I error—the error of rejecting the null hypothesis when it is true and should be retained.

Recall that the social researcher is generally willing to accept a 5% risk of making a Type I error (the .05 level of significance). He or she therefore expects that *by chance alone* 5 out of every 100 sample mean differences will be large enough to be considered as significant. The more statistical tests we conduct, however, the more likely we are to get statistically significant findings by sampling error (rather than by a true population difference) and hence to commit a Type I error. When we run a large number of such tests, the interpretation of our result becomes problematic. To take an extreme example: How would we interpret a significant *t* ratio out of 1,000 such comparisons made in a particular study? We know that at least a few large mean differences can be expected to occur simply on the basis of sampling error.

For a more typical example, suppose a researcher wished to survey and compare voters in eight regions of the country (New England, Middle Atlantic, South Atlantic, Midwest, South, Southwest, Mountain, and Pacific) on their opinions about the President. Comparing the regional samples would require 28 separate *t* ratios (New England versus Middle Atlantic, New England versus South Atlantic, Middle Atlantic versus South Atlantic, and so on). Out of 28 separate tests of difference between sample means, each with a .05 level of significance, 5% of the 28 tests—between one and two (1.4 to be exact)—would be expected to be significant due to chance or sampling error alone.

Suppose that from the 28 different *t* ratios, the researcher obtains two *t* ratios (New England versus South, and Middle Atlantic versus Mountain) that are significant. How should the researcher interpret these two significant differences? Should he go out on a limb and treat both as indicative of real population differences? Should he play it safe by maintaining that both could be the result of sampling error and go back to collect more data? Should he, based on the expectation that one *t* ratio will come out significant by chance, decide that only one of the two significant *t* ratios is valid? If so, in which of the two significant *t* ratios should he have faith? The larger one? The one that seems more plausible? Unfortunately, none of these solutions is particularly sound. The problem is that as the number of separate tests mounts, the likelihood of rejecting a true null hypothesis (Type I error) grows accordingly. Thus, while for each *t* ratio the probability of a

Type I error may be .05, overall the probability of rejecting *any* true null hypothesis is far greater than .05.

To overcome this problem and clarify the interpretation of our results, we need a statistical test that holds Type I error at a constant level (for example, .05) by making a *single* overall decision as to whether a significant difference is present among the three, four, eight, or however many sample means we seek to compare. Such a test is known as the *analysis of variance.*

The Logic of Analysis of Variance

To conduct an analysis of variance, we treat the total *variation* in a set of scores as being divisible into two components: the distance or deviation of raw scores from their group mean, known as *variation within groups,* and the distance or deviation of group means from one another, referred to as *variation between groups.*

To examine variation within groups, the achievement–motivation scores of members of four social classes—(1) lower, (2) working, (3) middle, and (4) upper—are graphically represented in Figure 8.1, where $X_1, X_2, X_3,$ and X_4 are any raw scores in their respective groups, and $\overline{X}_1, \overline{X}_2, \overline{X}_3,$ and $\overline{X}_4$ are the group means. In symbolic terms, we see that variation within groups refers to the deviations $X_1 - \overline{X}_1, X_2 - \overline{X}_2, X_3 - \overline{X}_3,$ and $X_4 - \overline{X}_4$.

We can also visualize variation between groups. With the aid of Figure 8.2, we see that degree of achievement motivation varies by social class: The upper-class group has greater achievement motivation than the middle-class group which, in turn, has greater achievement motivation than the working-class group which, in its turn, has greater achievement motivation than the lower-class group. More specifically, we determine the overall mean for the total sample of all groups combined, denoted here as $\overline{X}_T,$ and then compare each of the four group means to the total mean. In symbolic terms, we see that variation between groups focuses on the deviations $X_1 - X_T,$ $\overline{X}_2 - \overline{X}_T, \overline{X}_3 - \overline{X}_T,$ and $\overline{X}_4 - \overline{X}_T$.

The distinction between variation *within* groups and variation *between* groups is not peculiar to the analysis of variance. Although not named as such, we encountered a similar

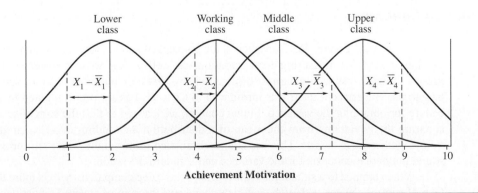

FIGURE 8.1 Graphical representation of variation within groups

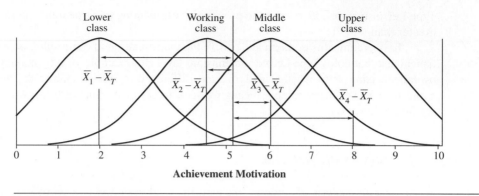

FIGURE 8.2   Graphical representation of variation between groups

distinction in the form of the *t* ratio, wherein a difference *between* $\overline{X}_1$ and $\overline{X}_2$ was compared against the standard error of the difference $S_{\overline{X}_1-\overline{X}_2}$, a combined estimate of differences *within* each group. That is,

$$t = \frac{\overline{X}_1 - \overline{X}_2}{s_{\overline{X}_1-\overline{X}_2}} \begin{array}{l} \leftarrow \text{ variation between groups} \\ \leftarrow \text{ variation within groups} \end{array}$$

In a similar way, the analysis of variance yields an *F ratio,* whose numerator represents variation between the groups being compared, and whose denominator represents variation within these groups. As we shall see, the *F* ratio indicates the size of the variation between groups *relative* to the size of the variation within each group. As was true of the *t* ratio, the larger the *F* ratio (the larger the variation between groups relative to the variation within groups), the greater the probability of rejecting the null hypothesis and accepting the research hypothesis.

The Sum of Squares

At the heart of the analysis of variance is the concept of *sum of squares,* which represents the initial step for measuring total variation, as well as variation between and within groups. It may come as a pleasant surprise to learn that only the label "sum of squares" is new to us. The concept itself was introduced in Chapter 4 as an important step in the procedure for obtaining the variance. In that context, we learned to find the sum of squares by squaring the deviations from the mean of a distribution and adding these squared deviations together $\Sigma(X - \overline{X})^2$. This procedure eliminated minus signs while still providing a sound mathematical basis for the variance and standard deviation.

When applied to a situation in which groups are being compared, there is more than one type of sum of squares, although each type represents the sum of squared deviations from a mean. Corresponding to the distinction between total variation and its two components, we have the *total* sum of squares (SS_{total}), *between-groups* sum of squares ($SS_{between}$), and *within-groups* sum of squares (SS_{within}).

Consider the hypothetical results shown in Figure 8.3. Note that only part of the data is shown to help us focus on the concepts of total, within-groups, and between-groups sums of squares.

The respondent with a 7 scored substantially higher than the total mean ($\overline{X}_{total} = 3$). His deviation from the total mean is ($X - \overline{X}_{total}$) = 4. Part of this elevated score represents, however, the fact that his group scored higher on average ($\overline{X}_{group} = 6$) than the overall or total mean ($\overline{X}_{total} = 3$). That is, the deviation of this respondent's group mean from the total mean is ($\overline{X}_{group} - \overline{X}_{total}$) = 3. After accounting for the group difference, this respondent's score remains higher than his own group mean. Within the group, his deviation from the group mean is ($X - \overline{X}_{group}$) = 1.

As we shall see very shortly, we can take these deviations (of scores from the total mean, deviations between group means and the total mean, and deviations of scores from their group means), square them, and then sum them to obtain SS_{total}, SS_{within}, and $SS_{between}$.

A Research Illustration

Let's consider a research situation in which each type of sum of squares might be calculated. Suppose a researcher is interested in comparing the degree of life satisfaction among adults with different marital statuses. She wants to know if single or married people are more satisfied with life and whether separated and divorced adults do in fact have a more negative view of life. She selects at random five middle-aged adults from each of the following four categories: widowed, divorced, never married, and married. The researcher then administers to each of the 20 respondents a 40-item checklist designed to measure satisfaction with various aspects of life. The scale ranges from 0 for dissatisfaction with all aspects of life to 40 for satisfaction with all aspects of life.

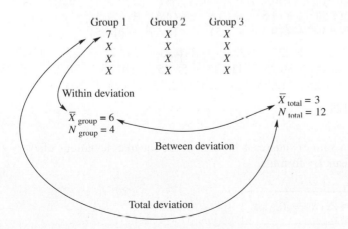

FIGURE 8.3 Analysis of variance

The researcher sets up her hypotheses as follows:

Null hypothesis: *Marital status has no effect on degree of satisfaction*
$(\mu_1 = \mu_2 = \mu_3 = \mu_4)$ *with life.*

Research hypothesis: *Marital status has an effect on degree of satisfaction*
(some $\mu_i \neq \mu_j$)   *with life.*

The total sum of squares is defined as the sum of the squared deviation of every raw score from the total mean. By formula,

$$SS_{total} = \Sigma(X - \overline{X}_{total})^2$$

where X = any raw score
       $\overline{X}_{total}$ = total mean for all groups combined

Using this formula, we subtract the total mean $(\overline{X}_{total})$ from each raw score (X), square the deviations that result, and then sum. Applying this formula to the data in Table 8.1, we obtain the following result:

$$
\begin{aligned}
SS_{total} &= (5 - 15.5)^2 + (6 - 15.5)^2 + (4 - 15.5)^2 + (5 - 15.5)^2 \\
&\quad + (0 - 15.5)^2 + (16 - 15.5)^2 + (5 - 15.5)^2 + (9 - 15.5)^2 \\
&\quad + (10 - 15.5)^2 + (5 - 15.5)^2 + (23 - 15.5)^2 + (30 - 15.5)^2 \\
&\quad + (20 - 15.5)^2 + (20 - 15.5)^2 + (27 - 15.5)^2 + (19 - 15.5)^2 \\
&\quad + (35 - 15.5)^2 + (15 - 15.5)^2 + (26 - 15.5)^2 + (30 - 15.5)^2 \\
&= (-10.5)^2 + (-9.5)^2 + (-11.5)^2 + (-10.5)^2 + (-15.5)^2 \\
&\quad + (.5)^2 + (-10.5)^2 + (-6.5)^2 + (-5.5)^2 + (-10.5)^2 \\
&\quad + (7.5)^2 + (14.5)^2 + (4.5)^2 + (4.5)^2 + (11.5)^2 \\
&\quad + (3.5)^2 + (19.5)^2 + (.5)^2 + (10.5)^2 + (14.5)^2 \\
&= 110.25 + 90.25 + 132.25 + 110.25 + 240.25 + .25 + 110.25 \\
&\quad + 42.25 + 30.25 + 110.25 + 56.25 + 210.25 + 20.25 + 20.25 \\
&\quad + 132.25 + 12.25 + 380.25 + .25 + 110.25 + 210.25 \\
&= 2{,}129
\end{aligned}
$$

The within-groups sum of squares is the sum of the squared deviations of every raw score from its group mean. By formula,

$$SS_{within} = \Sigma(X - \overline{X}_{group})^2$$

where X = any raw score
       $\overline{X}_{group}$ = mean of the group containing the raw score

TABLE 8.1 *Satisfaction with Life by Marital Status*

	Widowed			Divorced	
X_1	$X_1 - \overline{X}_1$	$(X_1 - \overline{X}_1)^2$	X_2	$X_2 - \overline{X}_2$	$(X_2 - \overline{X}_2)^2$
5	1	1	16	7	49
6	2	4	5	−4	16
4	0	0	9	0	0
5	1	1	10	1	1
0	−4	16	5	−4	16
$\Sigma X_1 = 20$	$\overline{X}_1 = \dfrac{20}{5} = 4$	$\Sigma(X_1 - \overline{X}_1)^2 = 22$	$\Sigma X_2 = 45$	$\overline{X}_2 = \dfrac{45}{5} = 9$	$\Sigma(X_2 - \overline{X}_2)^2 = 82$

	Never Married			Married	
X_3	$X_3 - \overline{X}_3$	$(X_3 - \overline{X}_3)^2$	X_4	$X_4 - \overline{X}_4$	$(X_4 - \overline{X}_4)^2$
23	−1	1	19	−6	36
30	6	36	35	10	100
20	−4	16	15	−10	100
20	−4	16	26	1	1
27	3	9	30	5	25
$\Sigma X_3 = 120$	$\overline{X}_3 = \dfrac{120}{5} = 24$	$\Sigma(X_3 - \overline{X}_3)^2 = 78$	$X_4 = 125$	$\overline{X}_4 = \dfrac{125}{5} = 25$	$\Sigma(X_4 - \overline{X}_4)^2 = 262$

$$\overline{X}_{\text{total}} = 15.5$$

Using this formula, we subtract the group mean $(\overline{X}_{\text{group}})$ from each raw score (X), square the deviations that result, and then sum. Applying this formula to the data in Table 8.1, we obtain

$$
\begin{aligned}
\text{SS}_{\text{within}} &= (5 - 4)^2 + (6 - 4)^2 + (4 - 4)^2 + (5 - 4)^2 + (0 - 4)^2 \\
&\quad + (16 - 9)^2 + (5 - 9)^2 + (9 - 9)^2 + (10 - 9)^2 + (5 - 9)^2 \\
&\quad + (23 - 24)^2 + (30 - 24)^2 + (20 - 24)^2 + (20 - 24)^2 + (27 - 24)^2 \\
&\quad + (19 - 25)^2 + (35 - 25)^2 + (15 - 25)^2 + (26 - 25)^2 + (30 - 25)^2 \\
&= (1)^2 + (2)^2 + (0)^2 + (-4)^2 + (7)^2 + (-4)^2 \\
&\quad + (0)^2 + (1)^2 + (-4)^2 + (-1)^2 + (6)^2 + (-4)^2 + (-4)^2 \\
&\quad + (3)^2 + (-6)^2 + (10)^2 + (-10)^2 + (1)^2 + (5)^2 \\
&= 1 + 4 + 0 + 1 + 16 + 49 + 16 + 0 + 1 + 16 \\
&\quad + 1 + 36 + 16 + 16 + 9 + 36 + 100 + 100 + 1 + 25 \\
&= 444
\end{aligned}
$$

Notice that the within-groups sum of squares could have been obtained simply by combining the sum of squares within each group. That is, with four groups,

$$\text{SS}_{\text{within}} = \Sigma(X_1 - \overline{X}_1)^2 + \Sigma(X_2 - \overline{X}_2)^2 + \Sigma(X_3 - \overline{X}_3)^2 + \Sigma(X_4 - \overline{X}_4)^2$$

From Table 8.1, we have

$$SS_{within} = 22 + 82 + 78 + 262 = 444$$

The between-groups sum of squares represents the *sum of the squared deviations of every group mean from the total mean.* Accordingly, we must determine the difference between each group mean and the total mean ($\bar{X}_{group} - \bar{X}_{total}$), square this deviation, multiply by the number of scores in that group, and add these quantities. Summing across groups, we obtain the following definitional formula for the between-groups sum of squares:

$$SS_{between} = \Sigma N_{group}(\bar{X}_{group} - \bar{X}_{total})^2$$

where N_{group} = number of scores in any group
$\quad\quad\bar{X}_{group}$ = mean of any group
$\quad\quad\bar{X}_{total}$ = mean of all groups combined

We apply the formula to the data in Table 8.1, and obtain

$$
\begin{aligned}
SS_{between} &= 5(4 - 15.5)^2 + 5(9 - 15.5)^2 + 5(24 - 15.5)^2 + 5(25 - 15.5)^2 \\
&= 5(-11.5)^2 + 5(-6.5)^2 + 5(8.5)^2 + 5(9.5)^2 \\
&= 5(132.25) + 5(42.25) + 5(72.25) + 5(90.25) \\
&= 661.25 + 211.25 + 361.25 + 451.25 \\
&= 1,685
\end{aligned}
$$

Thus, the sums of squares are

$$SS_{total} = 2,129$$

$$SS_{within} = 444$$

$$SS_{between} = 1,685$$

Notice that the total sum of squares is equal to the within-groups and between-groups sums of squares added together. This relationship among the three sums of squares can be used as a check on your work.

Computing Sums of Squares

The definitional formulas for total, within-groups, and between-groups sums of squares are based on the manipulation of deviation scores, a time-consuming and difficult process. Fortunately, we may instead employ the following much simpler computational formulas

to obtain results that are identical (except for rounding errors) to the lengthier definitional formulas:

$$SS_{total} = \Sigma X_{total}^2 - N_{total} \overline{X}_{total}^2$$

$$SS_{within} = \Sigma \overline{X}_{total}^2 - \Sigma N_{group} \overline{X}_{total}^2$$

$$SS_{between} = \Sigma N_{group} \overline{X}_{group}^2 - N_{total} \overline{X}_{total}^2$$

where ΣX_{total}^2 = all the scores squared and then summed

$\overline{X}_{total}$ = total mean of all groups combined

N_{total} = total number of scores in all groups combined

$\overline{X}_{group}$ = mean of any group

N_{group} = number of scores in any group

The raw scores in Table 8.1 have been set up in Table 8.2 for the purpose of illustrating the use of the computational sum-of-squares formulas. Note that before applying the

TABLE 8.2 *Computations for Life-Satisfaction Data*

Widowed		Divorced	
X_1	X_1^2	X_2	X_2^2
5	25	16	256
6	36	5	25
4	16	9	81
5	25	10	100
0	0	5	25
$\Sigma X_1 = 20$	$\Sigma X_1^2 = 102$	$\Sigma X_2 = 45$	$\Sigma X_2^2 = 487$
$\overline{X}_1 = \dfrac{20}{5} = 4$		$\overline{X}_2 = \dfrac{45}{5} = 9$	

Never Married		Married	
X_3	X_3^2	X_4	X_4^2
23	529	19	361
30	900	35	1,225
20	400	15	225
20	400	26	676
27	729	30	900
$\Sigma X_3 = 120$	$\Sigma X_3^2 = 2,958$	$\Sigma X_4 = 125$	$\Sigma X_4^2 = 3,387$
$\overline{X}_3 = \dfrac{120}{5} = 24$		$\overline{X}_4 = \dfrac{125}{5} = 25$	

$N_{total} = 20$	$\overline{X}_{total} = 15.5$	$\Sigma X_{total} = 310$	$\Sigma X_{total}^2 = 6,934$

formulas, we must first obtain the sum of scores (ΣX_{total}), sum of squared scores ($\Sigma X^2_{\text{total}}$), number of scores ($N_{\text{total}}$), and mean ($\overline{X}_{\text{total}}$) for all groups combined:

$$\begin{aligned}
\Sigma X_{\text{total}} &= \Sigma X_1 + \Sigma X_2 + \Sigma X_3 + \Sigma X_4 \\
&= 20 + 45 + 120 + 125 \\
&= 310
\end{aligned}$$

$$\begin{aligned}
\Sigma X^2_{\text{total}} &= \Sigma X_1^2 + \Sigma X_2^2 + \Sigma X_3^2 + \Sigma X_4^2 \\
&= 102 + 487 + 2{,}958 + 3{,}387 \\
&= 6{,}934
\end{aligned}$$

$$\begin{aligned}
N_{\text{total}} &= N_1 + N_2 + N_3 + N_4 \\
&= 5 + 5 + 5 + 5 \\
&= 20
\end{aligned}$$

$$\begin{aligned}
\overline{X}_{\text{total}} &= \frac{\Sigma X_{\text{total}}}{N_{\text{total}}} \\
&= \frac{310}{20} \\
&= 15.5
\end{aligned}$$

Next we move on to calculating the following sums of squares:

$$\begin{aligned}
\text{SS}_{\text{total}} &- \Sigma X^2_{\text{total}} - N_{\text{total}}\overline{X}^2_{\text{total}} \\
&- 6{,}934 - (20)(15.5)^? \\
&= 6{,}934 - (20)(240.25) \\
&= 6{,}934 - 4{,}805 \\
&= 2{,}129
\end{aligned}$$

$$\begin{aligned}
\text{SS}_{\text{within}} &= \Sigma X^2_{\text{total}} - \Sigma N_{\text{group}}\overline{X}^2_{\text{group}} \\
&= 6{,}934 - [(5)(4)^2 + (5)(9)^2 + (5)(24)^2 + (5)(25)^2] \\
&= 6{,}934 - [(5)(16) + (5)(81) + (5)(576) + (5)(625)] \\
&= 6{,}934 - (80 + 405 + 2880 + 3125) \\
&= 6{,}934 - 6{,}490 \\
&= 444
\end{aligned}$$

$$\begin{aligned}
\text{SS}_{\text{between}} &= \Sigma N_{\text{group}}\overline{X}^2_{\text{group}} - N_{\text{total}}\overline{X}^2_{\text{total}} \\
&= 6{,}490 - 4{,}805 \\
&= 1{,}685
\end{aligned}$$

These results agree with the computations obtained using the definitional formulas.

Mean Square

As we might expect from a measure of variation, the size of the sums of squares tends to become larger as variation increases. For example, SS = 10.9 probably designates greater variation than SS = 1.3. However, the sum of squares also gets larger with increasing sample size, so $N = 200$ will yield a larger SS than $N = 20$. As a result, the sum of squares cannot be regarded as an entirely satisfactory "pure" measure of variation, unless, of course, we can find a way to control for the number of scores involved.

Fortunately, such a method exists in a measure of variation known as the *mean square* (or *variance*), which we obtain by dividing $SS_{between}$ or SS_{within} by the appropriate degrees of freedom. Recall that in Chapter 4 we similarly divided $\Sigma(X - \overline{X})^2$ by N to obtain the variance. Therefore,

$$MS_{between} = \frac{SS_{between}}{df_{between}}$$

where $MS_{between}$ = between-groups mean square
$SS_{between}$ = between-groups sum of squares
$df_{between}$ = between-groups degrees of freedom

and

$$MS_{within} = \frac{SS_{within}}{df_{within}}$$

where MS_{within} = within-groups mean square
SS_{within} = within-groups sum of squares
df_{within} = between-groups degrees of freedom

But we must still obtain the appropriate degrees of freedom. For the between-groups mean square,

$$df_{between} = k - 1$$

where k = number of groups

To find degrees of freedom for within-groups mean square,

$$df_{within} = N_{total} - k$$

where N_{total} = total number of scores in all groups combined
k = number of groups

Illustrating with the data from Table 8.2, for which $SS_{between} = 1,685$ and $SS_{within} = 444$, we calculate our degrees of freedom as follows:

$$df_{between} = 4 - 1$$
$$= 3$$

and

$$df_{within} = 20 - 4$$
$$= 16$$

We are now prepared to obtain the following mean squares:

$$MS_{between} = \frac{1,685}{3}$$
$$= 561.67$$

and

$$MS_{within} = \frac{444}{16}$$
$$= 27.75$$

These then are the between and within variances, respectively.

The F Ratio

The analysis of variance yields an F ratio in which variation between groups and variation within groups are compared. We are now ready to specify the degree of each type of variation as measured by mean squares. Therefore, the F ratio can be regarded as indicating the size of the between-groups mean square relative to the size of the within-groups mean square, or

$$\boxed{F = \frac{MS_{between}}{MS_{within}}}$$

For Table 8.2,

$$F = \frac{561.67}{27.75}$$
$$= 20.24$$

Having obtained an F ratio, we must now determine whether it is large enough to reject the null hypothesis and accept the research hypothesis. Does satisfaction with life differ by marital status? The larger our calculated F ratio (the larger the $MS_{between}$ and the smaller the MS_{within}), the more likely we will obtain a statistically significant result.

But exactly how do we recognize a significant F ratio? Recall that in Chapter 7, our obtained t ratio was compared against a table t ratio for the .05 level of significance with the appropriate degrees of freedom. Similarly, we must now interpret our calculated F ratio with the aid of Table D in Appendix C. Table D contains a list of critical F ratios; these are the F ratios that we must obtain to reject the null hypothesis at the .05 and .01 levels of significance. As was the case with the t ratio, exactly which F value we must obtain depends on its associated degrees of freedom. Therefore, we enter Table D looking for the two df values, between-groups degrees of freedom and within-groups degrees of freedom. Degrees of freedom associated with the numerator ($df_{between}$) have been listed across the top of the page, and degrees of freedom associated with the denominator (df_{within}) have been placed down the left side of the table. The body of Table D presents critical F ratios at the .05 and .01 significance levels.

For the data in Table 8.2, we have found $df_{between} = 3$ and $df_{within} = 16$. Thus, we move in Table D to the column marked df = 3 and continue down the page from that point until we arrive at the row marked df = 16. By this procedure, we find that a significant F ratio at the $\alpha = .05$ level must exceed 3.24, and at the $\alpha = .01$ level it must exceed 5.29. Our calculated F ratio is 20.24. As a result, we reject the null hypothesis and accept the research hypothesis: Marital status appears to affect life satisfaction.

The results of our analysis of variance can be presented in a summary table such as the one shown in Table 8.3. It has become standard procedure to summarize an analysis of variance in this manner. The total sum of squares ($SS_{total} = 2,129$) is decomposed into two parts: the between-groups sum of squares ($SS_{between} = 1,685$) and the within-groups sum of squares ($SS_{within} = 444$). Each source of sum of squares is converted to mean square by dividing by the respective degrees of freedom. Finally, the F ratio (mean square *between* divided by mean square *within*) is calculated, which can be compared to the table critical value to determine significance.

To review some of the concepts presented thus far, consider Table 8.4, which shows two contrasting situations. Both Case 1 and Case 2 consist of three samples (A, B, and C) with sample means $\overline{X}_A = 3$, $\overline{X}_B = 7$, and $\overline{X}_C = 11$ and with $N = 3$ in each sample. Because the means are the same in both data sets, the between-groups sums of squares are identical $SS_{between} = 96$.

In Case 1, the three samples are clearly different. It would seem then that we should be able to infer that the population means are different. Relative to between-groups variation (the differences between the sample means), the within-groups variation is rather

TABLE 8.3 *Analysis of Variance Summary Table for the Data in Table 8.2*

Source of Variation	SS	df	MS	F
Between groups	1,685	3	561.67	20.24
Within groups	444	16	27.75	
Total	2,129	19		

TABLE 8.4 *Two Examples of Analysis of Variance*

Case 1 Data			Analysis-of-Variance Summary Table				
Sample A	*Sample B*	*Sample C*	*Source of Variation*	*SS*	*df*	*MS*	*F*
2	6	10	Between groups	96	2	48	48
3	7	11	Within groups	6	6	1	
4	8	12					
Mean 3	7	11	Total	102	8		

Distributions

```
              A   A   A       B   B   B       C   C   C
     |   |   |   |   |   |   |   |   |   |   |   |   |   |   |
     1   2   3   4   5   6   7   8   9  10  11  12  13  14  15
```

Case 2 Data			Analysis-of-Variance Summary Table				
Sample A	*Sample B*	*Sample C*	*Source of Variation*	*SS*	*df*	*MS*	*F*
1	3	7	Between groups	96	2	48	3.2
3	5	12	Within groups	90	6	15	
5	13	14					
Mean 3	7	11	Total	186	8		

Distributions

```
                  B       B
          A       A       A       C               C   B   C
     |   |   |   |   |   |   |   |   |   |   |   |   |   |   |
     1   2   3   4   5   6   7   8   9  10  11  12  13  14  15
```

small. Indeed, there is as much as a 48-to-1 ratio of between-groups mean square to within-groups mean square. Thus, $F = 48$ and is significant. Although the sample means and between-groups sum of squares are the same for Case 2, there is far more dispersion within groups, causing the samples to overlap quite a bit. The samples hardly appear as distinct as in Case 1, and so it would seem unlikely that we could generalize the differences between the sample means to differences between population means. The within-groups mean square is 15. The ratio of between to within mean square is then only 48-to-15, yielding a nonsignificant F ratio of 3.2.

Before moving to a step-by-step illustration, it will help to review the relationship between sums of squares, mean squares, and the F ratio. SS_{within} represents the sum of squared deviations of scores from their respective sample means, while the $SS_{between}$ involves the sum of squared deviations of the sample means from the total mean for all groups combined. The mean square is the sum of squares, within or between, divided by the corresponding degrees of freedom. In Chapter 6, we estimated the variance of a population using sample data through dividing the sum of squared deviations from the sample mean by the degrees of freedom (for one sample, $N - 1$). Now, we are confronted with two sources of variance, within and between, for each of which we divide the sum of squared deviations (within or between) by the appropriate degrees of freedom to obtain estimates of the variance (within or between) that exists in the population.

Finally, the F statistic is the ratio of two variances, for example between-groups variance (estimated by $MS_{between}$) divided by within-groups variance (estimated by MS_{within}). Even if the population means for all the groups were equal—as stipulated by the null hypothesis, the sample means would not necessarily be identical because of sampling error. Thus, even if the null hypothesis is true, $MS_{between}$ would still reflect sampling variability, but only sampling variability. On the other hand, MS_{within} reflects only sampling variability, whether the null hypothesis is true or false. Therefore, if the null hypothesis of no population mean differences holds, the F ratio should be approximately equal to 1, reflecting the ratio of two estimates of sampling variability. But, as the group means diverge, the numerator of the F ratio will grow, pushing the F ratio over 1. The F table then indicates how large an F ratio we require—that is, how many times larger $MS_{between}$ must be in comparison to MS_{within}, in order to reject the null hypothesis.

BOX 8.1 • *Step-by-Step Illustration: Analysis of Variance*

To provide a step-by-step illustration of an analysis of variance, suppose that a social researcher interested in employment discrimination issues conducts a simple experiment to assess whether male law school graduates are favored over females by large and well-established law firms. She asks the hiring partners at 15 major firms to rate a set of resumes in terms of background and potential on a scale of 0 to 10, with higher scores indicating stronger interest in the applicant. The set of resumes given to the 15 respondents are identical with one exception. On one of the resumes, the name is changed to reflect a male applicant (applicant name is "Jeremy Miller"), a gender-ambiguous applicant (applicant name is "Jordan Miller"), or a female applicant (applicant name is "Janice Miller"); everything else about the resume is the same in all conditions. The 15 respondents are randomly assigned one of these three resumes within their packets of resumes. The other resumes in the packets are the same for all of the hiring partners. These other resumes are included to provide a better context for the experiment, yet only the ratings of the Jeremy/Jordan/Janice Miller resumes are of interest.

The researcher establishes these hypotheses:

Null hypothesis: *Ratings of applicants do not differ based on gender*
$(\mu_1 = \mu_2 = \mu_3)$

Research hypothesis: *Ratings of applicants differ based on gender*
$(some \ \mu_i \neq \mu_j)$

(continued)

BOX 8.1   Continued

The rating scores given by the 15 hiring partners who had been randomly assigned to the three groups based on the apparent gender of the key resume are as follows:

Male ($N_1 = 5$)		Gender-Neutral ($N_2 = 5$)		Female ($N_3 = 5$)	
X_1	X_1^2	X_2	X_2^2	X_3	X_3^2
6	36	2	4	3	9
7	49	5	25	2	4
8	64	4	16	4	16
6	36	3	9	4	16
4	16	5	25	3	9
$\Sigma X_1 = 31$	$\Sigma X_1^2 = 201$	$\Sigma X_2 = 19$	$\Sigma X_2^2 = 79$	$\Sigma X_3 = 16$	$\Sigma X_3^2 = 54$

Step 1   Find the mean for each sample.

$$\bar{X}_1 = \frac{\Sigma X_1}{N_1}$$

$$= \frac{31}{5}$$

$$= 6.2$$

$$\bar{X}_2 = \frac{\Sigma X_2}{N_2}$$

$$= \frac{19}{5}$$

$$= 3.8$$

$$\bar{X}_3 = \frac{\Sigma X_3}{N_3}$$

$$= \frac{16}{5}$$

$$= 3.2$$

Notice that differences do exist, the tendency being for the apparently male applicant to be rated higher than the gender-neutral and apparently female applicant.

Step 2   Find the sum of scores, sum of squared scores, number of subjects, and mean for all groups combined.

$$\Sigma X_{\text{total}} = \Sigma X_1 + \Sigma X_2 + \Sigma X_3$$
$$= 31 + 19 + 16$$
$$= 66$$

$$\Sigma X_{total}^2 = \Sigma X_1^2 + \Sigma X_2^2 + \Sigma X_3^2$$
$$= 201 + 79 + 54$$
$$= 334$$

$$N_{total} = N_1 + N_2 + N_3$$
$$= 5 + 5 + 5$$
$$= 15$$

$$\overline{X}_{total} = \frac{\Sigma X_{total}}{N_{total}}$$
$$= \frac{66}{15}$$
$$= 4.4$$

Step 3   Find the total sum of squares.

$$SS_{total} = \Sigma X_{total}^2 - N_{total}\overline{X}_{total}^2$$
$$- 334 - (15)(4.4)^2$$
$$= 334 - (15)(19.36)$$
$$= 334 \quad 290.4$$
$$= 43.6$$

Step 4   Find the within-groups sum of squares.

$$SS_{within} = \Sigma X_{total}^2 \quad \Sigma N_{group}\overline{X}_{group}^2$$
$$= 334 - [(5)(6.2)^2 + (5)(3.8)^2 + (5)(3.2)^2]$$
$$= 334 - [(5)(38.44) + (5)(14.44) + (5)(10.24)]$$
$$= 334 - (192.2 + 72.2 + 51.2)$$
$$= 334 - 315.6$$
$$= 18.4$$

Step 5   Find the between-groups sum of squares.

$$SS_{between} = \Sigma N_{group}\overline{X}_{group}^2 - N_{total}\overline{X}_{total}^2$$
$$= [(5)(6.2)^2 + (5)(3.8)^2 + (5)(3.2)^2] - (15)(4.4)^2$$
$$= [(5)(38.44) + (5)(14.44) + (5)(10.24)] - (15)(19.36)$$
$$= (192.2 + 72.2 + 51.2) - 290.4$$
$$= 315.6 - 290.4$$
$$= 25.2$$

Step 6   Find the between-groups degrees of freedom.

$$df_{between} = k - 1$$
$$- 3 - 1$$
$$= 2$$

(continued)

BOX 8.1   Continued

Step 7   Find the within-groups degrees of freedom.

$$\begin{aligned} df_{within} &= N_{total} - k \\ &= 15 - 3 \\ &= 12 \end{aligned}$$

Step 8   Find the within-groups mean square.

$$\begin{aligned} MS_{within} &= \frac{SS_{within}}{df_{within}} \\ &= \frac{18.4}{12} \\ &= 1.53 \end{aligned}$$

Step 9   Find the between-groups mean square.

$$\begin{aligned} MS_{between} &= \frac{SS_{between}}{df_{between}} \\ &= \frac{25.2}{2} \\ &= 12.6 \end{aligned}$$

Step 10   Obtain the F ratio.

$$\begin{aligned} F &= \frac{MS_{between}}{MS_{within}} \\ &= \frac{12.6}{1.53} \\ &= 8.24 \end{aligned}$$

Step 11   Compare the obtained F ratio with the appropriate value found in Table D.

$$\text{obtained } F \text{ ratio} = 8.24$$
$$\text{table } F \text{ ratio} = 3.88$$
$$df = 2 \text{ and } 12$$
$$\alpha = .05$$

As shown in Step 11, to reject the null hypothesis at the .05 significance level with 2 and 12 degrees of freedom, our calculated F ratio must exceed 3.88. Because we have obtained an F ratio of 8.24, we can reject the null hypothesis and accept the research hypothesis. Specifically, we conclude that the ratings given law firm applicants tend to differ on the basis of their apparent gender.

A Multiple Comparison of Means

A significant F ratio informs us of an overall difference among the groups being studied. If we were investigating a difference between only two sample means, no additional analysis would be needed to interpret our result: In such a case, either the obtained difference is statistically significant or it is not. However, when we find a significant F for the differences among three or more means, it may be important to determine exactly where the significant differences lie. For example, in the foregoing step by step illustration, the social researcher uncovered statistically significant differences in the ratings of law firm applicants based on their gender.

Consider the possibilities raised by this significant F ratio: The ratings given the male applicant may differ significantly from those given the gender-neutral applicant, the ratings given the male applicant may differ significantly from those given the female applicant; and the ratings given the gender-neutral applicant may differ significantly from those given the female applicant. As explained earlier in this chapter, obtaining a t ratio for each comparison—$\bar{X}_1$ versus $\bar{X}_2$, $\bar{X}_2$ versus $\bar{X}_3$, and $\bar{X}_1$ versus $\bar{X}_3$—would entail a good deal of work and, more importantly, would increase the probability of Type I error. Fortunately, statisticians have developed a number of other tests for making multiple comparisons after a significant F ratio to pinpoint where the significant mean differences lie. We introduce *Tukey's HSD* (Honestly Significant Difference), one of the most useful tests for investigating the multiple comparison of means. Tukey's HSD is used only after a significant F ratio has been obtained. By Tukey's method, we compare the difference between any two mean scores against HSD. A mean difference is statistically significant only if it exceeds HSD. By formula,

$$\text{HSD} = q\sqrt{\frac{\text{MS}_{\text{within}}}{N_{\text{group}}}}$$

where
q = table value at a given level of significance for the total number of group means being compared
$\text{MS}_{\text{within}}$ = within-groups mean square (obtained from the analysis of variance)
N_{group} = number of subjects in each group (assumes the same number in each group)[1]

Unlike the t ratio, HSD takes into account that the likelihood of Type I error increases as the number of means being compared increases. The q value depends upon the number of group means, and the larger the number of group means, the more conservative HSD becomes with regard to rejecting the null hypothesis. As a result, fewer significant differences will be obtained with HSD than with the t ratio. Moreover, a mean difference is more likely to be significant in a multiple comparison of three means than in a multiple comparison of four or five means.

---

[1]Tukey's method can be used for comparison of groups of unequal size. In such cases, N is replaced by what is called the harmonic mean of the group sizes. The harmonic mean of the group sizes is the reciprocal of the mean of the reciprocal group sizes. That is, $N_{\text{group}} = \dfrac{k}{\dfrac{1}{N_1} + \dfrac{1}{N_2} + \cdots + \dfrac{1}{N_k}}$ where k is the number of groups being compared, and $N_1, N_2, \ldots N_k$ are the respective group sizes.

BOX 8.2 • *Step-by-Step Illustration: HSD for Analysis of Variance*

To illustrate the use of HSD, let us return to the previous example in which ratings of law school graduates by large law firm hiring partners differed significantly based on the apparent gender of the applicant. More specifically, a significant F ratio was obtained for differences among the mean ratings for the three groups of respondents—6.2 for those evaluating a resume from a male applicant, 3.8 for those assessing a gender neutral applicant, and 3.2 for those evaluating a resume from a female applicant, respectively. That is,

$$\overline{X}_1 = 6.2$$
$$\overline{X}_2 = 3.8$$
$$\overline{X}_3 = 3.2$$

Step 1   Construct a table of differences between ordered means.

For the present data, the rank order of means (from smallest to largest) is 3.2, 3.8, and 6.2. These mean scores are arranged in table form so that the difference between each pair of means is shown in a matrix. Thus, the difference between $\overline{X}_1$ and $\overline{X}_2$ is .6; the difference between $\overline{X}_1$ and $\overline{X}_3$ is 3.0; and the difference between $\overline{X}_2$ and $\overline{X}_3$ is 2.4. The subscripts for the group means should not change when arranged in order. Thus, for example, $\overline{X}_2$ represents the mean of the group originally designated as number 2, not the second highest group mean.

	$\overline{X}_1 = 3.2$	$\overline{X}_2 = 3.8$	$\overline{X}_3 = 6.2$
$\overline{X}_1 = 3.2$	—	.6	3.0
$\overline{X}_2 = 3.8$	—	—	2.4
$\overline{X}_3 = 6.2$	—	—	—

Step 2   Find q in Table H in Appendix C.

To find q from Table H, we must have (1) the degrees of freedom (df) for MS_{within}, (2) the number of group means k, and (3) a significance level of either .01 or .05. We already know from the analysis of variance that $df_{within} = 12$. Therefore, we look down the lefthand column of Table H until we arrive at 12 degrees of freedom. Second, because we are comparing three mean scores, we move across Table H to a number of group means (k) equal to three. Assuming a .05 level of significance, we find that $q = 3.77$.

Step 3   Find HSD.

$$HSD = q\sqrt{\frac{MS_{within}}{N_{group}}}$$
$$= 3.77\sqrt{\frac{1.53}{5}}$$
$$= 3.77\sqrt{.306}$$
$$= (3.77)(.553)$$
$$= 2.08$$

Step 4 Compare the HSD against the table of differences between means.

To be regarded as statistically significant, any obtained difference between means must exceed the HSD (2.08). Referring to the table of differences between means, we find that the mean rating difference of 3.0 between $\bar{X}_3$ (female applicant) and $\bar{X}_1$ (male applicant) and mean difference of 2.4 between $\bar{X}_1$ (male applicant) and $\bar{X}_2$ (gender-neutral applicant) are greater than HSD = 2.08. As a result, we conclude that these differences between means are statistically significant at the .05 level. Finally, the difference in mean rating of .6 between $\bar{X}_3$ (female applicant) and $\bar{X}_2$ (gender-neutral applicant) is not significant because it is less than HSD.

Two-Way Analysis of Variance

Suppose that a researcher is interested in the effect of particular drugs on heart rate. He first takes his subject's pulse and records the result as 60 beats per minute (bpm). He then administers Drug A to his subject and observes that the heart rate increases to 80 bpm. Thus, the effect produced by Drug A is 20 bpm ($80 - 60 = 20$). After waiting until the effect of Drug A wears off (until the heart rate returns to 60), the researcher then administers Drug B, observing that the heart rate increases to 70 bpm. Thus, the effect of Drug B is 10 bpm ($70 - 60 = 10$). Finally, the researcher is curious about what the result would be if he were to give his subject both drugs at the same time. The effect of Drug A is 20 and the effect of Drug B is 10. Would the combined effect of Drugs A and B be additive—that is, would the heart rate increase by $20 + 10 = 30$ bpm, reaching 90 bpm upon administration of both? Testing this out, the researcher observes his subject's heart rate actually skyrockets to 105 bpm, 15 bpm more than would have been expected by adding the separate effects of the drugs.

This simple example illustrates *main effects* and *interaction effects*. A main effect is the influence on heart rate of either Drug A or Drug B by itself. Here, Drug A has a main effect of 20 and Drug B has a main effect of 10. The interaction effect refers to the extra impact of the two drugs together, beyond the sum of their separate or main effects. Therefore, the interaction effect of Drugs A and B is 15. The main and interaction effects are illustrated in Figure 8.4. The heart rate is calibrated on the vertical axis, the absence or presence of Drug A is shown on the horizontal axis, and separate lines are employed for the absence and presence of Drug B. Had the effects of the two drugs been additive (non-interactive), then the lines would have been parallel, as shown.

There are many different possibilities for main and interaction effects, as illustrated in Figure 8.5. Graph (a) shows what happens when there are no effects of any kind. Under all four conditions (no drugs, Drug A alone, Drug B alone, and both drugs together), the heart rate is 60. Graph (b) shows what happens when only Drug A has an effect on heart rate. Here, the heart rate increases to 80, whether or not Drug B is present. Graph (c) shows what happens when only Drug B has an effect on heart rate. Here, the heart rate increases to 70, whether or not Drug A is present. Graph (d) shows what happens when both drugs have main effects but do not interact. Drug A increases the heart rate by 20, regardless of the presence of B, and Drug B increases the heart rate by 10, regardless of the presence of A. Together

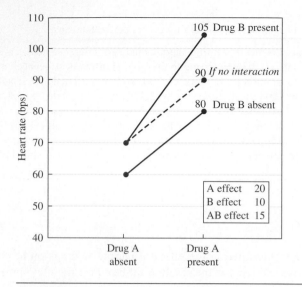

FIGURE 8.4 Effects of drugs A and B on heart rate

they increase the heart rate by 30, which is the sum total of the main effects. Graph (e) illustrates the situation that occurs when there is only an interaction but no main effects. By themselves the drugs have no effect, but together they cause the heart rate to increase to 75, an interaction effect of 15. Finally, graph (f) shows the presence of all three effects; a main effect for A of 20, a main effect for B of 10, and an interaction effect (AB) of 15.

The illustration in Figure 8.5 is designed to introduce the logic of determining main and interaction effects. However, the mechanics of two-way analysis of variance depart from our hypothetical example in two important respects. First, the main and interaction effects shown in the figure were derived by comparing particular treatment combinations against the control condition (no drugs) as a baseline. In the analysis of variance formulas to follow, treatment combinations are compared against the total mean of all groups as a baseline. Second, in actual research settings, a researcher would obviously observe more than one subject in order to have confidence in the results. In addition, it is often necessary to make comparisons between different groups of subjects—for example, a group that gets Drug A only, a group that gets Drug B only, a group that gets both Drugs A and B, and a control. The same way of thinking about main and interaction effects applies to groups of subjects, except that we focus on differences between group means. For example, we would analyze the mean heart rate for the group given only Drug A, for those given only Drug B, for those given both Drugs A and B, and for those given neither drug. In this chapter we only consider the situation of independent groups—different subjects for different conditions. More advanced books will extend this approach to repeated measures designs—the same subjects across different conditions.

As we saw earlier in the chapter, analysis of variance can be useful for examining group differences when three or more sample means are being compared. Actually, the analysis of variance discussed earlier is the simplest kind and is called a one-way analysis of variance, because it represents the effect of different categories or levels of a single factor or independent variable on a dependent variable. When we examine the influence of

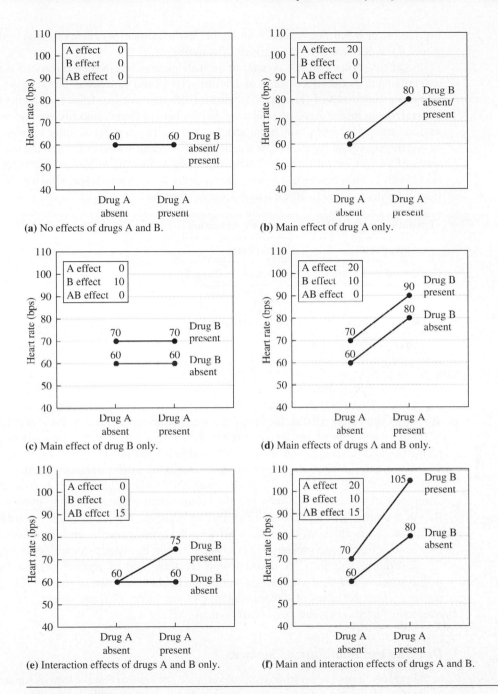

FIGURE 8.5   Main and interaction effects

two factors or independent variables together in the same experiment (such as with Drugs A and B), it is called two-way analysis of variance.

Let's consider a simple, yet far more realistic example. Suppose a social researcher specializing in media effects wants to determine if sex and violence in films impact the perceptions and attitudes of viewers. After gaining informed consent from 16 students from his undergraduate media class, he randomly divides his volunteers into four groups with four students each. One group watches a series of nonviolent, nonsexual films; another a series of violent, nonsexual films, another a series of nonviolent, sexual films; and the last group a series of violent, sexual films. Following the study period, the researcher/instructor administers to all 16 volunteers a test to measure empathy for victims of rape, with a scale ranging from 0 for no empathy to 10 for a high level of empathy.

Empathy Scores by Film Content Combination

Nonviolent		*Violent*	
Nonsexual	**Sexual**	**Nonsexual**	**Sexual**
8	9	6	2
10	5	4	1
7	7	8	1
9	7	6	2

One might be tempted to analyze these data using a one-way analysis of variance with four groups. However, this would not capture the fact that one of the groups combines the attributes of two others. That is, the group that watches sexually violent films is exposed to the effect of violent content, sexual content, and their interaction. Two-way analysis of variance permits us to disentangle main and interaction effects appropriately.

Figure 8.6 displays the group means to assess the main effects of violent and sexual content and their interaction on empathy levels. Not only do both content types appear to lower empathy levels of viewers, but the combination reduces empathy even more. Of course, as with any of the tests of differences covered thus far in the book, we still need to determine if the observed differences between the group means—and thus the main and interaction effects—are greater than one would obtain by chance. As with one-way analysis of variance, this involves decomposing the total sum of squares into between and within sources. But here, we shall further divide the between sum of squares into portions resulting from the effect of violent content (A), the effect of sexual content (B), and the interaction effect of violent and sexual content (AB).

Dividing the Total Sum of Squares

In our presentation of one-way analysis of variance, we learned that the sum of squares refers to the sum of squared deviations from a mean. For a one-way analysis of variance, the total sum of squares (SS_{total}) was divided into two components, the between-groups sum of squares ($SS_{between}$), representing variation between group means as a result of an independent variable, and the within-groups sum of squares (SS_{within}), representing random variation

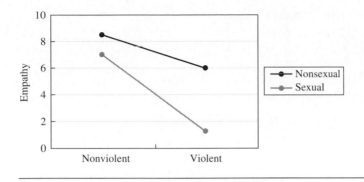

FIGURE 8.6 Plot of group means for film experiment

among the scores of the members of the same group. The total sum of squares was shown to be equal to a combination of its within- and between-groups components. Thus,

$$SS_{total} = SS_{between} + SS_{within}$$

In the case of a two-way analysis of variance, the total sum of squares can again be divided into its within-groups and between-groups components. This time, however, because more than one independent variable is involved, the between-groups sum of squares can be further broken down into

$$SS_{between} = SS_A + SS_B + SS_{AB}$$

where

SS_A is the sum of squares for the main effect A based on variation between levels of factor A

SS_B is the sum of squares for the main effect B based on variation between levels of factor B

SS_{AB} is the sum of squares for the interaction effect of A and B, based on variation between combinations of A and B

Consider the hypothetical experimental results shown in Figure 8.7. Note that in order to help us focus on the concepts of total, within-groups, main effect, and interaction sums of squares, only part of the data set is shown. For the sake of brevity, g represents the group, a is the level of the factor aligned in rows, b is the level of the factor aligned in columns, and t represents the total.

The subject indicated by the 14 scored substantially higher than the total mean for the study ($\overline{X}_{total} = 4$). His deviation from the total mean is ($X - \overline{X}_{total} = 10$). Part of his elevated score represents the fact that the groups in which his score is located (his level of Factor A) performed better than average; that is, ($\overline{X}_a - \overline{X}_{total}$) = 2. Part of his elevated score also represents the fact that the groups in which his score is located (his level of Factor B) performed better than average; that is, ($\overline{X}_b - \overline{X}_{total}$) = 3. Note that combining the difference due to Factor A (2) and Factor B (3) would suggest a group mean that is

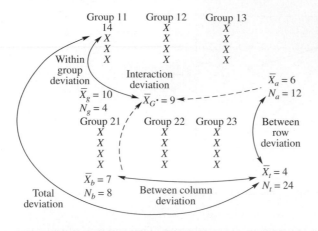

FIGURE 8.7   Illustration of two-way analysis of variance (g = group, a = row level, b = column level, t = total)

$2 + 3 = 5$ units above the total mean, that is; $\bar{X}^*_{\text{group}} = 4 + 5 = 9$. (The notation $\bar{X}^*_{\text{group}}$ is used to represent what the group mean would be in the absence of any interaction between Factors A and B.) The deviation of the group mean $(\bar{X}_{\text{group}})$ from that produced by adding the two main effects is a result of interaction, $\bar{X}_{\text{group}} - X_a - \bar{X}_b + \bar{X}_{\text{total}} = 1$. Finally, after accounting for deviations or differences due to main and interaction effects, the subject's score remains higher than the group mean. Within the group, his deviation is $X - \bar{X}_{\text{group}} = 4$.

All these deviations—the deviations of subjects from the total mean, deviations of subjects from their group means, deviations between Factor A mean and the total mean, deviations between Factor B mean and the total mean, and deviations between group means and those expected by combining main effects, can be squared, and then summed to obtain SS_{total}, SS_{within}, SS_A, SS_B, and SS_{AB}. Specifically,

$$SS_{\text{total}} = \Sigma(X - \bar{X}_{\text{total}})^2$$
$$SS_{\text{within}} = \Sigma(X - \bar{X}_{\text{group}})^2$$
$$SS_A = \Sigma N_a(\bar{X}_a - \bar{X}_{\text{total}})^2$$
$$SS_B = \Sigma N_b(\bar{X}_b - \bar{X}_{\text{total}})^2$$
$$SS_{AB} = \Sigma N_{\text{group}}(\bar{X}_{\text{group}} - \bar{X}_a - \bar{X}_b + \bar{X}_{\text{total}})^2$$

where

$\bar{X}_{\text{total}}$ = the total mean of all groups combined
$\bar{X}_{\text{group}}$ = the mean of any group
$\bar{X}_a$ = the mean of any level of Factor A
$\bar{X}_b$ = the mean of any level of Factor B
N_{group} = the number of cases in any group
N_a = the number of cases in any level of Factor A
N_b = the number of cases in any level of Factor B

As in the case of one-way analysis of variance, the sums of squares are generally more easily obtained using computational formulas based on raw scores rather than the previous definitional formulas based on deviations. For two-way analysis of variance, the computational formulas for calculating total, within, main effect, and interaction sums of squares are as follows:

$$SS_{total} = \Sigma X_{total}^2 - N_{total}\overline{X}_{total}^2$$
$$SS_{within} = \Sigma X_{total}^2 - \Sigma N_{group}\overline{X}_{group}^2$$
$$SS_A = \Sigma N_a\overline{X}_a^2 - N_{total}\overline{X}_{total}^2$$
$$SS_B = \Sigma N_b\overline{X}_b^2 - N_{total}\overline{X}_{total}^2$$
$$SS_{AB} = \Sigma N_{group}\overline{X}_{group}^2 - \Sigma N_a\overline{X}_a^2 - \Sigma N_b\overline{X}_b^2 + N_{total}\overline{X}_{total}^2$$

As with one-way analysis of variance, each of the sums of square is associated with degrees of freedom, which, in turn, are used to derive mean squares. Specifically,

$$df_{within} = N_{total} - ab$$
$$MS_{within} = \frac{SS_{within}}{df_{within}}$$
$$df_A = a - 1$$
$$MS_A = \frac{SS_A}{df_A}$$
$$df_B = b - 1$$
$$MS_B = \frac{SS_B}{df_B}$$
$$df_{AB} = (a - 1)(b - 1)$$
$$MS_B = \frac{SS_{AB}}{df_{AB}}$$

where a is the number of levels of Factor A, and b is the number of levels of Factor B.

Finally, as in the one-way case of analysis of variance, the mean squares for the main and interaction effects are tested for statistical significance by dividing each by the MS_{within}. This produces an F ratio for each effect which can in turn be compared to the value found in Table D in Appendix C with the appropriate degrees of freedom. That is,

$$F_A = \frac{MS_A}{MS_{within}} \text{ with } a - 1 \text{ and } N_{total} - ab \text{ degrees of freedom}$$
$$F_B = \frac{MS_B}{MS_{within}} \text{ with } b - 1 \text{ and } N_{total} - ab \text{ degrees of freedom}$$
$$F_{AB} = \frac{MS_{AB}}{MS_{within}} \text{ with } (a - 1)(b - 1) \text{ and } N_{total} - ab \text{ degrees of freedom}$$

If any of the F ratios exceed the value in Table D in Appendix C, we can conclude that the corresponding main or interaction effects are statistically significant, that is, larger than one would expect by chance alone.

BOX 8.3 • *Step-by-Step Illustration: Two-Way Analysis of Variance*

As a step-by-step illustration of two-way analysis of variance, let's return to the small experiment on the effects of violence and sex in films on the empathy of viewers toward rape victims. We have four groups of four subjects each, the results of which are displayed as follows in the form of a 2×2 table with various sums and means calculated for each group, for each level of Factor A (violent content) aligned in the rows, for each level of Factor B (sexual content) aligned in the columns and for the entire study.

Our objective is to test the null hypothesis that all groups are equal in terms of their population means with the research hypothesis being that not all population means are equal. That is,

Null hypothesis: *Violence and sex in films have no effect on viewer empathy*
$(\mu_1 = \mu_2 = \mu_3 = \mu_4)$

Research hypothesis: *Violence and sex in films have an effect on viewer empathy*
(some $\mu_i \neq \mu_j$)

		Sexual Content (B)				Total
		Nonsexual		**Sexual**		
		X	X^2	X	X^2	$N = 8$ $\overline{X} = 7.75$
		8	64	9	81	
		10	100	5	25	
	Nonviolent	7	49	7	49	
		9	81	7	49	
		34	294	28	204	
Violent		$N = 4$ $\overline{X} = 8.5$		$N = 4$ $\overline{X} = 7.0$		
Content (A)		X	X^2	X	X^2	$N = 8$ $\overline{X} = 3.75$
		6	36	2	4	
		4	16	1	1	
	Violent	8	64	1	1	
		6	36	2	4	
		24	152	6	10	
		$N = 4$ $\overline{X} = 6.0$		$N = 4$ $\overline{X} = 1.5$		
	Total	$N = 8$ $\overline{X} = 7.25$		$N = 8$ $\overline{X} = 4.25$		$N = 16$
						$\overline{X} = 5.75$

Step 1 Find the mean for each group.

$$\overline{X}_{group} = \frac{\Sigma X_{group}}{N_{group}}$$

$$= \frac{34}{4} = 8.5 \text{ (nonviolent/nonsexual)}$$

$$= \frac{28}{4} = 7.0 \text{ (nonviolent/sexual)}$$

$$= \frac{24}{4} = 6.0 \text{ (violent/nonsexual)}$$

$$= \frac{6}{4} = 1.5 \text{ (violent/sexual)}$$

Step 2 Find the total mean.

$$\overline{X}_{total} = \frac{\Sigma X_{total}}{N_{total}} = \frac{96}{16} = 5.75$$

Step 3 Find the mean for each level of Factor A.

$$\overline{X}_A = \frac{\Sigma X_A}{N_A}$$

$$= \frac{62}{8} = 7.75 \text{ (nonviolent)}$$

$$= \frac{30}{8} = 3.75 \text{ (violent)}$$

Step 4 Find the mean for each level of Factor B.

$$\overline{X}_B = \frac{\Sigma X_B}{N_B}$$

$$= \frac{58}{8} = 7.25 \text{ (nonviolent)}$$

$$= \frac{34}{8} = 4.25 \text{ (violent)}$$

Step 5 Calculate the total, within, main effect, and interaction sums of squares.

$$\begin{aligned}
SS_{total} &= \Sigma X_{total}^2 - N_{total}\overline{X}_{total}^2 \\
&= (294 + 204 + 152 + 10) - 16(5.75)^2 \\
&= 660 - 529 \\
&= 131
\end{aligned}$$

(continued)

BOX 8.3   Continued

$$\text{SS}_{\text{within}} = \Sigma X^2_{\text{total}} - \Sigma N_{\text{group}} \overline{X}^2_{\text{group}}$$
$$= (294 + 204 + 152 + 10) - [4(8.5)^2 + 4(6.0)^2 + 4(7.0)^2$$
$$+ 4(1.5)^2]$$
$$= 660 - (289 + 144 + 196 + 9)$$
$$= 660 - 638$$
$$= 22$$

$$\text{SS}_{\text{A}} = \Sigma N_a \overline{X}_a{}^2 - N_{\text{total}} \overline{X}^2_{\text{total}}$$
$$= [8(7.75)^2 + 8(3.75)^2] - 16(5.75)^2$$
$$= (480.5 + 112.5) - 529$$
$$= 593 - 529$$
$$= 64$$

$$\text{SS}_{\text{B}} = \Sigma N_b \overline{X}_b{}^2 - N_{\text{total}} \overline{X}^2_{\text{total}}$$
$$= [8(7.25)^2 + 8(4.25)^2] - 16(5.75)^2$$
$$= (420.5 + 144.5) - 529$$
$$= 565 - 529$$
$$= 36$$

$$\text{SS}_{\text{AB}} = \Sigma N_{\text{group}} \overline{X}^2_{\text{group}} - \Sigma N_a \overline{X}_a{}^2 - \Sigma N_b \overline{X}_b{}^2 + N_{\text{total}} \overline{X}^2_{\text{total}}$$
$$= [4(8.5)^2 + 4(6.0)^2 + 4(7.0)^2 + 4(1.5)^2] - [8(7.75)^2$$
$$+ 8(3.75)^2] - [8(7.25)^2 + 8(4.25)^2] + 16(5.75)^2$$
$$= 638 - 593 - 565 - 529$$
$$= 9$$

Step 6   Find the degrees of freedom for within, main effects, and interaction sums of squares.

$$\text{df}_{\text{within}} = N_{\text{total}} - ab = 16 - 2(2) = 12$$
$$\text{df}_{\text{A}} = a - 1 = 2 - 1 = 1$$
$$\text{df}_{\text{B}} = b - 1 = 2 - 1 = 1$$
$$\text{df}_{\text{AB}} = (a - 1)(b - 1) = (2 - 1)(2 - 1) = 1$$

Step 7   Find the mean squares for within, main effects, and interaction sums of squares.

$$\text{MS}_{\text{within}} = \frac{\text{SS}_{\text{within}}}{\text{df}_{\text{within}}} = \frac{22}{12} = 1.833$$

$$\text{MS}_{\text{A}} = \frac{\text{SS}_{\text{A}}}{\text{df}_{\text{A}}} = \frac{64}{1} = 64$$

$$MS_B = \frac{SS_B}{df_B} = \frac{36}{1} = 36$$

$$MS_{AB} = \frac{SS_{AB}}{df_{AB}} = \frac{9}{1} = 9$$

Step 8   Obtain the F ratios for the main effects and the interaction effect.

$$F_A = \frac{MS_A}{MS_{within}} = \frac{64}{1.833} = 34.909$$

$$F_B = \frac{MS_B}{MS_{within}} = \frac{36}{1.833} = 19.636$$

$$F_{AB} = \frac{MS_{AB}}{MS_{within}} = \frac{9}{1.833} = 4.909$$

Step 9   Compare the F ratios with the appropriate value found in Table D in Appendix C.

All three F ratios have 1 and 12 degrees of freedom and exceed 9.33, the F ratio from Table D for the .05 level of significance.

Step 10   Arrange the results in an analysis of variance summary table.

Source	SS	df	MS	*F*
Violent content (A)	64	1	64.0	34.909
Sexual content (B)	36	1	36.0	19.636
Interaction (AB)	9	1	9.0	4.909
Within group	22	12	1.833	
Total	131	15		

Based on these results, we can reject the null hypothesis that all group means are equal in the populations. Furthermore, we find that there are significant main effects of violent and sexual content as well as a significant interaction. In other words, not only do both forms of content tend to reduce viewer empathy, but in combination the effect is even greater than the sum of the main effects.

Requirements for Using the F Ratio

The analysis of variance should be made only after the researcher has considered the following requirements:

1. *A comparison between three or more independent means.* The F ratio is usually employed to make comparisons between three or more means from independent samples. It is possible, moreover, to obtain an F ratio rather than a t ratio when a two-sample comparison is made. For the two-sample case, $F = t^2$, identical results are obtained. However, a single sample arranged in a panel design (the same group

studied at several points in time) cannot be tested in this way. Thus, for example, one may not study improvement in class performance across three examinations during the term using this approach.

2. *Interval data.* To conduct an analysis of variance, we assume that we have achieved the interval level of measurement for the outcome or dependent variable. Categorized or ranked data should not be used. However, the groups can and are typically formed based on a categorical measure.

3. *Random sampling.* We should have taken our samples at random from a given population of scores.

4. *A normal distribution.* We assume the characteristic we are comparing between and within groups is normally distributed in the underlying populations. Alternatively, the normality assumption will hold if we draw large enough samples from each group.

5. *Equal variances.* The analysis of variance assumes that the population variances for the different groups are all equal. The sample variances, of course, may differ as a result of sampling error. Moderate differences among the sample variances do not invalidate the results of the F test. When such differences are extreme (for example, when one of the sample variances is many times larger than another), the F test presented here may not be appropriate.

Summary

Analysis of variance can be used to make comparisons among three or more sample means. Unlike the t ratio for comparing only two sample means, the analysis of variance yields an F ratio whose numerator represents variation between groups and whose denominator contains an estimate of variation within groups. The sum of squares represents the initial step for measuring variation, however, it is greatly affected by sample size. To overcome this problem and control for differences in sample size, we divide the between and within sums of squares by the appropriate degrees of freedom to obtain the mean square. The F ratio indicates the size of the between-groups mean square relative to the size of the within-groups mean square. The larger the F ratio (that is, the larger the between-groups mean square relative to its within-groups counterpart), the more likely we are to reject the null hypothesis and attribute our result to more than just sampling error. In the last chapter, we learned that in studying the difference between two sample means we must compare our calculated t ratio against the table t (Table B or C). For the purpose of studying differences among three or more means, we now interpret our calculated F ratio by comparing it against an appropriate F ratio in Table D. On this basis, we decide whether we have a significant difference—whether to retain or reject the null hypothesis. After obtaining a significant F, we can determine exactly where the significant differences lie by applying Tukey's method for the multiple comparison of means. Finally, analysis of variance may be extended to studies of more than one factor. In two-way analysis of variance, we distinguish and test for significant main effects of each factor as well as their interaction. Interaction exists when the effect of two factors combined differs from the sum of their separate (main) effects.

Terms to Remember _____

Analysis of variance
Sum of squares
   Within groups
   Between groups
   Total

Mean square
F ratio
Tukey's HSD
Main effect
Interaction effect

Questions and Problems _____

1. Rather than a series of *t* tests, analysis of variance is used because
   a. it holds Type 1 error at a constant level.
   b. it is too much work to do a series of *t* tests.
   c. it increases Type 1 error.
   d. it makes a number of decisions, whereas a series of *t* tests makes a single overall decision.

2. To find a significant difference with an analysis of variance, you hope to maximize
   a. the between-groups mean square.
   b. the within-groups mean square.
   c. the within-groups sum of square.
   d. variation within groups.

3. The *F* ratio is larger when
   a. the between-groups mean square is smaller.
   b. the within-groups mean square is smaller.
   c. the difference between means is smaller.
   d. None of the above

4. A multiple comparison of means is necessary when the analysis of variance results in
   a. a significant difference between two means.
   b. a nonsignificant difference between two means.
   c. a significant difference among three or more means.
   d. a nonsignificant difference among three or more means.

5. Which of the following is *not* a requirement of analysis of variance?
   a. A comparison of three or more independent means
   b. Random sampling
   c. A normal distribution
   d. Ordinal data

6. On the following random samples of social class, test the null hypothesis that neighborliness does not vary by social class. (*Note:* Higher scores indicate greater neighborliness.)

Lower	Working	Middle	Upper
8	7	6	5
4	3	5	2
7	2	5	1
8	8	4	3

7. A researcher is interested in the effect type of residence has on the personal happiness of college students. She selects samples of students who live in campus dorms, in off-campus apartments, and at home and asks the 12 respondents to rate their happiness on a scale of 1 (not happy) to 10 (happy). Test the null hypothesis that happiness does not differ by type of residence.

Dorms	Apartments	At Home
8	2	5
9	1	4
7	3	3
8	3	4

8. Construct a multiple comparison of means by Tukey's method to determine precisely where the significant differences occur in problem 7.

9. A pediatrician speculated that frequency of visits to his office may be influenced by type of medical insurance coverage. As an exploratory study, she randomly chose 15 patients: 5 whose parents belong to a health maintenance organization (HMO), 5 whose parents had traditional medical insurance, and 5 whose parents were uninsured. Using the frequency of visits per year from the following table, test the null hypothesis that type of insurance coverage has no effect on frequency of visits.

HMO	Traditional	None
12	6	3
6	5	2
8	7	5
7	5	3
6	1	1

10. Conduct a multiple comparison of means by Tukey's method to determine exactly where the significant differences occur in problem 9.

11. A high school guidance counselor hypothesized that peer group identity may have an effect on the number of schools to which her college-prep students apply. She draws eight students at random from the "geeks," the "punks," and the "jocks" of the senior class and compares the number of colleges to which the selected students had submitted an application.

"Geeks"	"Punks"	"Jocks"
8	4	2
6	4	2
4	3	3
5	6	2
3	5	4
7	6	1
6	3	2
9	3	2

Test the null hypothesis that peer group identity has no effect on the number of college applications.

12. Conduct a multiple comparison of means by Tukey's method to determine exactly where the significant differences occur in problem 11.

13. A health researcher is interested in comparing three methods of weight loss: low-calorie diet, low-fat diet, and low-carb diet. He selects 30 moderately overweight subjects and randomly assigns 10 to each weight-loss program. The following weight reductions (in pounds) were observed after a one-month period:

Low Calorie		Low Fat		Low Carb	
7	3	7	8	7	14
7	9	8	7	9	10
5	10	8	10	7	11
4	5	9	11	8	5
6	2	5	2	8	6

Test the null hypothesis that the extent of weight reduction does not differ by type of weight-loss program.

14. Consider an experiment to determine the effects of alcohol and of marijuana on driving. Five randomly selected subjects are given alcohol to produce legal drunkenness and then are given a simulated driving test (scored from a top score of 10 to a bottom score of 0). Five different randomly selected subjects are given marijuana and then the same driving test. Finally, a control group of five subjects is tested for driving while sober. Given the following driving test scores, test for the significance of differences among means of the following groups:

Alcohol	Drugs	Control
3	1	8
4	6	7
1	4	8
1	4	5
3	3	6

15. Conduct a multiple comparison of means by Tukey's method to determine exactly where the significant differences occur in problem 14.

16. Using Durkheim's theory of anomie (normlessness) as a basis, a sociologist obtained the following suicide rates (the number of suicides per 100,000 population), rounded to the nearest whole number, for five high-anomie, five moderate-anomie, and five low-anomie metropolitan areas (anomie was indicated by the extent to which newcomers and transients were present in the population):

	Anomie	
High	**Moderate**	**Low**
19	15	8
17	20	10
22	11	11
18	13	7
25	14	8

Test the null hypothesis that high-, moderate-, and low-anomie areas do not differ with respect to suicide rates.

17. Psychologists studied the relative efficacy of three different treatment programs—A, B, and C—on illicit drug abuse. The following data represent the number of days of drug abstinence accumulated by 15 patients (5 in each treatment program) for the 3 months after their treatment program ended. Thus, a larger number of days indicates a longer period free of drug use.

Treatment A	**Treatment B**	**Treatment C**
90	81	14
74	90	20
90	90	33
86	90	5
75	85	12

Test the null hypothesis that these drug-treatment programs do not differ in regard to their efficacy.

18. Conduct a multiple comparison of means by Tukey's method to determine exactly where the significant differences occur in problem 17.

19. Does a woman's chance of suffering from postpartum depression vary depending on the number of children she already has? To find out, a researcher collected random samples from four groups of women: the first group having just given birth to their first child, the second group having just given birth to their second child, and so on. He then rated their amount of postpartum depression on a scale from 1 to 5. Test the null hypothesis that the chances of developing postpartum depression do not differ with the number of children to which a woman has previously given birth.

First Child	**Second Child**	**Third Child**	**Fourth Child**
3	3	5	4
2	5	5	3
4	1	3	2
3	3	5	1
2	4	2	5

20. Studies have found that people find symmetrical faces more attractive than faces that are not symmetrical. To test this theory, a psychiatrist selected a random sample of people and showed them pictures of three different faces: a face that is perfectly symmetrical, a face that is slightly asymmetrical, and a face that is highly asymmetrical. She then asked them to rate the three faces in terms of their attractiveness on a scale from 1 to 7, with 7 being the most attractive. Test the null hypothesis that attractiveness does not differ with facial symmetry.

Symmetrical	Slightly Asymmetrical	Highly Asymmetrical
7	5	2
6	4	3
7	5	1
5	2	1
6	4	2
6	5	2

21. Conduct a multiple comparison of means by Tukey's method to determine exactly where the significant differences occur in problem 20.

22. Political theorist Karl Marx is known for his theory that the working class, to put an end to capitalism and establish a communist society, would eventually rise up and overthrow the upper-class members of society who exploit them. One reason for the capitalist workers' discontent, according to Marx's theory, is that these workers take no pride in their work because both the work they do and the products that result belong not to them but to the capitalists they work for. To test this insight, a researcher went to a large factory and interviewed people from three groups—the workers, the managers, and the owners—to see if there is a difference among them in terms of how much pride they take in their work. Given the following scores, with higher scores representing more pride in work, test the null hypothesis that pride in work does not differ by class.

Lower (Workers)	Middle (Managers)	Upper (Owners)
1	4	8
3	7	7
2	5	6
5	6	9
4	8	5
2	6	6
3	5	7

23. Conduct a multiple comparison of means by Tukey's method to determine exactly where the significant differences occur in problem 22.

24. A psychiatrist wonders if people with panic disorder benefit from one particular type of treatment over any others. She randomly selects patients who have used one of the following treatments: cognitive therapy, behavioral therapy, or medication. She asks them to rate on a scale from 1 to 10 how much the treatment has led to a decrease in symptoms (with a score of 10 being the greatest reduction in symptoms). Test the null hypothesis

that the different treatments for panic disorder did not differ in how much they helped these patients.

Cognitive Therapy	Behavior Therapy	Medication
4	6	8
2	3	6
5	4	5
3	8	9
7	6	3
5	4	4
3	7	5

25. Is there a relationship between a mother's education level and how long she breastfeeds her child? A curious researcher selects samples of mothers from three different education levels and determines their length of breastfeeding (measured in months). Test the null hypothesis that education level has no effect on how long a mother breastfeeds her child.

Less Than High School	High School Graduate	College Graduate
1.0	1.5	11.0
6.5	4.0	6.5
4.5	3.5	4.5
2.0	1.5	7.5
8.5	5.0	9.0

26. Conduct a multiple comparison of means by Tukey's method to determine exactly where the significant differences occur in problem 25.

27. A researcher collected samples of sexually active adolescents from various racial and ethnic backgrounds and asked them what percent of the time they use protection. Given the following data, test the null hypothesis that black, white, Hispanic, and Asian adolescents do not differ in terms of how often they use protection.

Black Adolescents	White Adolescents	Hispanic Adolescents	Asian Adolescents
50	75	65	80
55	60	55	65
65	55	60	80
75	60	50	60
45	70	40	75

28. A marriage counselor notices that first marriages seem to last longer than remarriages. To see if this is true, she selects samples of divorced couples from first, second, and third marriages and determines the number of years each couple was married before getting divorced.

Test the null hypothesis that first, second, and third marriages do not differ by marriage length before divorce.

First Marriage	Second Marriage	Third Marriage
8.50	7.50	2.75
9.00	4.75	4.00
6.75	3.75	1.50
8.50	6.50	3.75
9.50	5.00	3.50

29. Conduct a multiple comparison of means by Tukey's method to determine exactly where the significant differences occur in problem 28.

30. In recent years, a number of cases of high school teachers having sexual relationships with their students have made the national news. Interested in how gender combinations influence perceptions of impropriety, a social researcher asks 40 respondents in a survey of education issues about their reaction to a story of a 16-year-old student who is seduced by a 32-year-old teacher. Assigned at random, one-quarter of the respondents are told about a case involving a male teacher and a male student, one-quarter are given a scenario involving a male teacher and a female student, and one-quarter each are presented a situation of a female teacher with a male or female student, respectively. All respondents are asked to indicate the level of impropriety from 0 to 10, where 0 is not at all improper and 10 as improper as can be imagined. The results are as follows:

Male Teacher		Female Teacher	
Male Student	Female Student	Male Student	Female Student
10	10	6	5
10	9	7	8
10	9	5	7
9	10	5	9
9	8	2	7
9	6	4	7
10	7	5	5
9	10	7	6
7	9	2	6
9	8	3	10

Using two-way analysis of variance, test gender of teacher, gender of student, and their interaction impact on the level of perceived impropriety surrounding high school teacher–student sexual relationships.

31. How does gender and occupational prestige impact credibility? Graduate students in a public health program are asked to rate the strength of a paper concerning the health risks of childhood obesity. All 30 student raters are given the same paper to evaluate, except that the name and degrees associated with the author have been manipulated by a social researcher. The student raters are randomly assigned to one of six groups, with each group receiving a paper written by a combination of either a male name ("John Forrest") or female name ("Joan Forrest") followed by one of three degrees (M.D., R.N., or Ph.D.). The raters are asked to rate the paper from 1 to 5 on clarity, 1 to 5 on strength of argument, and 1 to 5 on thoroughness. The total rating scores (the sum of the three subscores) are given as follows for each student rater in each of the six groups

	M.D.	R.N.	Ph.D.
	12	10	10
	15	11	8
John Forrest	13	7	13
	15	8	12
	14	8	9
	15	11	11
	10	8	11
Joan Forrest	12	9	12
	14	11	8
	12	7	8

a. Plot the means for the six groups on a chart with degree on the horizontal axis, mean rating on the vertical axis, and separate lines for each gender.
b. Using two-way analysis of variance, test if gender of author, author's degree, and their interaction impact on the ratings.

SPSS Exercises

1. Using SPSS to analyze the Monitoring the Future Study, test whether lifetime use of cocaine (V124) varies by school region (V13). To do this test, assume that lifetime use of cocaine is an interval level variable. Hint: Select ANALYZE, COMPARE MEANS, ONE-WAY ANOVA. Note also that the Tukey test is selected in the Post Hoc button and that means and standard deviations are selected in the Options button.

2. Using SPSS to analyze the Best Places Study, test whether violent crime rates (CRIMEV) vary by region (REGION) using one-way ANOVA with the Tukey test.

3. Using SPSS to analyze the Best Places Study, determine whether average commute times vary by region using one-way ANOVA with the Tukey test.

4. Use SPSS to analyze two variables of your choice from the Monitoring the Future Study using one-way ANOVA. Remember the dependent variable needs to be an interval/ratio level variable and the independent variable needs to have three or more categories.

5. Use SPSS to reanalyze the example on film content that was presented in this chapter. Notice that the data set is structured with empathy scored on a zero to ten scale ranging

from none (0) to high (10). Sexual content is a categorical variable measured as nonsexual (1) or sexual (2). Violent content is also a categorical variable measure as nonviolent (1) or violent (2).

Film Content

Empathy	Sexual	Violent
2	1	1
1	1	1
1	1	1
2	1	1
9	1	2
5	1	2
7	1	2
7	1	2
6	2	1
4	2	1
8	2	1
6	2	1
8	2	2
10	2	2
7	2	2
9	2	2

Using the GLM (General Linear Model) procedure in SPSS, conduct a two-way ANOVA. (*Hint:* ANALYZE, GENERAL LINEAR MODEL, UNIVARIATE.) The dependent variable is EMPATHY and the independent variables are entered as fixed factors. The GLM procedure is the easiest method to calculate a two-way ANOVA. It will estimate all main and interaction effects. Compare these SPSS results to the hand calculations from the chapter. Are these results the same within rounding error?

6. Using the GLM procedure in SPSS to analyze the Best Places Study, test whether property crime rates vary by size of the metropolitan statistical area (POPSIZE) and southern geographical location (SOUTH).
   a. Are the main effects for population size and southern geographical location significant?
   b. Is the interaction effect significant?

9

Nonparametric Tests of Significance

As indicated in Chapters 7 and 8, we must ask a good deal of the social researcher who employs a t ratio or an analysis of variance with an F ratio to make comparisons between his or her groups of respondents. Each of these tests of significance has a list of requirements that includes the assumption that the characteristic studied is normally distributed in a specified population or that the sample is large enough that the sampling distribution of the sample mean approaches normality. In addition, each test asks for the interval level of measurement so that a score can be assigned to every case. When a test of significance, such as the t ratio or the F ratio, requires (1) normality in the population (or at least large samples so that the sampling distribution of means is normal) and (2) an interval-level measure, it is referred to as a *parametric test* because it makes assumptions about the nature of population parameters.

What about the social researcher who cannot employ a parametric test—that is, who either cannot honestly assume normality, does not work with large numbers of cases, or whose data are not measured at the interval level? Suppose, for example, that he or she is working with a skewed distribution or with data that have been categorized and counted (the nominal level) or ranked (the ordinal level). How does this researcher go about making comparisons between samples without violating the requirements of a particular test? Fortunately, statisticians have developed a number of *nonparametric tests of significance*—tests whose list of requirements does not include normality or the interval level of measurement. To understand the important position of nonparametric tests in social research, we must also understand the concept of the *power of a test*; that is, the probability of rejecting the null hypothesis when it is actually false and should be rejected.

Power varies from one test to another. The most powerful tests—those that are most likely to reject the null hypothesis when it is false—are tests that have the strongest or most difficult requirements to satisfy. Generally, these are parametric tests such as *t* or *F*, which assume that interval data are employed and that the characteristics being studied are normally distributed in their populations or are sampled in ample quantities. By contrast, the nonparametric alternatives make less stringent demands, but are less powerful tests of significance than their parametric counterparts. As a result, assuming that the null hypothesis is false (and holding constant such other factors as sample size), an investigator is more likely to reject the null hypothesis by the appropriate use of *t* or *F* than by a nonparametric alternative. In a statistical sense, you get what you pay for!

Understandably, social researchers are eager to reject the null hypothesis when it is false. As a result, many of them would ideally prefer to employ parametric tests of significance and might even be willing to "stretch the truth" a little bit to meet the assumptions. For example, if ordinal data are fairly evenly spaced and therefore approximate an interval scale, and if the data are not normal but also not terribly skewed, one can "get away with" using a parametric test.

As previously noted, however, it is often not possible—without deceiving yourself to the limit—to come even close to satisfying the requirements of parametric tests. In the first place, much of the data of social research are nowhere near the interval level. Second, we may know that certain variables or characteristics under study are severely skewed in the population and may not have large enough samples to compensate.

When its requirements have been severely violated, it is not possible to know the power of a statistical test. Therefore, the results of a parametric test whose requirements have gone unsatisfied may lack any meaningful interpretation. Under such conditions, social researchers wisely turn to nonparametric tests of significance.

This chapter introduces two of the best known nonparametric tests of significance for characteristics measured at the nominal or ordinal level: the chi-square test and the median test.

One-Way Chi-Square Test

Have you ever tried to psych out your instructor while taking a multiple-choice test? You may have reasoned, "The last two answers were both B; he wouldn't possibly have three in a row." Or you may have thought, "There haven't been very many D answers; maybe I should change a few of the ones I wasn't sure of to D." You are assuming, of course, that your instructor attempts to distribute his correct answers evenly across all categories, A through E.

Suppose your instructor returns the exam and hands out the answer key. You construct a frequency distribution of the correct responses to the 50-item test as follows:

A	12
B	14
C	9
D	5
E	10

Thus, 12 of the 50 items had a correct answer of A, 14 had a correct answer of B, and so on. These are called the *observed frequencies* (f_o). Observed frequencies refer to the set of frequencies obtained in an actual frequency distribution—that is, when we actually do research or conduct a study.

You can sense from this distribution that the instructor may favor putting the correct response near the top—that is, in the A and B positions. Conversely, he seems to shy away from the D response, for only five correct answers fell into that category.

Can we generalize about the tendencies of this professor observed from this one exam? Are the departures from an even distribution of correct responses large enough to indicate, for example, a real dislike for the category D? Or could chance variations account for these results; that is, if we had more tests constructed by this instructor, could we perhaps see the pattern even out in the long run?

These are the kinds of questions we asked in Chapters 7 and 8 concerning sample means, but now we need a test for frequencies. Chi-square (pronounced "ki-square" and written χ^2) is the most commonly used nonparametric test; not only is it relatively easy to follow, but it is applicable to a wide variety of research problems.

The *one-way chi-square* test can be used to determine whether the frequencies we observed previously differ significantly from an even distribution (or any other distribution we might hypothesize). In this example, our null and research hypotheses are as follows:

Null hypothesis:	*The instructor shows no tendency to assign any particular correct response from A to E.*
Research hypothesis:	*The instructor shows a tendency to assign particular correct responses from A to E.*

What should the frequency distribution of correct responses look like if the null hypothesis were true? Because there are five categories, 20% of the correct responses should fall in each. With 50 questions, we should expect 10 correct responses for each category.

The *expected frequencies* (f_e) are those frequencies that are expected to occur under the terms of the null hypothesis. The expected frequencies for the hypothesized even distribution are as follows:

A	10
B	10
C	10
D	10
E	10

Chi-square allows us to test the significance of the difference between a set of observed frequencies (f_o) and expected frequencies (f_e). We obtained 12 correct A answers, but we expected 10 under the null hypothesis; we obtained 14 correct B answers, but we expected 10 under the null hypothesis; and so on. Obviously, the greater the differences between the observed and the expected frequencies, the more likely we have a significant difference suggesting that the null hypothesis is unlikely to be true.

The chi-square statistic focuses directly on how close the observed frequencies are to what they are expected to be (represented by the expected frequencies) under the null hypothesis. Based on just the observed and expected frequencies, the formula for chi-square is

$$\chi^2 = \sum \frac{(f_o - f_e)^2}{f_e}$$

where f_o = observed frequency in any category
f_e = expected frequency in any category

According to the formula, we must subtract each expected frequency from its corresponding observed frequency, square the difference, divide by the expected frequency, and then add these quotients for all the categories to obtain the chi-square value.

Let's inspect how the chi-square statistic relates to the null hypothesis. If, in the extreme case, all the observed frequencies were equal to their respective expected frequencies (all $f_o = f_e$) as the null hypothesis suggests, chi-square would be zero. If all the observed frequencies were close to their expected frequencies, consistent with the null hypothesis except for sampling error, chi-square would be small. The more the set of observed frequencies deviates from the frequencies expected under the null hypothesis, the larger the chi-square value. At some point, the discrepancies of the observed frequencies from the expected frequencies become larger than could be attributed to sampling error alone. At that point, chi-square is so large that we are forced to reject the null hypothesis and accept the research hypothesis. Just how large chi-square must be before we reject the null hypothesis is something about which we will keep you in suspense, but only until we show you how chi-square is computed.

As stated earlier, we compute the difference between the observed and expected frequency, square the difference, and divide by the expected frequency for each category. Setting it up in tabular form,

Category	f_o	f_e	$f_o - f_e$	$(f_o - f_e)^2$	$\dfrac{(f_o - f_e)^2}{f_e}$
A	12	10	2	4	.4
B	14	10	4	16	1.6
C	9	10	−1	1	.1
D	5	10	−5	25	2.5
E	10	10	0	0	0.0
					$\chi^2 = 4.6$

By summing the last column (containing the quotients) over all the categories, we obtain $\chi^2 = 4.6$. To interpret this chi-square value, we must determine the appropriate number of degrees of freedom:

$$df = k - 1$$

where k = number of categories in the observed frequency distribution

In our example, there are five categories, and thus

$$df = 5 - 1 = 4$$

Turning to Table E in Appendix C, we find a list of chi-square values that are significant at the .05 and .01 levels. For the .05 significance level, we see that the critical value for chi-square with 4 degrees of freedom is 9.488. This is the value that we must exceed before we can reject the null hypothesis. Because our calculated chi-square is only 4.6, and therefore smaller than the table value, we must retain the null hypothesis and reject the research hypothesis. The observed frequencies do not differ enough from the frequencies expected under the null hypothesis of an equal distribution of correct responses. Thus, although we did not observe a perfectly even distribution (10 for each category), the degree of unevenness was not sufficiently large to conclude that the instructor had any underlying preference in designing his answer key.

BOX 9.1 • *Step-by-Step Illustration: One-Way Chi-Square*

To summarize the step-by-step procedure for calculating one-way chi-square, imagine that a social researcher is interested in surveying attitudes of high school students concerning the importance of getting a college degree. She questions a sample of 60 high school seniors about whether they believe that a college education is becoming more important, less important, or staying the same.

We might specify our hypotheses as follows:

Null hypothesis: *High school students are equally divided in their beliefs regarding the changing importance of a college education.*

Research hypothesis: *High school students are not equally divided in their beliefs regarding the changing importance of a college education.*

Let us say that of the 60 high school students surveyed, 35 feel that a college education is becoming more important, 10 feel that it is becoming less important, and 15 feel that the importance is about the same.

Step 1 Arrange the data in the form of a frequency distribution.

Category	Observed Frequency
More important	35
Less important	10
About the same	<u>15</u>
Total	60

Step 2 Obtain the expected frequency for each category.

The expected frequencies (f_e) are those frequencies expected to occur under the terms of the null hypothesis. Under the null hypothesis, we would expect the opinions to divide themselves equally across the three categories. Therefore, with three categories $(k = 3)$ and $N = 60$,

$$f_e = \frac{60}{3} = 20$$

Category	Observed Frequency (f_o)	Expected Frequency (f_e)
More important	35	20
Less important	10	20
About the same	<u>15</u>	<u>20</u>
Total	60	60

Step 3 Set up a summary table to calculate the chi-square value.

Category	f_o	f_e	$f_o - f_e$	$(f_o - f_e)^2$	$\dfrac{(f_o - f_e)^2}{f_e}$
More important	35	20	15	225	11.25
Less important	10	20	−10	100	5.00
About the same	15	20	−5	25	<u>1.25</u>
					$\chi^2 = 17.50$

Step 4 Find the degrees of freedom.

$$df = k - 1 = 3 - 1 = 2$$

Step 5 Compare the calculated chi-square value with the appropriate chi-square value from Table E.

Turning to Table E in Appendix C, we look up the chi-square value required for significance at the .05 level for 2 degrees of freedom and find that this critical

(continued)

BOX 9.1   Continued

value is 5.99. Because the calculated chi-square ($\chi^2 = 17.50$) is larger than the table value, we reject the null hypothesis. These findings suggest, therefore, that high school students are not equally divided about their views concerning the changing importance of pursuing a college education. In fact, the majority (35 out of 60) felt that it was becoming more important. More to the point, these findings cannot be passed off as merely the result of sampling error or chance.

Two-Way Chi-Square Test

Up to this point, we have considered testing whether the categories in an observed frequency distribution differ significantly from one another. To accomplish this, we used a one-way chi-square test to determine whether the observed set of frequencies is significantly different from a set of expected frequencies under the terms of the null hypothesis.

The chi-square test has a much broader use in social research: To test whether one observed frequency distribution significantly differs from another observed distribution. For example, the survey of high school students concerning the changing value of a college education discussed previously could be enhanced by testing to see whether male students respond differently from female students. Thus, rather than comparing a set of observed frequencies (more important, less important, about the same) against a hypothetically equal distribution, we might compare the observed frequencies (more important, less important, about the same) for both males and females.

As we saw in Chapter 2, nominal and ordinal variables are often presented in the form of a cross-tabulation. Specifically, cross-tabulations are used to compare the distribution of one variable, often called the dependent variable, across categories of some other variable, the independent variable. In a cross-tabulation, the focus is on the differences between groups—such as between males and females—in terms of the dependent variable; for example, opinion about the changing value of a college education.

We are now prepared to consider whether differences in a cross-tabulation, such as gender differences in beliefs regarding the importance of a college degree, are statistically significant. Recall that a one-way chi-square was used to test a single frequency distribution by comparing observed frequencies with expected frequencies under the null hypothesis. Similarly, a *two-way chi-square* can be used for testing a cross-tabulation, also by comparing observed frequencies with expected frequencies under the null hypothesis.

You will be happy to hear that the two forms of chi-square (one-way and two-way) are very similar in both logic and procedure. In fact, the only major difference is the basis for calculating the expected frequencies.

Computing the Two-Way Chi-Square

As in the case of the *t* ratio and analysis of variance, there is a sampling distribution for chi-square that can be used to estimate the probability of obtaining a significant chi-square value by chance alone rather than by actual population differences. Unlike these earlier tests of significance, however, chi-square is employed to make comparisons between frequencies rather than between mean scores. As a result, the null hypothesis for the chi-square test states that the populations do not differ with respect to the frequency of occurrence of a given characteristic, whereas the research hypothesis says that sample differences reflect actual population differences regarding the relative frequency of a given characteristic.

To illustrate the use of the two-way chi-square for frequency data (or for proportions that can be converted to frequencies), imagine that we have been asked to investigate the relationship between political orientation and child-rearing permissiveness. We might categorize our sample members on a strictly either–or basis; that is, we might decide that they are either permissive or not permissive. Therefore,

Null hypothesis: *The relative frequency (or percentage) of liberals who are permissive is the same as the relative frequency (or percentage) of conservatives who are permissive.*

Research hypothesis: *The relative frequency (or percentage) of liberals who are permissive is not the same as the relative frequency (or percentage) of conservatives who are permissive.*

Like its one-way counterpart, the chi-square test of significance for a two-way cross-tabulation is essentially concerned with the distinction between expected frequencies and observed frequencies. Once again, expected frequencies (f_e) refer to the terms of the null hypothesis, according to which the relative frequency (or proportion) is expected to be the same from one group to another. For example, if a certain percentage of the liberals is expected to be permissive, then we expect the same percentage of the conservatives to be permissive. By contrast, observed frequencies (f_o) refer to the results that we actually obtain when conducting a study and, therefore, may or may not vary from one group to another. Only if the difference between expected and observed frequencies is large enough do we reject the null hypothesis and decide that a true population difference exists.

Let's consider the simplest possible case in which we have equal numbers of liberals and conservatives as well as equal numbers of permissive and not permissive respondents. Assuming 40 respondents took part in the survey, the cross-tabulation showing the observed frequencies for each cell (f_o) might be as follows:

Child-Rearing Methods	Political Orientation		Total
	Liberals	*Conservatives*	
Permissive	13	7	20
Not Permissive	7	13	20
Total	20	20	$N = 40$

In this cross-tabulation, there are four cells and 40 respondents. Therefore, to calculate the expected frequencies (f_e), we might expect 10 cases per cell, as shown in the following:

Child-Rearing Methods	Political Orientation		Total
	Liberals	*Conservatives*	
Permissive	10	10	20
Not Permissive	10	10	20
Total	20	20	$N = 40$

This straightforward method of calculating expected frequencies works in this cross-tabulation, but *only* because the marginals—both row and column—are identical (they are all 20). Unfortunately, most research situations will not yield cross-tabulations in which both the row and column marginals are evenly split. By sampling technique, it may be possible to control the distribution of the independent variable—for example, to get exactly the same number of liberals and conservatives. But you cannot control the distribution of the dependent variable, for example, the number of permissive and not permissive respondents. Thus, we must consider a more general approach to calculating expected frequencies—one that can be used when either or both the row and column marginals are not evenly distributed.

Continuing with the present example in which we drew samples of 20 liberals and 20 conservatives, suppose that we observed more "permissive" respondents than "nonpermissive" respondents. Therefore, as shown in Table 9.1, the row marginals would not be equal.

The data in Table 9.1 indicate that permissive child-rearing methods were used by 15 out of 20 liberals and 10 out of 20 conservatives. To determine if these frequencies depart

TABLE 9.1   *Frequencies Observed in a Cross-Tabulation of Child-Rearing Methods by Political Orientation*

Child-Rearing Methods	Political Orientation		Total
	Liberals	*Conservatives*	
Permissive	15	10	25
Not Permissive	5	10	15
Total	20	20	$N = 40$

from what one would expect by chance alone, we need to determine the expected frequencies under the null hypothesis of no difference.

The observed and expected frequencies for each cell are displayed together in Table 9.2. The expected frequencies are derived purposely to be in line with the null hypothesis; that is, they represent the frequencies one would expect to see if the null hypothesis of no difference were true. Thus, 25 out of 40 of the respondents overall, or 62.5%, are permissive in their approach to child rearing. For there to be no difference between the liberals and the conservatives in this regard, as dictated by the null hypothesis, 62.5% of the liberals and 62.5% of the conservatives should be permissive. Translating into expected frequencies the fact that both groups should have the same percentage (or relative frequency) of permissive respondents, we expect 12.5 liberals (62.5% of 20, or .625 × 20 = 12.5) to be permissive and 12.5 conservatives (62.5% of 20, or .625 × 20 = 12.5) to be permissive, if the null hypothesis were true. Of course, the expected frequencies of respondents who are not permissive are 7.5 for both liberals and conservatives, because the expected frequencies must sum to the marginal totals (in this case, 12.5 + 7.5 = 20). Finally, it is important to note that the expected frequencies, as in the present example, do not have to be whole numbers.

As discussed earlier, chi-square focuses on how close the observed frequencies are to those expected under the null hypothesis. Based on the observed and expected frequencies, the chi-square formula is as follows:

$$\chi^2 = \sum \frac{(f_o - f_e)^2}{f_e}$$

where f_o = observed frequency in any cell
f_e = expected frequency in any cell

TABLE 9.2 *Frequencies Observed and Expected in a Cross-Tabulation of Child-Rearing Methods by Political Orientation*

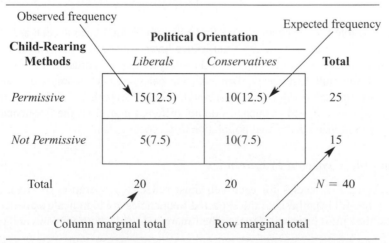

Observed frequency			Expected frequency
Child-Rearing Methods	**Political Orientation**		**Total**
	Liberals	*Conservatives*	
Permissive	15(12.5)	10(12.5)	25
Not Permissive	5(7.5)	10(7.5)	15
Total	20	20	N = 40
	Column marginal total	Row marginal total	

We subtract each expected frequency from its corresponding observed frequency, square the difference, divide by the expected frequency, and then add up these quotients for all the cells to obtain the chi-square value.

By applying the chi-square formula to the case at hand,

Cell	f_o	f_e	$f_o - f_e$	$(f_o - f_e)^2$	$\dfrac{(f_o - f_e)^2}{f_e}$
Upper left	15	12.5	2.5	6.25	.50
Upper right	10	12.5	−2.5	6.25	.50
Lower left	5	7.5	−2.5	6.25	.83
Lower right	10	7.5	2.5	6.25	.83
					$\chi^2 = 2.66$

Thus, we learn that $\chi^2 = 2.66$. To interpret this chi-square value, we must still determine the appropriate number of degrees of freedom. This can be done for tables that have any number of rows and columns by employing the formula

$$df = (r - 1)(c - 1)$$

where r = number of rows in the table of observed frequencies
c = number of columns in the table of observed frequencies
df = degrees of freedom

Because the observed frequencies in Table 9.2 form two rows and two columns (2×2),

$$df = (2 - 1)(2 - 1)$$
$$= (1)(1)$$
$$= 1$$

Turning to Table E in Appendix C, we find a list of chi-square values that are significant at the .05 and .01 levels. For the .05 significance level, we see that the critical value for chi-square with 1 degree of freedom is 3.84. This is the value that must be exceeded before we can reject the null hypothesis. Because our calculated χ^2 is only 2.66 and therefore smaller than the table value, we must retain the null hypothesis and reject the research hypothesis. The observed frequencies do not differ enough from the frequencies expected by chance to indicate that actual population differences exist.

Finding the Expected Frequencies

The expected frequencies for each cell must reflect the operation of chance under the terms of the null hypothesis. If the expected frequencies are to indicate sameness across all samples, they must be proportional to their marginal totals, both for rows and columns.

To obtain the expected frequency for any cell, we multiply together the column and row marginal totals for a particular cell and divide the product by N. Therefore,

$$f_e = \frac{(\text{row marginal total})(\text{column marginal total})}{N}$$

For the upper-left cell in Table 9.2 (permissive liberals),

$$f_e = \frac{(25)(20)}{40}$$

$$= \frac{500}{40}$$

$$= 12.5$$

Likewise, for the upper-right cell in Table 9.2 (permissive conservatives),

$$f_e = \frac{(25)(20)}{40}$$

$$= \frac{500}{40}$$

$$= 12.5$$

For the lower-left cell in Table 9.2 (not permissive liberals),

$$f_e = \frac{(15)(20)}{40}$$

$$= \frac{300}{40}$$

$$= 7.5$$

For the lower-right cell in Table 9.2 (not permissive conservatives),

$$f_e = \frac{(15)(20)}{40}$$

$$= \frac{300}{40}$$

$$= 7.5$$

As we will see, the foregoing method for determining f_e can be applied to any chi-square problem for which the expected frequencies must be obtained.

BOX 9.2 • *Step-by-Step Illustration: Two-Way Chi-Square Test of Significance*

To summarize the step-by-step procedure for obtaining chi-square for a cross-tabulation, let us consider a study in which the effectiveness of hypnosis as a means of improving the memory of eyewitnesses to a crime is examined. The hypotheses might be specified as follows:

Null hypothesis: *Hypnosis does not affect the recognition memory of eyewitnesses to a crime.*

Research hypothesis: *Hypnosis does affect the recognition memory of eyewitnesses to a crime.*

To test the null hypothesis at the $\alpha = .05$ level of significance, all subjects first view a videotape of a pickpocket plying his trade. One week later, subjects are randomly assigned to one of two conditions. The subjects in the experimental group are hypnotized and then asked to pick the thief out of a lineup. Subjects in the control group are not hypnotized and attempt the same lineup identification. Suppose that the results are as shown in Table 9.3. We can see from the results that the hypnotized group actually did worse in attempting to identify the culprit. Only 7 of the 40 subjects in the experimental group were correct, whereas 17 of the 40 control subjects made the right choice. This difference may suggest that hypnosis does have an effect (although not the kind of effect that one might desire); but is the difference significant?

TABLE 9.3 *Hypnosis and Accuracy of Eyewitness Identification*

	Hypnotized	**Control**
Correct identification	7	17
Incorrect identification	33	23
Total	40	40

Step 1 Rearrange the data in the form of a 2×2 table containing the observed frequencies for each cell.

	Hypnotized	**Control**	
Correct Identification	7	17	24
Incorrect Identification	33	23	56
	40	40	$N = 80$

Step 2   Obtain the expected frequency for each cell.

	Hypnotized	**Control**	
Correct Identification	7(12)	17(12)	24
Incorrect Identification	33(28)	23(28)	56
	40	40	$N = 80$

Upper left:

$$f_e = \frac{(24)(40)}{80}$$

$$= \frac{960}{80}$$

$$= 12$$

Upper right:

$$f_e = \frac{(24)(40)}{80}$$

$$= \frac{960}{80}$$

$$= 12$$

Lower left:

$$f_e = \frac{(56)(40)}{80}$$

$$= \frac{2,240}{80}$$

$$= 28$$

Lower right:

$$f_e = \frac{(56)(40)}{80}$$

$$= \frac{2,240}{80}$$

$$= 28$$

Step 3   Construct a summary table in which, for each cell, you report the observed and expected frequencies, subtract the expected from the observed frequency, square

(continued)

BOX 9.2   Continued

this difference, divide by the expected frequency, and sum these quotients to obtain the chi-square value.

Cell	f_o	f_e	$f_o - f_e$	$(f_o - f_e)^2$	$\dfrac{(f_o - f_e)^2}{f_e}$
Upper left	7	12	−5	25	2.08
Upper right	17	12	5	25	2.08
Lower left	33	28	5	25	.89
Lower right	23	28	−5	25	.89
					$\chi^2 = 5.94$

Step 4   Find the degrees of freedom.

$$\begin{aligned}
df &= (r - 1)(c - 1) \\
&= (2 - 1)(2 - 1) \\
&= (1)(1) \\
&= 1
\end{aligned}$$

Step 5   Compare the obtained chi-square value with the appropriate chi-square value in Table E.

$$\text{Obtained } \chi^2 = 5.94$$

$$\text{Table } \chi^2 = 3.84$$

$$df = 1$$

$$\alpha = .05$$

As indicated in Step 5, to reject the null hypothesis at the .05 significance level with 1 degree of freedom, our calculated chi-square value would have to be larger than 3.84. Because we have obtained a chi-square value of 5.94, we can reject the null hypothesis and accept the research hypothesis. Our results suggest a significant difference in the ability of hypnotized and control subjects to identify a thief. Before you recommend that all crime witnesses be hypnotized, however, take a second look at how the data themselves line up. The hypnotized subjects were *less*, not more, accurate in identifying the thief. Perhaps under hypnosis, subjects felt compelled to use their imaginations to fill in gaps in their memories.

Comparing Several Groups

Until now we have limited our illustrations to the widely employed 2 × 2 problem. However, chi-square is frequently calculated for tables that have more than two rows or columns, for example, when several groups or categories are to be compared or where the outcome variable has more than two values. The step-by-step procedure for comparing several groups is essentially the same as its 2×2 counterpart. Let us illustrate with a 3×3 problem (3 rows by 3 columns), although any number of rows and columns could be used.

Suppose that a sociologist is interested in how child-discipline practices differ among men based on their type of parental role. Specifically, he compares samples of biological fathers, stepfathers, and male live-in partners in terms of how strict they tend to be in dealing with their own or their wives'/partners' children. He draws three random samples—32 fathers, 30 stepfathers, and 27 live-in partners—and categorizes their reported child-discipline practices as permissive, moderate, or authoritarian. Therefore,

Null hypothesis: *The relative frequency of permissive, moderate, and authoritarian child discipline is the same for fathers, stepfathers, and live-in partners.*

Research hypothesis: *The relative frequency of permissive, moderate, and authoritarian child discipline is not the same for fathers, stepfathers, and live-in partners.*

Suppose the sociologist obtains the the sample differences in child-discipline practices shown in Table 9.4 on page 324. We see, for example, that 7 out of 32 fathers, 9 out of 30 stepfathers, and 14 out of 27 live-in partners could be regarded as permissive in their approach.

To determine whether there is a significant difference in Table 9.4, we must apply the original chi-square formula as introduced earlier in the chapter:

$$\chi^2 = \sum \frac{(f_o - f_e)^2}{f_e}$$

BOX 9.3 • *Step-by-Step Illustration: Comparing Several Groups*

The chi-square formula can be applied to the 3×3 problem using the following step-by-step procedure:

Step 1   Arrange the data in the form of a 3×3 table.

Discipline Practices	Type of Relationship			Total
	Father	*Step-father*	*Live-in Partner*	
Permissive	7	9	14	30
Moderate	10	10	8	28
Authoritarian	15	11	5	31
Total	32	30	27	89

Column marginal total   Observed frequency   Row marginal total

(continued)

BOX 9.3   Continued

Step 2   Obtain the expected frequency for each cell.

Upper left:

$$f_e = \frac{(30)(32)}{89}$$

$$= \frac{960}{89}$$

$$= 10.79$$

Upper middle:

$$f_e = \frac{(30)(30)}{89}$$

$$= \frac{900}{89}$$

$$= 10.11$$

Upper right:

$$f_e = \frac{(30)(27)}{89}$$

$$= \frac{810}{89}$$

$$= 9.10$$

Middle left:

$$f_e = \frac{(28)(32)}{89}$$

$$= \frac{896}{89}$$

$$= 10.07$$

Middle middle:

$$f_e = \frac{(28)(30)}{89}$$

$$= \frac{840}{89}$$

$$= 9.44$$

Middle right:

$$f_e = \frac{(28)(27)}{89}$$

$$= \frac{756}{89}$$

$$= 8.49$$

Lower left:

$$f_e = \frac{(31)(32)}{89}$$

$$= \frac{992}{89}$$

$$= 11.15$$

Lower middle:

$$f_e = \frac{(31)(30)}{89}$$

$$= \frac{930}{89}$$

$$= 10.45$$

Lower right:

$$f_e = \frac{(31)(27)}{89}$$

$$= \frac{837}{89}$$

$$= 9.40$$

Step 3 Construct a summary table in which, for each cell, you report the observed and expected frequencies, subtract the expected from the observed frequency, square this difference, divide by the expected frequency, and sum these quotients to obtain the chi-square value.

Cell	f_o	f_e	$f_o - f_e$	$(f_o - f_e)^2$	$\dfrac{(f_o - f_e)^2}{f_e}$
Upper left	7	10.79	−3.79	14.36	1.33
Upper middle	9	10.11	−1.11	1.23	.12
Upper right	14	9.10	4.90	24.01	2.64
Middle left	10	10.07	−.07	.00	.00
Middle middle	10	9.44	.56	.31	.03
Middle right	8	8.49	−.49	.24	.03
Lower left	15	11.15	3.85	14.82	1.33
Lower middle	11	10.45	.55	.30	.03
Lower right	5	9.40	−4.40	19.36	2.06
					$\chi^2 = 7.57$

(continued)

BOX 9.3 Continued

Step 4 Find the number of degrees of freedom.

$$df = (r - 1)(c - 1)$$
$$= (3 - 1)(3 - 1)$$
$$= (2)(2)$$
$$= 4$$

Step 5 Compare the obtained chi-square value with the appropriate chi-square value in Table E.

$$\text{Obtained } \chi^2 = 7.57$$
$$\text{Table } \chi^2 = 9.49$$
$$df = 4$$
$$\alpha = .05$$

Therefore, we need a chi-square value above 9.49 to reject the null hypothesis. Because our obtained chi-square is only 7.57, we must retain the null hypothesis and attribute our sample differences to the operation of chance alone. We have not found statistically significant evidence to indicate that the relative frequency of child-discipline methods differs due to type of relationship.

Correcting for Small Expected Frequencies

One of the primary reasons why the chi-square test is so popular among social researchers is that it makes very few demands on the data. That is, the numerous assumptions associated with the t ratio and the analysis of variance are absent from the chi-square alternative. Despite this relative freedom from assumptions, however, chi-square cannot be used indiscriminately. In particular, chi-square does impose some rather modest requirements on sample size. Although chi-square does not require the same large samples as some of the

TABLE 9.4 *Child-Discipline Practices by Type of Relationship*

| Discipline Practices | Type of Relationship | | |
	Father	*Stepfather*	*Live-in Partner*
Permissive	7	9	14
Moderate	10	10	8
Authoritarian	15	11	5
Total	32	30	27

parametric tests, an extremely small sample can sometimes yield misleading results, as we will see in what follows.

Generally, chi-square should be used with great care whenever some of the expected frequencies are below 5. There is no hard-and-fast rule concerning just how many expected frequencies below 5 will render an erroneous result. Some researchers contend that all expected frequencies should be at least 5, and others relax this restriction somewhat and only insist that most of the expected frequencies be at least 5. The decision concerning whether to proceed with the test depends on what impact the cells with the small expected frequencies have on the value of chi-square.

Consider, for example, the cross-tabulation in Table 9.5 of murder weapon and gender of offender for 200 homicide cases. The "female/other weapon" cell has an expected frequency of only 2. For this cell, the observed frequency is 6 (females tend to use poison far more than men), and so its contribution to the chi-square statistic is

$$\frac{(f_o - f_e)^2}{f_e} = \frac{(6 - 2)^2}{2} = 8$$

No matter what happens with the seven other cells, this value of 8 for the female/other cell will cause the null hypothesis to be rejected. That is, for a 4×2 table (df = 3) the critical chi-square value from Table E is 7.815, which is already surpassed because of this one cell alone. One should feel uncomfortable indeed about rejecting the null hypothesis just because there were four more women than expected who used an "other" weapon, such as poison. The problem here is that the expected frequency of 2 in the denominator causes the fraction to be unstable. With even a modest difference between observed and expected frequencies in the numerator, the quotient explodes because of the small divisor. For this reason, small expected frequencies are a concern.

In instances like this in which expected frequencies of less than 5 create such problems, you should collapse or merge together some categories, but only if it is logical to do so. One would not want to merge together categories that are substantively

TABLE 9.5 *Cross-Tabulation of Murder Weapon and Gender of Offender (Expected Frequencies in Parentheses)*

	Male	**Female**	**Total**
Gun	100	20	120
	(90)	(30)	
Knife	39	21	60
	(45)	(15)	
Blunt object	9	3	12
	(9)	(3)	
Other	2	6	8
	(6)	(2)	
Total	150	50	$N = 200$

TABLE 9.6 *Revised Cross-Tabulation of Murder Weapon and Gender of Offender (Expected Frequencies in Parentheses)*

	Male	Female	Total
Gun	100	20	120
	(90)	(30)	
Knife	39	21	60
	(45)	(15)	
Other	11	9	20
	(15)	(5)	
Total	150	50	$N = 200$

very different. But in this instance, we can reasonably combine the "blunt object" and "other" categories into a new category that we can still label "other." The revised cross-tabulation is shown in Table 9.6. Note that now none of the cells has problematically low expected frequencies.

In 2×2 tables, the requirement for having all expected frequencies at least equal to 5 is particularly important. In addition, for 2×2 tables, distortions can also occur if expected frequencies are under 10. Fortunately, however, there is a simple solution for 2×2 tables with any expected frequency less than 10 but greater than 5, known as *Yates's correction.*[1] By using Yates's correction, the difference between observed and expected frequencies is reduced by .5. Because chi-square depends on the size of that difference, we also reduce the size of our calculated chi-square value. The following is the corrected chi-square formula for small expected frequencies:

$$\chi^2 = \sum \frac{(|f_o - f_e| - .5)^2}{f_e}$$

In the corrected chi-square formula, the vertical lines surrounding $f_o - f_e$ indicate that we must reduce the absolute value (ignoring minus signs) of each $f_o - f_e$ by .5.

To illustrate, suppose that an instructor at a U.S. university close to the Canadian border suspects that his Canadian students are more likely than his U.S. students to be cigarette smokers. To test his hypothesis, he questions the 36 students in one of his classes about their smoking status and nationality. The results are shown in Table 9.7.

If we were to use the original chi-square formula for a 2×2 problem ($\chi^2 = 5.13$), we would conclude that the difference between U.S. and Canadian students is significant. Before we make much of this result, however, we must be concerned about the potential effects of small expected frequencies and compute Yates's corrected formula.

---

[1]Some researchers recommend that Yates's corrections be used for all 2×2 tables, not just those with deficient expected frequencies. Although technically correct, it makes little practical difference when all the expected frequencies are fairly large. That is, the corrected and uncorrected chi-square values are very similar with large expected frequencies.

TABLE 9.7 *Cross-Tabulation of Smoking Status and Nationality*

	Nationality	
Smoking Status	*Americans*	*Canadians*
Nonsmokers	15	5
Smokers	6	10
Total	21	15

The procedure for applying the corrected 2×2 chi-square formula can be summarized in tabular form:

f_o	f_e	$|f_o - f_e|$	$|f_o - f_e| - .5$	$(|f_o - f_e| - .5)^2$	$\dfrac{(|f_o - f_e| - .5)^2}{f_e}$
15	11.67	3.33	2.83	8.01	.69
5	8.33	3.33	2.83	8.01	.96
6	9.33	3.33	2.83	8.01	.86
10	6.67	3.33	2.83	8.01	1.20
					$\chi^2 = 3.71$

As shown, Yates's correction yields a smaller chi-square value ($\chi^2 = 3.71$) than was obtained by means of the uncorrected formula ($\chi^2 = 5.13$). In the present example, our decision regarding the null hypothesis would depend on whether we had used Yates's correction. With the corrected formula, we retain the null hypothesis; without it, we reject the null hypothesis. Given these very different results, one should go with the more conservative formula that uses Yates's correction.

Requirements for the Use of Chi-Square

The chi-square test of significance has few requirements for its use, which might explain in part why it is applied so frequently. Unlike the *t* ratio, for example, it does not assume a normal distribution in the population nor interval-level data. The following requirements still need to be considered before using chi-square:

1. *A comparison between two or more samples.* As illustrated and described in the present chapter, the chi-square test is employed to make comparisons between two or more *independent* samples. This requires that we have at least a 2×2 table (at least 2 rows and at least 2 columns). The assumption of independence indicates that chi-square cannot be applied to a single sample that has been studied in a before–after panel design or matched samples. At least two samples of respondents must be obtained.

2. *Nominal data.* Chi-square does not require data that are ranked or scored. Only frequencies are required.

3. *Random-sampling.* We should have drawn our samples at random from a particular population.

4. *The expected cell frequencies should not be too small.* Exactly how large f_e must be depends on the nature of the problem. For a 2×2 problem, no expected frequency should be smaller than 5. In addition, Yates's corrected formula should be used for a 2×2 problem in which an expected cell frequency is smaller than 10. For a situation wherein several groups are being compared (say, a 3×3 or 4×5 problem), there is no hard-and-fast rule regarding minimum cell frequencies, although we should be careful to see that few cells contain fewer than five cases. In such instances, categories with small numbers of cases should be merged together if at all possible.

The Median Test

For ordinal data, the *median test* is a simple nonparametric procedure for determining the likelihood that two or more random samples have been taken from populations with the same median. Essentially, the median test involves performing a chi-square test of significance on a cross-tabulation in which one of the dimensions is whether the scores fall above or below the median of the two groups combined. Just as before, Yates's correction is used for a 2×2 problem (comparing two samples) having small expected frequencies.

BOX 9.4 • *Step-by-Step Illustration: Median Test*

To illustrate the procedure for carrying out the median test, suppose an investigator wanted to study male–female reactions to a socially embarrassing situation. To create the embarrassment, the investigator asked 15 men and 12 women with only average singing ability to sing individually several difficult songs, including the national anthem, in front of an audience of "experts." The following table shows the order in which subjects quit performing (1 = first to quit; 27 = last to quit; a shorter period of time singing indicates greater embarrassment):

Men	Women
4	1 (first to quit)
6	2
9	3
11	5
12	7
15	8
16	10
18	13
19	14
21	17
23	20
24	22
25	
26	
27 (last to quit)	

Step 1   Find the median of the two samples combined.

$$\text{Position of median} = \frac{N + 1}{2}$$

$$= \frac{27 + 1}{2}$$

$$= 14\text{th}$$

Step 2   Count the number in each sample falling above the median and not above the median (Mdn = 14).

	Men	Women
Above median	10	3
Not above median	5	9
	$N = 27$	

As shown, the numbers above and not above the median singing time from each sample are represented in a 2×2 frequency table. In this table, we see that 10 of the 15 men but only 3 of the 12 women continued singing for a period of time that was greater than the median singing time for the group as a whole.

Step 3   Perform a chi-square test of significance. If no gender differences exist with respect to singing time (and therefore social embarrassment), we would expect the same median split within each sample so that half of the men and half of the women fall above the median. To find out whether the gender differences obtained are statistically significant or merely a product of sampling error, we conduct a chi-square test (using Yates's correction). The following table shows the observed and expected frequencies.

	Men	Women
Above median	10 (7.22)	3 (5.78)
Not above median	5 (7.78)	9 (6.22)
	$N = 27$	

$$\chi^2 = \sum \frac{(|f_o - f_e| - .5)^2}{f_e}$$

(continued)

BOX 9.4   Continued

Set up the calculations in tabular form:

f_o	f_e	$|f_o - f_e|$	$|f_o - f_e| - .5$	$(|f_o - f_e| - .5)^2$	$\dfrac{(|f_o - f_e| - .5)^2}{f_e}$
10	7.22	2.78	2.28	5.20	.72
3	5.78	2.78	2.28	5.20	.90
5	7.78	2.78	2.28	5.20	.67
9	6.22	2.78	2.28	5.20	.84
					$\chi^2 = 3.13$

Referring to Table E in Appendix C, we learn that chi-square must exceed 3.84 (df = 1) to be regarded as significant at the .05 level. Because our obtained $\chi^2 = 3.13$, we cannot reject the null hypothesis. There is insufficient evidence to conclude on the basis of our results that men differ from women with respect to their reactions to a socially embarrassing situation.

In the previous step-by-step example, we were interested in comparing two groups on an ordinal-level variable. This was accomplished by constructing a chi-square test on the 2×2 cross-tabulation of placement above versus not above the median by group membership (in this case, gender).

If we were instead interested in comparing three groups on an ordinal-level variable, we would need to apply a chi-square for comparing several groups. That is, we would first compute the median for all three groups combined, then construct a 2×3 cross-tabulation of placement above versus not above the median by group membership, and finally calculate the chi-square test of significance.

Requirements for the Use of the Median Test

The following conditions must be satisfied to appropriately apply the median test to a research problem:

1. *A comparison between two or more medians.* The median test is employed to make comparisons between two or more medians from independent samples.

2. *Ordinal data.* To perform a median test, we assume at least the ordinal level of measurement. Nominal data cannot be used.

3. *Random sampling.* We should draw our samples on a random basis from a given population.

Summary

It is not always possible to meet the requirements of parametric tests of significance such as the *t* ratio or analysis of variance. Fortunately, statisticians have developed a number of non-parametric alternatives—tests of significance whose requirements do not include a normal distribution or the interval level of measurement. Though less powerful than their parametric counterparts *t* and *F*, nonparametric techniques can be applied to a wider range of research situations. They are useful when a researcher works with ordinal or nominal data or with a small number of cases representing a highly asymmetrical underlying distribution. The most popular nonparametric test of significance, the chi-square test, is widely used to make comparisons between frequencies rather than between mean scores. In a one-way chi-square, the frequencies observed among the categories of a variable are tested to determine whether they differ from a set of hypothetical frequencies. But chi-square can also be applied to cross-tabulations of two variables. In a two-way chi-square, when the differences between expected frequencies (expected under the terms of the null hypothesis) and observed frequencies (those we actually obtain when we do research) are large enough, we reject the null hypothesis and accept the validity of a true population difference. This is the requirement for a significant chi-square value. The chi-square test of significance assumes that the expected frequencies are at least equal to 5. When several groups are being compared, it may be possible to collapse or merge together some categories. In 2 × 2 tables, Yates's correction for small expected frequencies is often used. The chi-square test requires only nominal (frequency) data. Finally, the median test, which is based on the chi-square analysis, is used to determine whether there is a significant difference between the medians of two or more independent variables. Ordinal or interval data are required for the median test.

Terms to Remember

Parametric test	Observed frequencies
Nonparametric test	Expected frequencies
Power of a test	Yates's correction
One-way chi-square	Median test
Two-way chi-square	

Questions and Problems

1. Which is *not* true of parametric tests?
   a. They require interval data.
   b. They require normality in the population.
   c. They are less powerful than nonparametric tests.
   d. None of the above is true of parametric tests.

2. In a chi-square test, the expected frequencies
   a. are expected to occur under the terms of the null hypothesis.
   b. are expected to occur under the terms of the research hypothesis.
   c. refer to those frequencies actually observed from the results of conducting research.
   d. are never known by the researcher.

3. The two-way chi-square might be used as a nonparametric alternative instead of _____ when comparing two groups.
   a. confidence intervals
   b. standard deviation
   c. *t* ratio
   d. analysis of variance

4. In a chi-square analysis, the larger the difference between expected and observed frequencies, the more likely you are to
   a. retain the null hypothesis.
   b. reject the null hypothesis.
   c. instead use a *t* ratio or some other parametric test.

5. Which of the following is *not* a requirement of two-way chi square?
   a. A comparison between two or more mean scores
   b. Nominal data
   c. Random sampling
   d. The expected cell frequencies should not be too small

6. To employ the median test, you must be able to
   a. compare the means.
   b. assume a normal distribution.
   c. compare three or more independent samples.
   d. rank order a set of cases.

7. A researcher was interested in studying the phenomenon known as social distance, the reluctance of people to associate with members of different ethnic and racial groups. She designed an experiment in which students enrolled in a lecture course were asked to choose a discussion group (all meeting at the same time in the same building) based only on the ethnic–racial stereotype associated with the names of the teaching assistant:

Group	Teaching Assistant	Room	Enrollment
A	Cheng	106	10
B	Schultz	108	24
C	Goldberg	110	10
D	Rodriguez	112	16

Based on the enrollments, use the one-way chi-square to test the null hypothesis that the ethnic–racial name made no difference in students' selection of a discussion group.

8. Some politicians have been known to complain about the liberal press. To determine if in fact the press is dominated by left-wing writers, a researcher assesses the political leanings of a random sample of 60 journalists. He found that 15 were conservative, 18 were moderate, and 27 were liberal. Test the null hypothesis that all three political positions are equally represented in the print media.

9. A researcher is interested in determining if the pattern of traffic stops by the local police indicates discriminatory treatment against minorities. She takes a random sample of 250

records of traffic stops by the police, observing the racial distribution of the drivers shown in the following table. She also determines from census data that the city's resident population is 68% white, 22% black, and 10% other races. Using a one-way chi-square, test the null hypothesis that the observed distribution of driver race in the sample of traffic stops matches the resident population distribution in the community.

Race of Driver	f	%
White	154	61.6%
Black	64	25.6%
Other	32	12.8%
Total	250	100%

10. Some recent studies have suggested that the chances of a baby being born prematurely are increased when the mother suffers from chronic oral infections such as periodontal disease. An interested researcher collected the following data. Applying Yates's correction, conduct a chi-square analysis to test the null hypothesis that pregnant women who suffer from chronic oral infections are no more likely to give birth prematurely than women who do not suffer from chronic oral infections.

Baby Born Premature	Suffers from Chronic Oral Infections	
	Yes	*No*
Yes	44	8
No	6	57

11. A researcher interested in suicide created the following 2×2 cross-tabulation, which crosses gender with whether a suicide attempt was successfully completed. Applying Yates's correction, test the null hypothesis that the relative frequency of women who successfully commit suicide is the same as the relative frequency of men who successfully commit suicide. What do your results indicate?

Gender	Suicide Completed?	
	Yes	*No*
Male	22	7
Female	6	19

12. The following is a 2 × 2 cross-tabulation of gender of schizophrenic patients by their responsiveness to medication. Applying Yates's correction, conduct a chi-square test of significance.

Responds Well to Medication?	Gender	
	Female	*Male*
Yes	16	5
No	7	24

13. The following is a 2 × 2 cross-tabulation of gender of juvenile violent offenders by whether they were victims of abuse before they began offending. Applying Yates's correction, conduct a chi-square test of significance.

Victim of Abuse?	Gender of Offender	
	Male	*Female*
Yes	25	13
No	11	6

14. The following is a 2 × 2 cross-tabulation represents whether politicians are Republican or Democrat by whether they favor or oppose stricter gun control laws. Applying Yates's correction, conduct a chi-square test of significance.

Gun Control	Political Party	
	Democrat	*Republican*
Favor	26	5
Oppose	7	19

15. The following 2 × 2 cross-tabulation represents whether high school students passed the road test for their driver's license on the first attempt by whether they took a driver's education course. Applying Yates's correction, conduct a chi-square test of significance.

Driver's Education Course

Test Results	Yes	No
Pass	16	8
Fail	7	11

16. The following is a 2 × 2 cross-tabulation of preference for slasher films such as *Halloween* or *Friday the 13th* by gender of respondent. Applying Yates's correction, conduct a chi-square test of significance.

Preference for Films	Gender of Respondent	
	Female	Male
Like	8	12
Dislike	10	15

17. A computer company conducted a "new and improved" course designed to train its service representatives in learning to repair personal computers. One hundred trainees were split into two groups on a random basis: 50 took the customary course and 50 took the new course. At the end of 6 weeks, all 100 trainees were given the same final examination.

    Using chi-square, test the null hypothesis that the new course was no better than the customary course in terms of teaching service representatives to repair personal computers. What do your results indicate?

Skills	Customary	"New and Improved"
Above average	15	19
Average	25	23
Below average	10	8

18. Does voting behavior vary by social class? To find out, a political scientist questioned a random sample of 80 registered voters about the candidate for office, A or B, they intended to support in an upcoming election. The researcher also questioned members of her sample

concerning their social class membership—whether upper, middle, working, or lower. Her results are as follows:

Vote for	Social Class			
	Upper	*Middle*	*Working*	*Lower*
Candidate A	14	9	8	6
Candidate B	10	9	11	13

Using chi-square, test the null hypothesis that voting behavior does not differ by social class. What do your results indicate?

19. Do single-parent families tend to be more impoverished than families with two parents? A family researcher studied a sample of 35 one-parent and 65 two-parent families in a particular city to determine whether their total family income fell below the poverty level. Applying chi-square to the following data, test the null hypothesis that one- and two-parent families do not differ with respect to poverty.

Total Family Income	Family Structure	
	One Parent	*Two Parents*
Below poverty level	24	20
Not below poverty level	11	45

20. Should the Internet be censored on school computers? These days, computers and the Internet are being increasingly used in schools for all ages, and a common issue that must be addressed is whether students should have access to everything the Internet has to offer. Wondering if teachers view this issue differently depending on the level at which they teach, a researcher asked 65 teachers from four different levels of education whether they believed the Internet should be censored on their school computers. The results were as follows:

Internet Censorship	Level of Teaching			
	Grades 1–4	*Grades 5–8*	*High School*	*College or University*
For	16	12	9	4
Against	1	4	5	14

Using chi-square, test the null hypothesis that teachers' views of Internet censorship do not vary with the level of education at which they teach. What do your results indicate?

21. Does the rate of serious mental illness among substance abusers vary according to the type of substance abused? To find out, a psychiatrist collected data on the mental health of a random

sample of substance abusers: drug abusers, alcohol abusers, and abusers of both drugs and alcohol. His results were as follows:

Substance Abused

Mental Illness?	Drugs	Alcohol	Drugs and Alcohol
Yes	12	8	16
No	31	33	35

Using chi-square, test the null hypothesis that the prevalence of serious mental illness does not differ by type of substance abuse.

22. A sample of 118 college students is asked whether they are involved in campus activities. Using the following cross tabulation depicting student responses by the region in which their colleges are located, conduct a chi-square test of significance for regional differences.

Campus Activity Participation

Region	Involved	Uninvolved
East	19	10
South	25	6
Midwest	15	15
West	8	20

23. Returning to the data on the racial distribution of traffic stops from Problem 9, the researcher cross-tabulates the race of the driver with the race of the police officer for the random sample of 250 recorded stops. Note that there were too few stops by officers who are neither white nor black to maintain an "Other" category. Using the chi-square test of independence, test the null hypothesis of no difference between white and nonwhite officers in teams of the race of drivers they stop for traffic violations.

Race of Driver	Race of Officer White	Nonwhite	Total
White	96	58	154
Black	48	16	64
Other	16	16	32
Total	160	90	250

24. A radio executive considering a switch in his station's format collects data on the radio preferences of various age groups of listeners. Using the following cross-tabulation, test the null hypothesis that radio format preference does not differ by age group.

	Age		
Radio Format Preference	*Young Adult*	*Middle Age*	*Older Adult*
Music	14	10	3
News–talk	4	15	11
Sports	7	9	5

25. Conduct chi-square tests of significance for the choice of murder weapon by gender of offender cross-tabulations shown in Tables 9.5 and 9.6. What is the effect of collapsing categories?

26. Two samples of students were asked to read and then evaluate a short story written by a new author. One-half of the students were told that the author was a woman, and the other half were told that the author was a man. The following evaluations were obtained (higher scores indicating more favorable evaluations):

Told Author Was a Woman	**Told Author Was a Man**
6	6
5	8
1	8
1	2
3	5
4	6
3	3
6	8
5	6
5	8
1	2
3	2
5	6
6	8
6	4
3	3

Applying the median test, determine whether there is a significant difference between the medians of these groups. Were student evaluations of the short story influenced by the attributed gender of its author?

27. A researcher concerned with racial justice suspects that jurors might perceive the severity of a crime as greater when the victim is white rather than black. She provides mock jurors with videotapes of trials (in which the victim is not shown) and arbitrarily describes the victim as either white or black. Using the following severity scores (1 for less serious to 9 for most serious), apply the median test to determine if there is a significant difference between crimes against white and black victims.

White Victim		Black Victim	
7	9	4	3
8	5	7	2
7	9	3	2
6	8	2	6
7	9	3	4
7	7	4	5
8	9	7	4
9	9	4	4
7		5	4
6		6	3
9		2	

28. How many registered voters would vote to elect a female president? A curious political scientist selected two equivalent samples of voters. One group was given the description of a hypothetical male candidate who takes the most popular stance on most major issues, while the other group was given this same description while being told that the candidate was female. The two groups were then asked to rate on a scale from 1 to 5 the likelihood that they would vote for their respective candidate (with 1 being least likely and 5 being most likely). The following scores were obtained:

Told Candidate Was a Man	Told Candidate was a Woman
5	3
4	2
3	1
5	3
5	1
4	2
3	2
4	4
3	3
5	4
4	2
3	3

Applying the median test, determine whether there is a significant difference between the medians of these groups. Was the likelihood of this hypothetical candidate being voted for influenced by his or her gender?

SPSS Exercises

1. Use SPSS with the Monitoring the Future Study to calculate a chi-square statistic to test the null hypothesis that lifetime marijuana or hashish use (V115) does not differ by regional location (V13). Hint: ANALYZE, DESCRIPTIVE STATISTICS, CROSSTABS, select row and column variables, choose column or row percentages in options, and select chi-square in statistics.

2. Use the Monitoring the Future Study to calculate a chi-square statistic to test the following hypothesis: lifetime cocaine use (V124) will differ by sex (V150).

3. In Chapter 7, you used SPSS with the Monitoring the Future Study to find out if boys or girls (V150) were more likely to have smoked marijuana or hashish in their lifetime (v115). Use SPSS now to conduct a chi-square test of the null hypothesis for the same variables. What do these results indicate? Which statistic provides better information? Why?

4. Using the General Social Survey, calculate a chi-square statistic to test the null hypothesis that attitudes toward abortion for any reason (ABANY) do not vary by age (AGE).
   a. How large will the cross-tabulation be? (r × c = # cells) _____ × _____ = _____ (Remember that age goes from 18 through 89)
   b. What percentage of cells have small expected frequencies?_____
   c. Is this acceptable? Yes/No Explain.
   d. What could be done to improve it? Explain.

5. Recode age into three groups (18 to 39 = 1)(40 to 59 = 2)(60 and above = 3) by using TRANSFORM then RECODE INTO DIFFERENT VARIABLES and naming the new variable AGEGRP3 (for age grouped in three categories). Actually, the new name is up to you, but this name is fairly clear. Use SPSS to recalculate chi-square from the previous problem. What is the level of measurement for grouped age? Apply a chi-square statistic to test null hypothesis that attitudes toward abortion for any reason (ABANY) do not vary by grouped age (AGEGRP3).

Looking at the Larger Picture: Testing for Differences

Until this point, we have flirted with the idea that smoking and drinking might differ by gender. At the end of Part I we examined bar charts for gender differences in the percentage distribution for smoking and drinking, and at the end of Part II we constructed confidence intervals for the percentage who smoke and the mean daily use of cigarettes, overall but also separately for males and females. There seemed to be some differences between male and female students, but now we can determine whether they are statistically significant.

Our null hypothesis is that there are no differences between male and female urban public high school students in terms of smoking percentage and mean cigarette usage as well as mean drinking frequency. The following table summarizes the calculations needed for a z test of proportions (for the percentage who smoke) and the t test of means (for daily cigarettes among smokers and drinking occasions for all students).

We see quite clearly that the difference in the percentage of male and female smokers within the sample is large enough to reject the null hypothesis of no difference in population proportions. Thus, there is indeed a significant gender difference in terms of whether a student is a smoker. Turning to the daily consumption of cigarettes among the smokers, we can see that the difference between sample means for males and females is not large enough to reject the null hypothesis. Therefore, there is not enough evidence of a true gender difference in the population. The t test for differences between sample means for drinking frequency does, however, lead us to reject the null hypothesis of no difference in population means for males and females. Specifically, males and females were found significantly different in their alcohol consumption.

Other background characteristics are divided into more than two groups. Suppose that we want to compare whites, blacks, Latinos, and

Tests of Differences between Groups

Variable	Statistic	Group	
		Males	*Females*
If smoker			
	N	127	123
	%	52.8	71.5
		SE = 6.14	
		$z = 3.06$	
		(significant)	
Daily smoking			
	N	67	88
	Mean	15.9	17.7
		SE = 1.68	
		$z = -1.08$	
		(not significant)	
Occasions drinking			
	N	127	123
	Mean	1.80	1.36
		SE = 0.14	
		$z = 3.09$	
		(significant)	

others (with so few Asians in the sample, we should collapse this group into the "other" category). Using the analysis of variance, we determine that the race differences in smoking are not significant (although whites appeared to smoke more heavily); that is, we cannot quite reject the null hypothesis of no differences between population means for the various races. The differences in sample mean drinking frequency by race appear rather minimal and to be the result of sampling variability alone. With a tabled critical value of 2.68, the F test is not quite large enough to indicate significance for racial differences by alcohol consumption.

Note that we cannot assess race differences in the percentage of students who smoke using an analysis of variance, because the variable

Continued

Testing Differences between Groups with Analysis of Variance

Variable	Statistic	Group			
		White	*Black*	*Latino*	*Other*
Daily smoking					
	N	103	24	18	10
	Mean	18.5	13.8	14.2	13.0
			df = 3,151		
			F = 2.38 (not significant)		
Occasions drinking					
	N	156	44	32	18
	Mean	1.53	1.68	1.75	1.50
			df = 3,246		
			F = 0.45 (not significant)		

smoke–not smoke is nominal. Instead, we can construct a cross-tabulation of smoking by race and calculate a chi-square test. Although the sample percentages who smoke may differ among the races, these differences are relatively small and turn out not to be statistically significant.

Cross-Tabulation of Race and Smoking

	Smoke	**Not Smoke**
White	103 (66.0%)	53 (34.0%)
Black	24 (54.5%)	20 (45.5%)
Latino	18 (56.2%)	14 (43.8%)
Other	10 (55.6%)	8 (44.4%)

χ^2 = 2.88, 3 df (not significant)

Finally, we might also examine a cross-tabulation comparing whether a student smokes to whether one or both parents smoke. Over three-quarters of the students with a smoker parent

smoke, whereas about 57% of those with non-smoker parents smoke. Using the chi-square test, we determine that this difference is large enough to lead us to reject the null hypothesis.

Cross-Tabulation of Parental Smoking and Student Smoking

	Student Smokes	**Student Does Not Smoke**
Parent smokes	49 (75.4%)	16 (24.6%)
Parents do not smoke	106 (57.3%)	79 (42.7%)

χ^2 = 6.68, 1 df (significant)

In Part IV we will look at the relationship between variables. Not only can we determine the strength of association between parents and respondents smoking, but also between age and smoking or drinking.

From Decision Making to Association

10

Correlation

Characteristics such as age, intelligence, and educational attainment vary from one person to another and are therefore referred to as variables. In earlier chapters we have been concerned with establishing the presence or absence of a relationship between any two variables, which we will now label X and Y; for example, between age (X) and frequency of Internet use (Y), between intelligence (X) and achievement-motivation (Y), or between educational attainment (X) and income (Y). Aided by the t ratio, analysis of variance, or nonparametric tests such as chi-square, we previously sought to discover whether a difference between two or more samples could be regarded as statistically significant—reflective of a true population difference—and not merely the product of sampling error. Now we turn our attention to the existence and strength of relationship between two variables—that is, on their co-relationship or, simply, their correlation.

Strength of Correlation

Finding that a relationship exists does not indicate much about the degree of association, or *correlation*, between two variables. Many relationships are statistically significant—that is, stronger than you would expect to obtain just as a result of sampling error alone, yet rather few express *perfect* correlation. To illustrate, we know that height and weight are associated, since the taller a person is, the more he or she tends to weigh. There are numerous exceptions to the rule, however. Some tall people weigh very little; some short people weigh a lot. In the same way, a relationship between age and net worth does not preclude the possibility of finding many young adults who have accumulated a greater net worth in just a few years than some older adults have over decades.

Correlations actually vary with respect to their *strength*. We can visualize differences in the strength of correlation by means of a *scatter plot* or *scatter diagram*, a graph that shows the way scores on any two variables, X and Y, are scattered throughout the range of possible score values. In the conventional arrangement, a scatter plot is set up so that the X variable is located along the horizontal base line, and the Y variable is measured on the vertical line.

Turning to Figure 10.1, we find two scatter plots, each representing the relationship between years of education (X) and income (Y). Figure 10.1(a) depicts this relationship for males, and Figure 10.1(b) represents the relationship for females. Note that every point in these scatter plots depicts *two* scores, education and income, obtained by *one* respondent. In Figure 10.1(a), for example, we see that a male having 9 years of education earned just below $40,000, whereas a male with 17 years of education made about $60,000.

We can say that the strength of the correlation between X and Y increases as the points in a scatter plot more closely form an imaginary diagonal line through the center of the scatter of points in the graph. Although not necessarily included in a scatter plot, we have drawn these lines with light gray dashes to illustrate how the strength of correlation translates to closeness of points to a straight line. Actually, we will discuss this line in great detail in the next chapter.

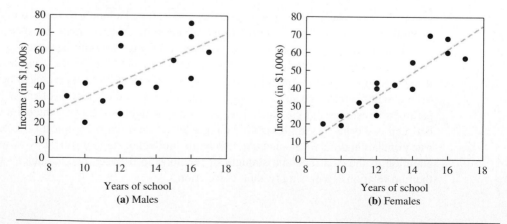

FIGURE 10.1   Strength of relationship

As should be clear, Figure 10.1(a) represents a weaker correlation than does Figure 10.1(b), although both scatter plots indicate that income tends to increase with greater education. Such data would indeed support the view that the income of women (relative to that of men) is more related to the level of education they attain. Alternatively, this suggests that income levels for men is more a result of factors other than education than is the case for women.

Direction of Correlation

Correlation can often be described with respect to direction as either positive or negative. A *positive correlation* indicates that respondents getting *high* scores on the *X* variable also tend to get *high* scores on the *Y* variable. Conversely, respondents who get *low* scores on *X* also tend to get *low* scores on *Y*. Positive correlation can be illustrated by the relationship between education and income. As we have previously seen, respondents completing many years of school tend to make large annual incomes, whereas those who complete only a few years of school tend to earn very little annually. The overall relationship between education and income for males and females combined is shown in Figure 10.2(a), again with an imaginary straight line through the scatter of points drawn to show the direction of the relationship.

A *negative correlation* exists if respondents who obtain *high* scores on the *X* variable tend to obtain *low* scores on the *Y* variable. Conversely, respondents achieving *low* scores on *X* tend to achieve *high* scores on *Y*. The relationship between education and income would certainly *not* represent a negative correlation, because respondents completing many years of school *do not* tend to make small annual incomes. A more likely example of negative correlation is the relationship between education and the degree of prejudice against minority groups. Prejudice tends to diminish as the level of education increases. Therefore, as shown in Figure 10.2(b), individuals having little formal education tend to hold strong prejudices, whereas individuals completing many years of education tend to be low with respect to prejudice.

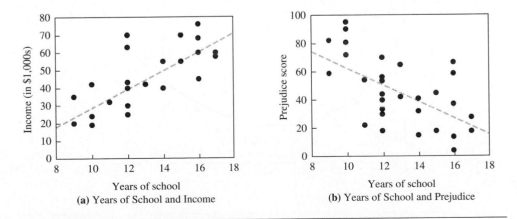

(a) Years of School and Income

(b) Years of School and Prejudice

FIGURE 10.2 Direction of relationship

A positive or negative correlation represents a type of *straight-line* relationship. Depicted graphically, the points in a scatter plot tend to form an imaginary straight line through the center of the points in the graph. If a positive correlation exists, then the points in the scatter plot will cluster around the imaginary straight line, as drawn in Figure 10.2(a). In contrast, if a negative correlation is present, the points in the scatter plot will surround the imaginary straight line as shown in Figure 10.2(b).

Curvilinear Correlation

For the most part, social researchers seek to establish a straight-line correlation, whether positive or negative. It is important to note, however, that not all relationships between X and Y can be regarded as forming a straight line. There are many curvilinear relationships, indicating, for example, that one variable increases as the other variable increases until the relationship reverses itself, so that one variable finally decreases while the other continues to increase. That is, a relationship between X and Y that begins as positive becomes negative; a relationship that starts as negative becomes positive. To illustrate a curvilinear correlation, consider the relationship between age and hours of television viewing. As shown in Figure 10.3, the points in the scatter plot tend to form a U-shaped curve rather than a straight line. Thus, television viewing tends to decrease with age, until the thirties, after which viewing tends to increase with age.

The Correlation Coefficient

The process for finding curvilinear correlation lies beyond the scope of this text. Instead, we turn our attention to *correlation coefficients*, which numerically express both strength

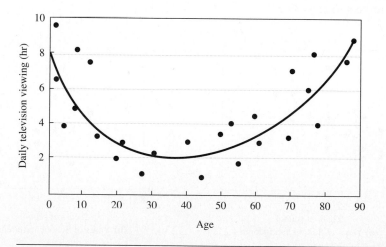

FIGURE 10.3   The relationship between age (X) and television viewing (Y): A curvilinear correlation

and direction of straight-line correlation. Such correlation coefficients range between -1.00 and $+1.00$ as follows:

$-1.00 \leftarrow$ perfect negative correlation
⋮
$-.60 \leftarrow$ strong negative correlation
⋮
$-.30 \leftarrow$ moderate negative correlation
⋮
$-.10 \leftarrow$ weak negative correlation
⋮
$.00 <$ no correlation
⋮
$+.10 \leftarrow$ weak positive correlation
⋮
$+.30 \leftarrow$ moderate positive correlation
⋮
$+.60 \leftarrow$ strong positive correlation
⋮
$+1.00 \leftarrow$ perfect positive correlation

We see, then, that negative numerical values such as -1.00, $-.60$, $-.30$, and $-.10$ signify negative correlation, whereas positive numerical values such as $+1.00$, $+.60$, $+.30$, and $+.10$ indicate positive correlation. Regarding degree of association, the closer to 1.00 in either direction, the greater the strength of the correlation. Because the strength of a correlation is independent of its direction, we can say that $-.10$ and $+.10$ are equal in strength (both are weak); $-.80$ and $+.80$ have equal strength (both are very strong).

Pearson's Correlation Coefficient

With the aid of *Pearson's correlation coefficient* (r), we can determine the strength and the direction of the relationship between X and Y variables, both of which have been measured at the interval level. For example, we might be interested in examining the association between height and weight for the following sample of eight children:

Child	Height (in.) (X)	Weight (lb) (Y)
A	49	81
B	50	88
C	53	87
D	55	99
E	60	91
F	55	89
G	60	95
H	50	90

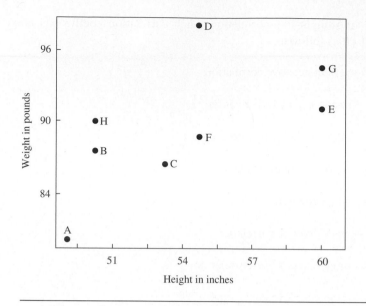

FIGURE 10.4   Scatter plot of height and weight

In the scatter plot in Figure 10.4, the positive association that one would anticipate between height (X) and weight (Y) in fact appears. But note that there are some exceptions. Child C is taller but weighs less than child H; child D is shorter but weighs more than child E. These exceptions should not surprise us because the relationship between height and weight is not perfect. Overall, nonetheless, the general rule that "the taller one is the heavier one is" holds true: F is taller and heavier than A, as is G compared to H.

Pearson's r does more than just consider if subjects are simply taller or heavier than other subjects; it considers precisely how much heavier and how much taller. The quantity that Pearson's r focuses on is the product of the X and Y deviations from their respective means. The deviation ($X - \overline{X}$) tells how much taller or shorter than average a particular child is; the deviation ($Y - \overline{Y}$) tells how much heavier or lighter than average a particular child is.

With Pearson's r, we add the products of the deviations to see if the positive products or negative products are more abundant and sizable. Positive products indicate cases in which the variables go in the same direction (that is, both taller and heavier than average or both shorter and lighter than average); negative products indicate cases in which the variables go in opposite directions (that is, taller but lighter than average or shorter but heavier than average).

In Figure 10.5, dashed lines are added to the scatter plot of X and Y to indicate the location of the mean height ($\overline{X} = 54$ inches) and the mean weight ($\overline{Y} = 90$ pounds). Child G is apparently much taller and much heavier than average. His deviations on the two variables are ($X - \overline{X}$) = 60 − 54 = 6 (inches) and ($Y - \overline{Y}$) = 95 − 90 = 5 (pounds), which when multiplied yield +30. Child A is much shorter and much lighter than average. Her deviations (−5 and −9) multiply to +45. On the other hand, child C is only slightly shorter and lighter than average; her product of deviations (−1 × −3 = 3) is far less dramatic. This is as it would seem intuitively: The more dramatically a child demonstrates the

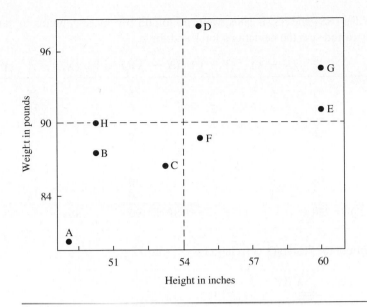

FIGURE 10.5   Scatter plot of height and weight with mean axes

rule "the taller, the heavier," the larger the product of the X and Y deviations. Finally, child F is a slight exception: He is a little taller than average yet lighter than average. As a result, his $+1$ deviation on X and -1 deviation on Y produce a negative product (-1).

We can compute the sum of the products for the data as follows. In the following table, columns 2 and 3 reproduce the heights and weights for the eight children in the sample. Columns 4 and 5 give the deviations from the means for the X and Y values. In column 6, these deviations are multiplied and then summed.

Child	X	Y	$(X - \overline{X})$	$(Y - \overline{Y})$	$(X - \overline{X})(Y - \overline{Y})$	
A	49	81	-5	-9	45	$N = 8$
B	50	88	-4	-2	8	$\overline{X} = 54$
C	53	87	-1	-3	3	$\overline{Y} = 90$
D	55	99	1	9	9	
E	60	91	6	1	6	
F	55	89	1	-1	-1	
G	60	95	6	5	30	
H	50	90	-4	0	0	
	$\Sigma X = 432$	$\Sigma Y = 720$			SP $= 100$	

The sum of the final column (denoted SP, for sum of products) is positive—indicating a positive association between X and Y. But, as we have learned, correlation coefficients are constrained to range from -1 to $+1$ to aid in their interpretation. The formula for r accomplishes this by dividing the SP value by the square root of the product of the sum of squares of both variables (SS_X and SS_Y). Thus, we need to add two more columns to our table in which we square and sum the deviations for X and for Y.

X	Y	$(X - \bar{X})$	$(Y - \bar{Y})$	$(X - \bar{X})(Y - \bar{Y})$	$(X - \bar{X})^2$	$(Y - \bar{Y})^2$
49	81	−5	−9	45	25	81
50	88	−4	−2	8	16	4
53	87	−1	−3	3	1	9
55	99	1	9	9	1	81
60	91	6	1	6	36	1
55	89	1	−1	−1	1	1
60	95	6	5	30	36	25
50	90	−4	0	0	16	0
$\Sigma X = 432$	$\Sigma Y = 720$			SP = 100	$SS_X = 132$	$SS_Y = 202$

The formula for Pearson's correlation is as follows:

$$r = \frac{\Sigma(X - \bar{X})(Y - \bar{Y})}{\sqrt{\Sigma(X - \bar{X})^2 \Sigma(Y - \bar{Y})^2}} = \frac{SP}{\sqrt{SS_X SS_Y}}$$

$$= \frac{100}{\sqrt{(132)(202)}}$$

$$= \frac{100}{\sqrt{26,664}}$$

$$= \frac{100}{\sqrt{163.3}}$$

$$= +.61$$

Therefore, Pearson's correlation indicates, as suggested by the scatter plot, that height and weight are fairly strongly correlated in the positive direction.

A Computational Formula for Pearson's r

Computing Pearson's r from deviations helps relate the topic of correlation to our earlier discussion. However, the previous formula for Pearson's r requires lengthy and time-consuming calculations. Fortunately, there is an alternative formula for Pearson's r that works directly with raw scores, thereby eliminating the need to obtain deviations for the X and Y variables. Similar to the computational formulas for variance and standard deviation in Chapter 4, there are raw-score formulas for SP, SS_X, and SS_Y.

$$SP = \Sigma XY - N\bar{X}\bar{Y}$$

$$SS_X = \Sigma X^2 - N\bar{X}^2$$

$$SS_Y = \Sigma Y^2 - N\bar{Y}^2$$

Using these expressions in our formula for Pearson's correlation, we obtain the following computational formula for r:

$$r = \frac{\Sigma XY - N\overline{X}\overline{Y}}{\sqrt{(\Sigma X^2 - N\overline{X}^2)(\Sigma Y^2 - N\overline{Y}^2)}}$$

To illustrate the use of Pearson's r computational formula, consider the following data on the number of years of school completed by the father (X) and the number of years of school completed by the child (Y). To apply our formula, we must obtain the sums of X and Y (to calculate the means) and of X^2, Y^2, and XY:

X	Y	X^2	Y^2	XY	
12	12	144	144	144	$N = 8$
10	8	100	64	80	$\Sigma X = 84$
6	12	36	144	72	$\Sigma Y = 92$
16	11	256	121	176	$\overline{X} = \dfrac{\Sigma X}{N} = \dfrac{84}{8} = 10.5$
8	10	64	100	80	
9	8	81	64	72	$\overline{Y} = \dfrac{\Sigma Y}{N} = \dfrac{92}{8} = 11.5$
12	16	144	256	192	
11	15	121	225	165	$\Sigma X^2 = 946$
84	92	946	1,118	981	$\Sigma Y^2 = 1,118$
					$\Sigma XY = 981$

The Pearson's correlation is then equal to

$$r = \frac{\Sigma XY - N\overline{X}\overline{Y}}{\sqrt{(\Sigma X^2 - N\overline{X}^2)(\Sigma Y^2 - N\overline{Y}^2)}}$$

$$= \frac{981 - 8(10.5)(11.5)}{\sqrt{[946 - 8(10.5)^2][1,118 - 8(11.5)^2]}}$$

$$= \frac{981 - 966}{\sqrt{(946 - 882)(1,118 - 1,058)}}$$

$$= \frac{15}{\sqrt{(64)(60)}}$$

$$= \frac{15}{\sqrt{3,840}}$$

$$= \frac{15}{61.97}$$

$$= +.24$$

Testing the Significance of Pearson's r

Pearson's r gives us a precise measure of the strength and direction of the correlation in the sample being studied. If we have taken a random sample from a specified population, we may still seek to determine whether the obtained association between X and Y exists in the *population* and is not due merely to sampling error.

To test the significance of a measure of correlation, we usually set up the null hypothesis that no correlation exists in the population. With respect to the Pearson correlation coefficient, the null hypothesis states that the population correlation ρ (rho) is zero. That is,

$$\rho = 0$$

whereas the research hypothesis says that

$$\rho \neq 0$$

As was the case in earlier chapters, we test the null hypothesis by selecting the alpha level of .05 or .01 and computing an appropriate test of significance. To test the significance of Pearson's r, we can compute a t ratio with the degrees of freedom equal to $N - 2$ (N equals the number of pairs of scores). For this purpose, the t ratio can be computed by the formula,

$$t = \frac{r\sqrt{N - 2}}{\sqrt{1 - r^2}}$$

where t = t ratio for testing the statistical significance of Pearson's r
N = number of pairs of scores X and Y
r = obtained Pearson's correlation coefficient

Returning to the previous example, we can test the significance of a correlation coefficient equal to +.24 between a father's educational level (X) and that of his son (Y):

$$
\begin{aligned}
t &= \frac{.24\sqrt{8 - 2}}{\sqrt{1 - (.24)^2}} \\
&= \frac{(.24)(2.45)}{\sqrt{1 - .0576}} \\
&= \frac{.59}{\sqrt{.9424}} \\
&= \frac{.59}{.97} \\
&= .61
\end{aligned}
$$

When we turn to Table C in Appendix C, we find that the critical value of t with 6 degrees of freedom and $\alpha = .05$ is 2.447. Because our calculated t value does not even come close to exceeding this critical value, we cannot reject the null hypothesis that $\rho = 0$. Although a correlation of $+.24$ is not especially weak, with a sample size of only 8, it is not nearly statistically significant. That is, given a small sample size of 8, it is very possible that the obtained r of $+.24$ is a result of sampling error. Thus, we are forced to retain the null hypothesis that the population correlation (ρ) is zero, at least until we have more data bearing on the relationship between father's and child's educational attainment.

A Simplified Method for Testing the Significance of r

Fortunately, the process of testing the significance of Pearson's r as previously illustrated has been simplified, so it becomes unnecessary actually to compute a t ratio. Instead, we turn to Table F in Appendix C, where we find a list of significant values of Pearson's r for the .05 and .01 levels of significance, with the number of degrees of freedom ranging from 1 to 90. Directly comparing our calculated value of r with the appropriate table value yields the same result as though we had actually computed a t ratio. If the calculated Pearson's correlation coefficient does not exceed the appropriate table value, we must retain the null hypothesis that $\rho = 0$; if, on the other hand, the calculated r is greater than the table critical value, we reject the null hypothesis and accept the research hypothesis that a correlation exists in the population.

For illustrative purposes, let us return to our previous example in which a correlation coefficient equal to $+.24$ was tested by means of a t ratio and found not to be statistically significant. Turning to Table F in Appendix C, we now find that the value of r must be at least .7067 to reject the null hypothesis at the .05 level of significance with 6 degrees of freedom. Hence, this simplified method leads us to the same conclusion as the longer procedure of computing a t ratio.

BOX 10.1 • *Step-by-Step Illustration: Pearson's Correlation Coefficient*

To illustrate the step-by-step procedure for obtaining a Pearson's correlation coefficient (r), let us examine the relationship between years of school completed (X) and prejudice (Y) as found in the following sample of 10 immigrants:

Respondent	Years of School (X)	Prejudice (Y)[a]
A	10	1
B	3	7
C	12	2
D	11	3
E	6	5
F	8	4
G	14	1
H	9	2
I	10	3
J	2	10

[a]Higher scores on the measure of prejudice (from 1 to 10) indicate greater prejudice.

(continued)

BOX 10.1   Continued

To obtain Pearson's r, we must proceed through the following steps:

Step 1   Find the values of ΣX, ΣY, ΣX^2, ΣY^2, and ΣXY, as well as $\overline{X}$ and $\overline{Y}$.

X	Y	X^2	Y^2	XY	
10	1	100	1	10	$N = 10$
3	7	9	49	21	$\Sigma X = 85$
12	2	144	4	24	$\Sigma Y = 38$
11	3	121	9	33	$\overline{X} = \dfrac{\Sigma X}{N} = \dfrac{85}{10} = 8.5$
6	5	36	25	30	
8	4	64	16	32	$\overline{Y} = \dfrac{\Sigma Y}{N} = \dfrac{38}{10} = 3.8$
14	1	196	1	14	
9	2	81	4	18	$\Sigma X^2 = 855$
10	3	100	9	30	$\Sigma Y^2 = 218$
2	10	4	100	20	$\Sigma XY = 232$
85	38	855	218	232	

Step 2   Plug the values from Step 1 into Pearson's correlation formula.

$$r = \frac{\Sigma XY - N\overline{X}\,\overline{Y}}{\sqrt{(\Sigma X^2 - N\overline{X}^2)(\Sigma Y^2 - N\overline{Y}^2)}}$$

$$= \frac{232 - (10)(8.5)(3.8)}{\sqrt{[855 - (10)(8.5)^2][218 - (10)(3.8)^2]}}$$

$$= \frac{232 - 323}{\sqrt{(855 - 722.5)(218 - 144.4)}}$$

$$= \frac{-91}{\sqrt{(1325)(73.6)}}$$

$$= \frac{-91}{\sqrt{9,752}}$$

$$= \frac{-91}{98.75}$$

$$= -.92$$

Our result indicates a rather strong negative correlation between education and prejudice.

Step 3   Find the degrees of freedom.

$$df = N - 2$$

$$= 10 - 2$$
$$= 8$$

Step 4   Compare the obtained Pearson's r with the appropriate value of Pearson's r in Table F.

obtained $r = -.92$

table $r = .6319$

df $= 8$

$\alpha = .05$

As indicated, to reject the null hypothesis that $\rho = 0$ at the .05 level of significance with 8 degrees of freedom, our calculated value of Pearson's r must exceed .6319. Because our obtained r equals $-.92$, we reject the null hypothesis and accept the research hypothesis. That is, our result suggests that a negative correlation between education and prejudice is present in the immigrant population from which our sample was taken.

Requirements for the Use of Pearson's r Correlation Coefficient

To employ Pearson's correlation coefficient correctly as a measure of association between X and Y variables, the following requirements must be taken into account:

1. *A straight-line relationship.* Pearson's r is only useful for detecting a straight-line correlation between X and Y.

2. *Interval data.* Both X and Y variables must be measured at the interval level so that scores may be assigned to the respondents.

3. *Random sampling.* Sample members must have been drawn at random from a specified population to apply a test of significance.

4. *Normally distributed characteristics.* Testing the significance of Pearson's r requires both X and Y variables to be normally distributed in the population. In small samples, failure to meet the requirement of normally distributed characteristics may seriously impair the validity of the test. However, this requirement is of minor importance when the sample size equals or exceeds 30 cases.

The Importance of Scatter Plots

It seems instinctive to look for shortcuts and time-saving devices in our lives. For social researchers, the development of high-speed computers and simple statistical software has become what the advent of the automatic washer and liquid detergent was for the housekeeper. Unfortunately, these statistical programs have been used too often without sufficient concern for their appropriateness. This is particularly true in correlational analysis.

The correlation coefficient is a very powerful statistical measure. Moreover, for a data set containing several variables, with a very short computer run, one can obtain in just seconds a correlation matrix, such as that in Table 10.1.

A correlation matrix displays in compact form the interrelationships of several variables simultaneously. Along the diagonal from the upper-left corner to the bottom-right corner is a series of 1.00s. These represent the correlation of each variable with itself, and so they are necessarily perfect and therefore equal to 1. The off-diagonal entries are the intercorrelations. The entry in the second row, fourth column (.78) gives the correlation of $X2$ and $X4$ (respondent's and spouse's education). The matrix is symmetrical—that is, the triangular portion above the diagonal is identical to that below the diagonal. Thus, the entry for the fourth row, second column is .78 as well.

The value of computer programs that produce results like this is that the researcher can quickly glance at the intercorrelations of a large number of variables—say, 10—and quickly pick out the strong and interesting correlations. One immediate problem, as we discussed earlier in reference to analysis of variance, is that such a fishing expedition for a large number of correlations will tend to pick up correlations that are significant by chance. An even greater pitfall, however, is that correlations may gloss over some major violations of the assumptions of Pearson's r. That is, a correlation matrix only provides (linear) correlation coefficients; it does not tell if the relationships are linear in the first place or whether there are peculiarities in the data that are worth noting. To prevent falling victim to data peculiarities, one really should inspect scatter plots before jumping to conclusions about what is related to what.

It is a far more tedious task to look at scatter plots in conjunction with the correlation matrix, because they must be examined one pair at a time. For example, to inspect scatter plots for all pairs of 10 variables would require 45 plots and a great deal of time and effort. As a result, far too many students and researchers skip over this step, often with misleading or disastrous results. Sometimes, as we shall see, what seems like a strong association on the basis of the correlation coefficient may be proved illusory after seeing the scatter plot. Conversely, truly important associations may be misrepresented by the single summary value of Pearson's r.

TABLE 10.1 *A Correlation Matrix*

	Respondent's Age $X1$	Respondent's Education $X2$	Family Income $X3$	Spouse's Education $X4$
$X1$	1.00	−.48	.35	−.30
$X2$	−.48	1.00	.67	.78
$X3$	.35	.67	1.00	.61
$X4$	−.30	.78	.61	1.00

Consider, for example, the following data on homicide and suicide rates (per 100,000 population) for the six New England states:

State	Homicide Rate	Suicide Rate
Maine	3.2	14.3
New Hampshire	2.9	11.3
Vermont	4.3	17.8
Massachusetts	3.6	8.9
Rhode Island	4.2	12.3
Connecticut	5.4	8.6

The correlation coefficient is −.17, suggesting a weak to moderate negative relationship. This would seem to support the contention of some sociologists that these two forms of violence (other-directed and self-directed) are trade-offs; when one rate is high, the other rate is low.

Before we get too excited about this result, however, let's inspect the scatter plot in Figure 10.6. Although the scatter plot appears to show a slight negative association, the lower-right point deserves further consideration. This corresponds to Connecticut. There is some justification for suspecting that Connecticut is in fact systematically different from the rest of the New England states. Suppose, for the sake of argument, we exclude Connecticut and recalculate the correlation. By using only the five other states, $r = .44$. Indeed, Connecticut has both the lowest suicide rate and the highest homicide rate in New England, which seems to have distorted the initial correlation coefficient.

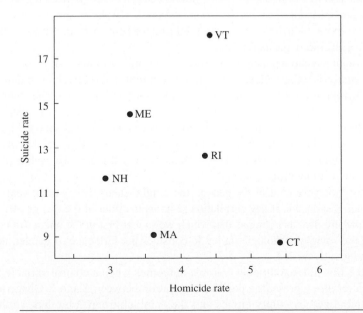

FIGURE 10.6   Scatter plot of homicide and suicide rates in New England

There are statistical procedures for determining if this or any other data point should be excluded; they are, however, beyond the scope of this book. Nevertheless, the importance of inspecting for these outliers is a lesson well worth learning. It can be distressing to promote a particular correlation as substantively meaningful, only to find later that the exclusion of one or two observations radically alters the results and interpretation.

Partial Correlation

In this chapter we have considered a powerful method for studying the association or relationship between two interval-level variables. It is important to consider if a correlation between two measures holds up when controlling for additional variables. That is, does our interpretation of the relationship between two variables change in any way when looking at the broader context of other related factors?

To see this most easily, we will focus again on scatter plots. A scatter plot visually displays all the information contained in a correlation coefficient—both its direction (by the trend underlying the points) and its strength (by the closeness of the points to a straight line). We can construct separate scatter plots for different subgroups of a sample to see if the correlation observed for the full sample holds when controlling for the subgroup or control variable. For example, there has been some research by social psychologists in recent years on the relationship between physical characteristics (such as attractiveness) and professional attainment (for example, salary or goal fulfillment). Suppose that within the context of studying the relationship between personal attributes and salary, a social psychologist stumbles on a strong positive association between height and salary, as shown in Figure 10.7. This would make sense to the social psychologist; he or she reasons that taller people tend to be more assertive and are afforded greater respect from others, which pays off in being successful in requests for raises.

But this social psychologist could be misled—in total or in part—if he or she fails to bring into the analysis other relevant factors that might alternatively account for the height–salary correlation. Gender of employee is one such possible variable. Men tend to be taller than women, and, for a variety of reasons, tend to be paid more. Perhaps this could explain all or part of the strong correlation between height and salary. Figure 10.7 also provides scatter plots of height and salary separately for males and females in the sample. It is important to note, first, that if we superimposed these two scatter plots they would produce the original plot.

Apparently, when we control for gender, the height–salary correlation weakens substantially—in fact, disappears. If any correlation remains in either of the two gender-specific subplots, it is nowhere near as strong as that which we saw at first in the uncontrolled scatter plot. Thus, had the social psychologist failed to consider the influence of gender, he or she would have been greatly misled.

Figure 10.8 illustrates additional possible outcomes when a control variable is introduced. Each scatter plot represents a positive correlation between *X* and *Y*. Observations in group 1 are symbolized by empty circles and those in subgroup 2 by dark circles. This

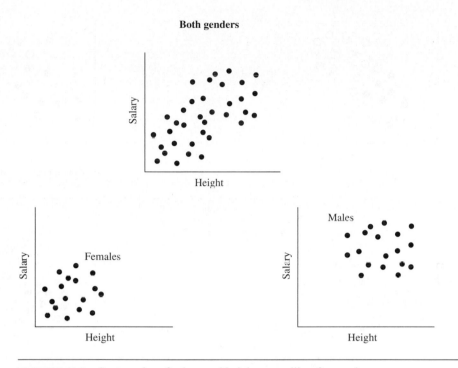

FIGURE 10.7   Scatter plot of salary and height controlling for gender

allows us to see the *X–Y* relationship within the two subgroups separately. Note that these are prototypes—in practice, one may not observe such clear-cut situations.

In scatter plot (a), we see that the *X–Y* association observed overall holds for each subgroup as well. Group 1 tends to exceed group 2 on both *X* and *Y*, and within these two groups *X* and *Y* are still related positively and strongly. That is, controlling for the grouping variable does not alter the *X–Y* relationship. For example, the positive relationship between education and income holds both for whites and non-whites. If one observes this kind of outcome when testing for a range of control variables (for example, race, sex, and age), one develops confidence in interpreting the association (for example, between education and income) as causal.

Scatter plot (b) shows a conditional relationship. Again, there is a strong relationship between *X* and *Y* for one group, but no relationship for the other. If the grouping variable is ignored, the correlation between *X* and *Y* misrepresents the more accurate picture within the subgroups.

Scatter plot (c) illustrates a spurious or misleading correlation. Within both subgroups, *X* and *Y* are unrelated. Overall, group 1 tends to be higher on both variables. As a result, when ignoring the subgroup distinction, it appears as if *X* and *Y* are related. Our association noted previously between height and salary is an example of a spurious correlation. Spurious correlations frequently occur in practice, and one should always be wary that two variables are related only because of their having a common cause.

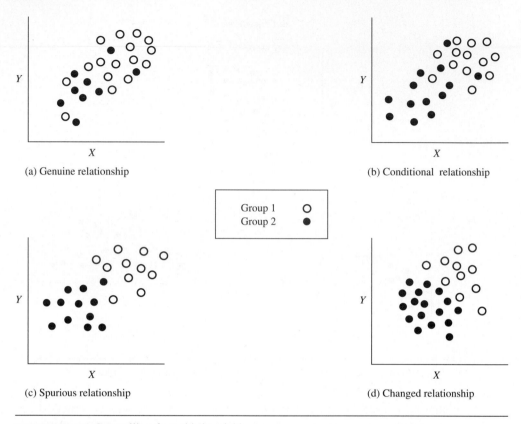

(a) Genuine relationship

(b) Conditional relationship

Group 1 ○
Group 2 ●

(c) Spurious relationship

(d) Changed relationship

FIGURE 10.8 Controlling for a third variable

Finally, scatter plot (d) shows a relationship that changes direction when a third variable is controlled. That is, the original positive association between *X* and *Y* becomes negative within the two subgroups. That group 1 was so much greater than group 2 on both *X* and *Y* overshadowed the negative relationship within each subgroup. This type of situation occurs rarely in practice, but one still should be aware that an apparent finding could be just the opposite of what it should be.

All the comparisons we have considered thus far involve dichotomous (two-category) control variables. The same approach applies to control variables having three or more levels or categories. For example, one could investigate the influence of religion on the relationship between two variables by computing Pearson's *r* separately for Protestants, Catholics, and Jews.

How would one handle an interval-level control variable like age? There is a temptation to categorize age into a number of subgroups (for example, under 18, 18–34, 35–49, 50 and over) and then to plot the *X–Y* association separately for each age category. However, this would be both inefficient and a waste of information (for example, the distinction between 18-year-olds and 34-year-olds is lost because these two ages are within the same category). Perhaps, then, we could use narrower age groups, but we still are being less precise than we could be. Fortunately, a simple method exists for adjusting a correlation between two variables for the influence of a third variable when all three are interval level. That is, we do not have to categorize any variables artificially.

The *partial correlation coefficient* is the correlation between two variables, after removing (or partialing out) the common effects of a third variable. Like simple correlations, a partial correlation can range from −1 to +1 and is interpreted exactly the same way as a simple correlation. The formula for the partial correlation of X and Y controlling for Z is

$$r_{XY.Z} = \frac{r_{XY} - r_{XZ}r_{YZ}}{\sqrt{1 - r_{XZ}^2}\sqrt{1 - r_{YZ}^2}}$$

In the notation $r_{XY.Z}$ the variables before the period are those being correlated, and the variable after the period is the control variable. The partial correlation is computed exclusively on the basis of three quantities: the correlations between X and Y, X and Z, and Y and Z.

For example, consider the following correlation matrix for height *(X)*, weight *(Y)*, and age *(Z)*. Not only are height and weight positively correlated, but both increase with age. One might wonder, then, how much of the correlation between height and weight $(r_{XY} = .90)$ is due to the common influence of age and how much remains after the influence of age is controlled:

	Height	**Weight**	**Age**
Height *(X)*	1.00	.90	.80
Weight *(Y)*		1.00	.85
Age *(Z)*			1.00

$$
\begin{aligned}
r_{XY.Z} &= \frac{r_{XY} - r_{XZ}r_{YZ}}{\sqrt{1 - r_{XZ}^2}\sqrt{1 - r_{YZ}^2}} \\[2mm]
&= \frac{.90 - (.80)(.85)}{\sqrt{1 - (.80)^2}\sqrt{1 - (.85)^2}} \\[2mm]
&= \frac{.90 - .68}{\sqrt{1 - .64}\sqrt{1 - .7225}} \\[2mm]
&= \frac{.22}{\sqrt{.36}\sqrt{.2775}} \\[2mm]
&= \frac{.22}{(.60)(.5268)} \\[2mm]
&= \frac{.22}{.3161} \\[2mm]
&= +.70
\end{aligned}
$$

Thus, the strong initial correlation between height and weight (r_{XY} = .90) weakens somewhat when the effects of age are removed ($r_{XY.Z}$ = .70).

We saw in Figure 10.8 that there are many possible patterns when controlling for a third variable. Similarly, partial correlations can be smaller, equal to, or greater than the two-variable simple correlation. Consider, for example, the following correlation matrix for education *(X)*, salary *(Y)*, and age *(Z)*.

	Education	**Salary**	**Age**
Education *(X)*	1.00	.40	−.30
Salary *(Y)*		1.00	.50
Age *(Z)*			1.00

The simple correlation between education and salary is .40, but the partial correlation between education and salary controlling for age is even higher:

$$r_{XY.Z} = \frac{r_{XY} - r_{XZ}r_{YZ}}{\sqrt{1 - r_{XZ}^2}\sqrt{1 - r_{YZ}^2}}$$

$$= \frac{.40 - (-.30)(.50)}{\sqrt{1 - (-.30)^2}\sqrt{1 - (.50)^2}}$$

$$= \frac{.55}{\sqrt{.91}\sqrt{.75}}$$

$$= \frac{.55}{(.9539)(.8660)}$$

$$= \frac{.55}{.8261}$$

$$= +.67$$

Thus, ignoring age suppresses the observed association between education and salary. Younger employees, because of their low seniority, have lower salaries, despite their higher level of educational attainment. As a result, the influence of education is dwarfed in the simple correlation because highly educated employees, who you would think should be paid more than they are, do not have the salary expected because they tend to be younger and newer employees. By controlling for age, we isolate the effect of education on salary, absent of the influence of age.

The partial correlation coefficient is a very useful statistic for finding spurious relationships, as is demonstrated in this classic case of a "vanishing" correlation.[1] The correlation between the rate of forcible rape (per 100,000) in 1982 and the circulation of *Playboy* (per 100,000) in 1979 for 49 U.S. states (Alaska is an outlier on rape and is excluded) is

---

[1]We thank Rodney Stark and Cognitive Development, Inc., for this fine illustration and for these data.

$r = +40$. Because of this substantial correlation, many observers have asked, if *Playboy* has this kind of effect on sex crimes, imagine what harm may be caused by truly hard-core pornography?

This concern requires the unjustified assumption that the correlation implies cause. Before making such a leap, however, we need to consider whether the two variables have a third variable as a common cause, thereby producing a spurious result.

As it turns out, both the rape and the *Playboy* subscription rate are related to the rate of homes without an adult female (per 1,000 households): For the rape rate *(Y)* and the rate of homes without an adult female *(Z)*, $r_{YZ} = +.48$; for the rate of subscription to *Playboy* *(X)* and the rate of homes without an adult female *(Z)*, $r_{YZ} = +.85$. Apparently, both types of sexual outlet (one illegal and one legal) sometimes stem from the absence of adult females in the home.

To determine the correlation of *Playboy (X)* with rape *(Y)*, controlling for homes without adult females *(Z)*, we calculate the partial correlation:

$$
\begin{aligned}
r_{XY.Z} &= \frac{r_{XY} - r_{XZ}r_{YZ}}{\sqrt{1 - r_{XZ}^2}\sqrt{1 - r_{YZ}^2}} \\[2mm]
&= \frac{.40 - (.85)(.48)}{\sqrt{1 - (.85)^2}\sqrt{1 - (.48)^2}} \\[2mm]
&= \frac{.40 - .41}{\sqrt{1 - .7225}\sqrt{1 - .2304}} \\[2mm]
&= \frac{-.01}{\sqrt{.2775}\sqrt{.7696}} \\[2mm]
&= \frac{-.01}{(.53)(.88)} \\[2mm]
&= \frac{-.01}{.47} \\[2mm]
&= -.02
\end{aligned}
$$

As a result, after controlling for one common variable, the original correlation disappears.

Summary

In this chapter we went beyond the task of establishing the presence or absence of a relationship between two variables. In correlation, the social researcher is interested in the degree of association between two variables. With the aid of the correlation coefficient known as Pearson's *r*, it is possible to obtain a precise measure of both the strength—from 0.0 to 1.0—and direction—positive versus negative—of a relationship between two variables that have been measured at the interval level. Moreover, if a researcher has taken a random sample of scores, he or she may also compute a *t* ratio to determine whether the obtained relationship between *X* and *Y* exists in the population and is not due merely to sampling error. In addition, the partial correlation coefficient allows the researcher to control a two-variable relationship for the impact of a third variable.

Terms to Remember

Variable
Correlation
  Strength
  Direction (positive versus negative)
  Curvilinear versus straight line

Correlation coefficient
Scatter plot
Pearson's correlation coefficient
Partial correlation coefficient

Questions and Problems

1. When the points in a scatter plot cluster closely around the regression line, the correlation can be said to be
   a. weak.
   b. strong.
   c. positive.
   d. negative.

2. When respondents who score high on the SATs tend also to get high grades in college, whereas those who score low on the SATs do poorly in college, there is reason to posit
   a. a positive correlation between SATs and college grades.
   b. a negative correlation between SATs and college grades.
   c. a zero correlation between SATs and college grades.
   d. None of the above

3. A correlation coefficient expresses in a single number
   a. the strength of a correlation.
   b. the direction of a correlation.
   c. both the strength and the direction of a correlation.
   d. None of the above

4. $r = -.17$ represents
   a. a correlation that is strong and negative.
   b. a correlation that is strong and positive.
   c. a correlation that is weak and negative.
   d. a correlation that is weak and positive.

5. Which of the following is *not* required by Pearson's *r*?
   a. A straight-line relationship
   b. Nominal data
   c. Random sampling
   d. Normally distributed characteristics

6. A partial correlation coefficient gives us the correlation between two variables
   a. after removing the common effect of a third variable.
   b. at the interval level of measurement.
   c. after introducing the common effect of a third variable.
   d. when we are partially certain that no relationship between them actually exists.

7. The following six students were questioned regarding *(X)* their attitudes toward the legalization of prostitution and *(Y)* their attitudes toward the legalization of marijuana. Compute a Pearson's correlation coefficient for these data and determine whether the correlation is significant.

Student	X	Y
A	1	2
B	6	5
C	4	3
D	3	3
E	2	1
F	7	4

8. A high school guidance counselor is interested in the relationship between proximity to school and participation in extracurricular activities. He collects the data on distance from home to school (in miles) and number of clubs joined for a sample of 10 juniors. Using the following data, compute a Pearson's correlation coefficient and indicate whether the correlation is significant.

	Distance to School (miles)	Number of Clubs Joined
Lee	4	3
Ronda	2	1
Jess	7	5
Evelyn	1	2
Mohammed	4	1
Steve	6	1
George	9	9
Juan	7	6
Chi	7	5
David	10	8

9. An urban sociologist interested in neighborliness collected data for a sample of 10 adults on *(X)* how many years they have lived in their neighborhood and *(Y)* how many of their neighbors they regard as friends. Compute a Pearson's correlation coefficient for these data and determine whether the correlation is significant.

X	Y	X	Y
1	1	2	1
5	4	5	2
6	2	9	6
1	3	4	7
8	5	2	0

10. An economist is interested in studying the relationship between length of unemployment and job-seeking activity among white-collar workers. He interviews a sample of 12 unemployed accountants as to the number of weeks they have been unemployed *(X)* and seeking a job during the past year *(Y)*. Compute a Pearson's correlation coefficient for these data and determine whether the correlation is significant.

Accountant	X	Y
A	2	8
B	7	3
C	5	4
D	12	2
E	1	5
F	10	2
G	8	1
H	6	5
I	5	4
J	2	6
K	3	7
L	4	1

11. A psychiatrist is concerned about his daughter, who has suffered from extremely low self-esteem and high anxiety since entering high school last year. Wondering if many of his daughter's peers have the same troubles, he collects a random sample of high school girls and anonymously asks them how much they agree or disagree (on a scale from 1 to 7, with 1 being "strongly disagree" and 7 being "strongly agree") with the following statements: *(X)* "I felt better about myself before I started high school" and *(Y)* "I have felt very anxious since I started high school." Compute a Pearson's correlation coefficient for the following data and indicate whether the correlation is significant.

X	Y
7	5
6	4
4	3
5	6
3	2
6	7
5	5

12. A new high school special education teacher wonders if there really is a correlation between reading disabilities and attention disorders. She collects data from six of her students on their reading abilities *(X)* and their attention abilities *(Y)*, with a higher score indicating greater ability for both variables. Compute a Pearson's correlation coefficient for the following data and indicate whether the correlation is significant.

X	Y
2	3
1	2
5	3
4	2
2	4
1	3

13. A researcher wonders if there is a correlation between *(X)* people's opinions of bilingual education and *(Y)* their opinions about whether foreign-born citizens should be allowed to run for president. She collects the following data, with both variables being measured on a scale from 1 to 9 (1 being strongly opposed and 9 being strongly in favor).

X	Y
2	1
5	4
8	9
6	5
1	1
2	1
8	7
3	2

Calculate a Pearson's correlation coefficient and determine whether the correlation is significant.

14. Obesity in children is a major concern because it puts them at risk for several serious medical problems. Some researchers believe that a major issue related to this is that children these days spend too much time watching television and not enough time being active. Based on a sample of boys of roughly the same age and height, data were collected regarding hours of television watched per day and weight. Compute a Pearson's correlation coefficient and indicate whether the correlation is significant.

TV Watching (hr)	Weight (lb)
1.5	79
5.0	105
3.5	96
2.5	83
4.0	99
1.0	78
0.5	68

15. Is there a relationship between *(X)* rate of poverty (measured as percent of population below poverty level) and *(Y)* rates of teen pregnancy (measured per 1,000 females aged 15 to 17)? A researcher selected random states and collected the following data. Compute a Pearson's correlation coefficient and determine whether the correlation is significant.

State	X	Y
A	10.4	41.7
B	8.9	38.6
C	13.3	43.2
D	6.9	35.7
E	16.0	46.9
F	5.2	33.5
G	14.5	43.3
H	15.3	44.8

16. In preparing for an examination, some students in a class studied more than others. Each student's grade on the 10-point exam and the number of hours studied were as follows:

	Hours Studied	Exam Grade
Barbara	4	5
Bob	1	2
Deidra	3	1
Owen	5	5
Charles	8	9
Emma	2	7
Sanford	7	6
Luis	6	8

Calculate a Pearson's correlation coefficient and determine whether the correlation is significant.

17. A researcher set out to determine whether suicide and homicide rates in metropolitan areas around the country are correlated and, if so, whether they vary inversely (negative correlation) or together (positive correlation). Using available data for a recent year, he compared the following sample of 10 metropolitan areas with respect to their rates (number per 100,000), of suicide and homicide:

Metropolitan Area	Suicide Rate	Homicide Rate	Metropolitan Area	Suicide Rate	Homicide Rate
A	20.2	22.5	F	21.4	19.5
B	22.6	28.0	G	9.8	13.2
C	23.7	15.4	H	13.7	16.0
D	10.9	12.3	I	15.5	17.7
E	14.0	12.6	J	18.2	20.8

What is the strength and direction of correlation between suicide and homicide rates among the 10 metropolitan areas sampled? Test the null hypothesis that rates of suicide and homicide are not correlated in the population.

18. An educational researcher interested in the consistency of school absenteeism over time studied a sample of eight high school students for whom complete school records were available. The researcher counted the number of days each student had missed while in the sixth grade and then in the tenth grade. He obtained the following results:

Student	Days Missed (6th)	Days Missed (10th)
A	4	10
B	2	4
C	21	11
D	1	3
E	3	1
F	5	5
G	4	9
H	8	5

What is the strength and direction of the relationship between the number of days these students were absent from elementary school (sixth grade) and how many days they missed when they reached high school (tenth grade)? Can the correlation be generalized to a larger population of students?

19. Do reading and television viewing compete for leisure time? To find out, a communication specialist interviewed a sample of 10 children regarding the number of books they had read during the last year and the number of hours they had spent watching television on a daily basis. Her results are as follows:

Number of Books	Hours of TV Viewing
0	3
7	1
2	2
1	2
5	0
4	1
3	3
3	2
0	7
1	4

What is the strength and direction of the correlation between number of books read and hours of television viewing daily? Is the correlation significant?

20. In addition to job-seeking activity, the age of a white-collar worker may be related to his or her length of unemployment. Suppose then that age *(Z)* is added to the two variables in problem 10.

Accountant	Weeks Unemployed *(X)*	Weeks Seeking *(Y)*	Age *(Z)*
A	2	8	30
B	7	3	42
C	5	4	36
D	12	2	47
E	1	5	29
F	10	2	56
G	8	1	52
H	6	5	40
I	5	4	27
J	2	6	31
K	3	7	36
L	4	1	33

Find the partial correlation of weeks unemployed and weeks seeking a job, holding the age of the worker constant.

21. Besides studying time, intelligence itself may be related to test performance. Suppose then that IQ *(Z)* is added to the two variables in problem 16.

	Hours Studied *(X)*	Exam Grade *(Y)*	IQ *(Z)*
Barbara	4	5	100
Bob	1	2	95
Deidra	3	1	95
Owen	5	5	108
Charles	8	9	110
Emma	2	7	117
Sanford	7	6	110
Luis	6	8	115

Find the partial correlation of studying time and exam grade, holding IQ constant.

22. The following is a correlation matrix among family size *(X)*, weekly grocery bill *(Y)*, and income *(Z)* for a random sample of 50 families.

	X	Y	Z
X	1.00	.60	.20
Y	.60	1.00	.30
Z	.20	.30	1.00

a. Which of the correlations are significant at the .05 level?

b. What is the partial correlation between family size and grocery bill, holding income constant? Discuss the difference between the simple correlation r_{XY} and the partial correlation $r_{XY.Z}$.

SPSS Exercises

1. Using SPSS to analyze the Monitoring the Future Study, generate a correlation matrix which will allow you to test the following null hypotheses:

   Null hypothesis 1: There is no relationship between lifetime reported use of marijuana or hashish (V115) and lifetime reported use of cocaine (V124).

   Null hypothesis 2: There is no relationship between self-reported lifetime use of marijuana or hashish (V115) and the likelihood of getting a speeding ticket or a warning from police for a traffic violation (V197).

   Create a correlation matrix of the three variables. Report the strength and direction of the Pearson's *r* correlation coefficients. Hint: ANALYZE, CORRELATE, BIVARIATE and choose the variables.

2. Use SPSS to generate a single correlation matrix which will allow you to test the null hypothesis of no relationship for all of the following pairs of variables:

   Grades (V179) and income from a job after school (V192);
   Lifetime marijuana use (V115) and income from a job after school (V192);
   Grades (V179) and lifetime marijuana use (V115); and
   Satisfaction with school (V1682) and lifetime marijuana use (V115).

   a. Create the correlation matrix.
   b. Report the strength and direction of the Pearson's *r* correlation coefficients for each pair of variables.
   c. What other pairs of variables could be tested using this same correlation matrix?

3. Using SPSS to analyze the Best Places Study, generate a single correlation matrix to test the null hypothesis of no relationship for all of the following pairs of variables:

   Suicide rates (SUICIDE) and divorce rates (DIVORCE);
   Suicide rates (SUICIDE) and violent crime (CRIMEV);
   Property crime (CRIMEP) and violent crime (CRIMEV); and
   Another variable of your choice to correlate with divorce rates (DIVORCE).

   a. Create the correlation matrix.
   b. Report the strength and direction of the Pearson's *r* correlation coefficients for each pair of variables.

4. Using the General Social Survey, calculate Pearson's *r* to test the following null hypotheses:

Null hypothesis 1:   *Respondent's highest degree (DEGREE) is not related to their fathers' highest degree (PADEG).*

Null hypothesis 2:   *Respondent's highest degree (DEGREE) is not related to level of personal income (RINCOM98).*

   a. Create the correlation matrix.
   b. Report the strength and direction of the Pearson's *r* correlation coefficients for each pair of variables.

5. Choose two variables from the General Social Survey so that you may generate and interpret a Pearson's *r* correlation to test the null hypothesis of no correlation between the variables.

11

Regression Analysis

Certain concepts of statistics, such as percentages and means, are so commonplace that you may have understood them long before taking a statistics course. Other concepts are new, and in the process of learning statistics thoroughly, you'll begin to see the usefulness of measures that initially you may have learned to calculate by just "plugging into" a formula. It is analogous to becoming fluent in a foreign language: Initially, one needs a dictionary to translate words, but later the context of the words also becomes meaningful.

In Chapter 4 we learned a new concept, that of variance. We also saw that in some instances it was an even more important concept than the mean. But still you may not yet have a feel for what the variance signifies and its fundamental role in statistics. In the context of *regression analysis*, the important notion of variance should become clearer.

Let's reconsider the problem in Chapter 4 concerning the length of sentences given criminal defendants. Suppose a criminologist collected data on sentences given defendants convicted under a gun control law, which mandates a 1-year prison term for the illegal possession of a firearm. Obviously, the entire data set would consist of 12-month sentences. Although knowing how many people had been sentenced under the law might be mildly interesting, the mean sentence length of 12 would be of no analytic value: All sentences would be 12 months, and so the variance would be zero.

To learn something about sentencing patterns and the tendencies of judges in handing out their sentences, the criminologist would be far better off focusing on a crime for which the sentences vary. If nothing varies, there is nothing to explain. That is, the goal of research is to explain why variables vary. For example, why do certain defendants obtain short sentences and others long sentences? Can we identify characteristics of defendants or of the nature of their crimes that account for this variation in sentence length? Or are sentence lengths to some degree random and thus unpredictable?

Suppose that a particular judge, known to be moody at times, has given these sentences (in months) to 10 defendants convicted of assault: 12, 13, 15, 19, 26, 27, 29, 31, 40, and 48. The immediate questions are these: Why did certain defendants receive only a term of or near 12 months? Why did one defendant receive as much as 48 months? Was it deserved? Is it because this defendant had a long criminal record or because the crime was particularly vicious? Or was it just a result of how the judge felt that day? Worse, could it have something to do with the defendant's race and the race of the victim?

The mean of these data is 26 months. There is nothing apparently unreasonable (not too harsh or too lenient) about the average sentence given by this judge. But our concern is with his consistency in sentencing—that is, with how disparate his sentences seem. We find that the variance of these data is 125 (you should verify this yourself). This then is a measure of the amount of dissimilarity among the sentences. It may seem large, but, more to the point, is it justified? How much of this 125 value is a result of, say, the number of prior convictions the defendants had?

Regression analysis is used to answer these questions. Each of us might have a theory about which factors encourage stiff prison sentences and which encourage leniency. But regression analysis gives us the ability to quantify precisely the relative importance of any proposed factor or variable. There is little question, therefore, why the regression technique is used more than any other in the social sciences.

The Regression Model

Regression is closely allied with correlation in that we are still interested in the strength of association between two variables—for example, length of sentence and number of prior convictions. In regression, however, we are also concerned with specifying the nature of this relationship. We specify one variable as dependent and one as independent. That is, one variable is believed to influence the other. In our case, sentence length is dependent and number of prior convictions is independent.

In regression analysis, a mathematical equation is used to predict the value of the dependent variable (denoted Y) on the basis of the independent variable (denoted X):

$$Y = a + bX + e$$

Mathematically, this equation states that the sentence length (Y) received by a given defendant is the sum of three components: (1) a base-line amount given to all defendants (denoted *a*); (2) an additional amount given for each prior conviction (denoted *b*); and (3) a residual value (denoted *e*) that is unpredictable and unique to that individual's case.

The term *a* is called the *Y-intercept.* It refers to the expected level of *Y* when $X = 0$ (no priors). This is the base-line amount because it is what *Y* should be before we take the level of *X* into account.

The term *b* is called the *slope* (or the *regression coefficient*) for *X*. This represents the amount that *Y* is expected to change (increase or decrease) for each increase of one unit in *X*. Thus, for example, the difference in sentence length between a defendant with $X = 0$ (no priors) and $X = 1$ (one prior) is expected to be *b*; the difference in expected sentence length between an offender with $X = 0$ (no priors) and a defendant with $X = 2$ (two priors) is $b + b = 2b$.

Finally, *e* is called the *error term* or *disturbance term.* It represents the amount of the sentence that cannot be accounted for by *a* and *bX*. In other words, *e* represents the departure of a given defendant's sentence from that which would be expected on the basis of his number of priors *(X)*.

Let's consider this model geometrically. Figure 11.1 gives the scatter plot of the sentence length and prior conviction data of Table 11.1 for the 10 defendants.

Clearly, there is a strong positive association between *X* and *Y*. Regression involves placing or fitting a line through the scatter of points: If the line is drawn accurately, the value of *a* (the *Y*-intercept) would be the location where the line crosses the *Y* axis. The value of *b* (the slope) would correspond to the incline or rise of the line for a unit increase in *X*. We learned in Chapter 10 that Pearson's correlation coefficient *(r)* measures the degree to which the points lie close to a straight line; the line to which we were referring is this *regression line.* It falls closest to all the points in a scatter plot.

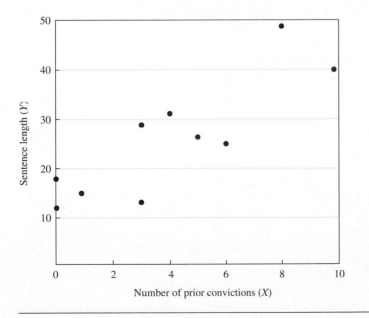

FIGURE 11.1   Scatter plot of sentencing data in Table 11.1

TABLE 11.1 *Sentence Length and Prior Convictions for 10 Defendants*

Priors (X)	Sentence (in months) (Y)
0	12
3	13
1	15
0	19
6	26
5	27
3	29
4	31
10	40
8	48

Because regression is closely allied with correlation, it should not be surprising that the calculations are similar. In fact, almost all the computational steps for regression are the same as those for correlation.

The values of a and b that most closely fit the data are given by

$$b = \frac{\text{SP}}{\text{SS}_X}$$

or

$$b = \frac{\Sigma(X - \bar{X})(X - \bar{Y})}{\Sigma(X - \bar{X})^2} \quad \text{for deviations}$$

or

$$b = \frac{\Sigma XY - N\bar{X}\bar{Y}}{\Sigma X^2 - N\bar{X}^2} \quad \text{for raw scores}$$

and the intercept

$$a = \bar{Y} - b\bar{X}$$

The necessary calculations for data given in the previous table are shown in Table 11.2.

TABLE 11.2 *Regression Calculations for Data in Table 11.1*

X	Y	X − $\overline{X}$	Y−Y	(X − $\overline{X}$)(Y − $\overline{Y}$)	(X − X)²	(Y − $\overline{Y}$)²	
0	12	−4	−14	56	16	196	$\overline{X}$ = 4
3	13	−1	−13	13	1	169	$\overline{Y}$ = 26
1	15	−3	−11	33	9	121	SP = 300
0	19	−4	−7	28	16	49	SP$_X$ = 100
6	26	2	0	0	4	0	SS$_Y$ = 250
5	27	1	1	1	1	1	r = +.85
3	29	−1	3	−3	1	9	b = 3
4	31	0	5	0	0	25	a = 14
10	40	6	14	84	36	196	
8	48	4	22	88	16	484	
ΣX = 40	ΣY = 260			SP = 300	SS$_X$ = 100	SS$_Y$ = 1,250	

First, we compute the means of the two variables ($\overline{X}$ — mean number of priors; $\overline{Y}$ = mean sentence length in months):

$$\overline{X} = \frac{\Sigma X}{N}$$

$$= \frac{40}{10}$$

$$= 4$$

$$\overline{Y} = \frac{\Sigma Y}{N}$$

$$= \frac{260}{10}$$

$$= 26$$

Next we compute the sum of products and sums of squares:

$$SP = \Sigma(X − \overline{X})(Y − \overline{Y})$$
$$= 300$$

$$SS_X = \Sigma(X − \overline{X})^2$$
$$= 100$$

$$SS_Y = \Sigma(Y − \overline{Y})^2$$
$$= 1,250$$

Using these calculations, we now compute *b* and then *a*:

$$b = \frac{SP}{SS_X}$$

$$= \frac{300}{100}$$

$$= 3$$

$$a = \overline{Y} - b\overline{X}$$

$$= 26 - (3)(4)$$

$$= 26 - 12$$

$$= 14$$

The slope *(b)* and the *Y*-intercept *(a)* form the equation for the regression line. Because the regression line represents expected or predicted sentences, rather than the actual sentences, we use $\hat{Y}$ (the caret symbol ˆ means *predicted*) on the left side to represent predicted sentence, as opposed to actual sentence *(Y)*:

$$\hat{Y} = a + bX$$

$$\hat{Y} = 14 + 3X$$

The next step is to plot the regression line on the scatter plot. To do this, we only need to find two points on the line and then connect the points (two points uniquely determine a line).

The easiest point to plot is the *Y*-intercept. That is, the value of *a* is where the regression line crosses the *Y* axis. In other words, the regression line always passes through the point $(X = 0, Y = a)$. The next easiest point to determine is the intersection of the two means. It makes intuitive sense that a case average on *X* can be predicted to be average on *Y*. In short, the regression line always passes through the point $(X = \overline{X}, Y = \overline{Y})$.

In our example, then, we know immediately that the regression line passes through the points (0, 14) and (4, 26). And in Figure 11.2 we can connect these points to form this line.

At times, the two easiest points to plot—the *Y*-intercept and the intersection of the two means—are too close together to allow us to draw the regression line with accuracy. That is, if the points are too near, there is considerable room for error in setting down a ruler between them. In such cases, one needs to select a different point than the intersection of means. One should select a large value of *X* (so that the point is far from the *Y*-intercept) and plug it into the equation. In our example, we could select $X = 10$, and so

$$\hat{Y} = a + b(10)$$

$$= 14 + (3)(10)$$

$$= 14 + 30$$

$$= 44$$

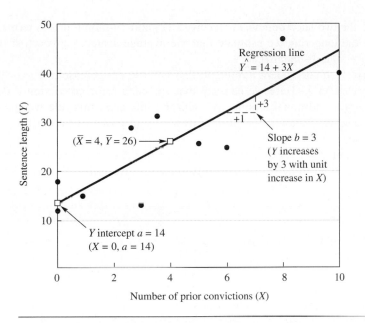

FIGURE 11.2   Regression line for sentencing data in Table 11.1

Point (10, 44) could be used with point (0, 14) to plot the line. Besides the advantage of selecting points as far apart as possible to make the drawing simpler, it does not matter which points you choose—all the predicted points will lie on the same line.

Requirements for Regression

The assumptions underlying regression are the same as those for Pearson's r. In particular:

1. It is assumed that both variables are measured at the interval level.
2. Regression assumes a straight-line relationship. If this is not the case, there are various transformations (which are more advanced than this presentation) that can be used to make the relationship into a straight line. Also, if extremely deviant cases are observed in a scatter plot, these should be removed from the analysis.
3. Sample members must be chosen randomly to employ tests of significance.
4. To test the significance of the regression line, one must also assume normality for both variables or else have a large sample.

Interpreting the Regression Line

Let's consider what the values of a and b mean in substantive terms. The Y-intercept corresponds to the expected or predicted value of Y when X is zero. In our case, then, we can expect that first offenders (that is, those without prior convictions) will be sentenced to $a = 14$ months. Of course, not all first offenders will receive a 14-month sentence—and in our sample the two such defendants received 12- and 19-month prison terms, respectively. But in the long run, we estimate that the average sentence given first offenders (those with $X = 0$) is 14 months.

The regression coefficient b refers to the increase or decrease in Y expected with each unit increase in X. Here we can say that for each prior conviction a defendant tends to get $b = 3$ additional months. As with the intercept, this rule will not hold in every case; however, 3 months is the long-run cost in terms of prison time for each prior conviction.

With this notion, we can also make predictions of a defendant's sentence on the basis of his number of prior convictions. If a defendant has five priors, for example, we can expect or predict

$$\hat{Y} = a + b(5)$$
$$= 14 + (3)(5)$$
$$= 14 + 15$$
$$= 29$$

In other words, this defendant should or can be expected to receive the base line of 14 months plus 3 additional months for each of his five priors. Note as well that this point (5, 29) also lies on the regression line drawn in Figure 11.2. Thus, we could simply use this line to make predictions of the sentence length for any defendant, even defendants outside this sample, as long as they are a part of the population from which this sample was drawn (that is, defendants convicted of assault in the same jurisdiction). Although we cannot expect to predict sentences exactly, the regression line will make the best prediction possible on the basis of just the number of priors.

Unfortunately, the interpretation of the regression line is not always as direct and meaningful as in this example, particularly with regard to the Y-intercept. In our example we could interpret the Y-intercept because a value of $X = 0$ was realistic. If, however, we were regressing weight on height (that is, predicting weight from height), the Y-intercept would represent the predicted weight of an individual 0 inches tall. The interpretation would be as foolish as the thought of such a person.

Meaningful or not, the Y-intercept is nevertheless an important part of the regression equation, but never as substantively important as the slope. In the height–weight instance, the slope refers to the expected weight increase for each inch of height. Indeed, the old adage of 5 pounds for every inch of growth is actually a regression slope.

Regression equations are frequently used to project the impact of the independent variable (X) beyond its range in the sample. There were no defendants with more than 10 priors, but we could still predict the sentence given by the judge to a hypothetical defendant with 13 priors (see Figure 11.3):

$$\hat{Y} = a + b(13)$$
$$= 14 + (3)(13)$$
$$= 14 + 39$$
$$= 53$$

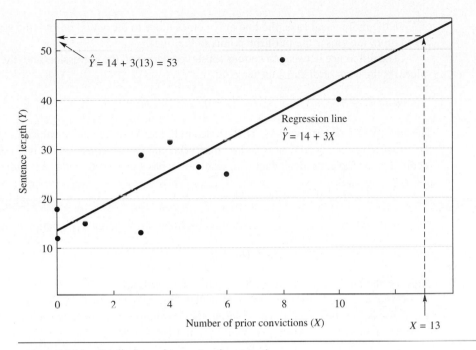

FIGURE 11.3 Prediction using sentencing equation

One has to be cautious, however, about making predictions that fall far afield from the sample of data points. It would be farfetched to use a height–weight regression to predict the weight of a 10-foot-tall man. Similarly, in our sentencing example, it would be mathematically possible to predict the sentence that would be awarded a defendant with 100 priors:

$$\hat{Y} = a + b(100)$$
$$= 14 + (3)(100)$$
$$= 14 + 300$$
$$= 314$$

A 314-month sentence (over 26 years) for assault is absurd, but then so is the idea of a defendant with 100 prior convictions. Once you exceed the sample range of values too far, the ability to generalize the regression line breaks down. Because the largest value of X in the sample is 10, good sense would dictate against predicting sentences of defendants with more than, say, 15 priors. It would be quite unlikely that the mathematical rule of $b = 3$ months per prior would be applied to someone with as long a criminal record as this. Other considerations would surely intervene that would invalidate such a farfetched prediction.

Prediction Errors

In the special case in which the correlation is perfect ($r = +1$ or -1), all the points lie precisely on the regression line and all the Y values can be predicted perfectly on the basis of *X*. In the more usual case, the line only comes close to the actual points (the stronger the correlation, the closer the fit of the points to the line).

The difference between the points (observed data) and the regression line (the predicted values) is the error or disturbance term *(e):*

$$e = Y - \hat{Y}$$

The concept of a disturbance term is illustrated in Figure 11.4. A positive value of *e* means that the sentence given a defendant is greater than what you would expect on the basis of his prior record. For example, the defendant with eight priors has a predicted sentence length of

$$\hat{Y} = 14 + (3)(8) = 14 + 24 = 38$$

His actual sentence was as much as 48 months, however, yielding a prediction error of

$$e = Y - \hat{Y} = 48 - 38 = 10$$

Thus, on the basis of priors alone, this sentence is underpredicted by 10 months.

Negative prediction errors occur when the data points lie below the regression line. That is, on the basis of the *X* value, the *Y* value is overpredicted. For example, for the defendant with six priors, we would predict his sentence to be

$$\hat{Y} = 14 + (3)(6) = 14 + 18 = 32$$

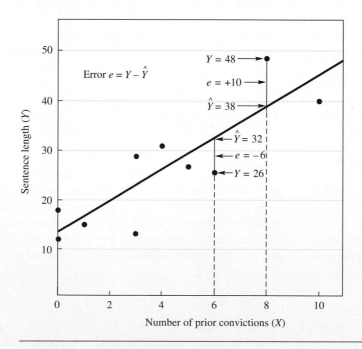

FIGURE 11.4   Prediction error in regression

This defendant received only 26 months, producing a prediction error of

$$e = Y - \hat{Y} = 26 - 32 = -6$$

The predictive value of a regression line (say, for predicting sentences on the basis of priors) can be assessed by the magnitude of these error terms. The larger the error, the poorer is the regression line as a prediction device.

It would seem intuitively logical to add the error terms to obtain a measure of predictive ability. However, the negative and positive errors cancel out. That is, $\Sigma e = 0$. To prevent this, we can square the errors before we sum. The so-called *error sum of squares* (or *residual sum of squares*), denoted by SS_{error}, is

$$SS_{error} = \Sigma e^2 = \Sigma(Y - \hat{Y})^2$$

The usefulness of the error sum of squares lies in comparing it to the magnitude of error that would have resulted had one not used X in making predictions. Without knowing anything about a defendant, what would you guess his sentence to be? The best guess would be the average sentence, or $\overline{Y}$. If we guess or predict $\overline{Y}$ for every defendant, the errors would simply be the deviations from the mean:

$$\text{Error without knowing } X = Y - \overline{Y}$$

The sum of squared prediction errors or deviations without using X is called the *total sum of squares:*

$$SS_{total} = \Sigma(Y - \overline{Y})^2$$

The predictive value of the regression equation is in its ability to reduce prediction error—that is, the extent that SS_{error} is smaller than SS_{total}. The difference between the two is the sum of squares that X can explain, and this is called the *regression sum of squares* (or *explained sum of squares*). The regression sum of squares is then

$$SS_{reg} = SS_{total} - SS_{error}$$

To summarize:

	Not Knowing X	**Knowing X**
Actual value	Y	Y
Predicted value	$\overline{Y}$	$\hat{Y} = a + bX$
Prediction error	$Y - \overline{Y}$	$Y - \hat{Y}$
Sum of squares	$SS_{total} = \Sigma(Y - \overline{Y})^2$	$SS_{error} = \Sigma(Y - \hat{Y})^2$
Difference	$SS_{reg} = SS_{total} - SS_{error}$	

Let's now calculate these sums of squares for the sentencing data. We already have from Table 11.2 the total sum of squares (previously called SS_Y):

$$SS_{total} = \Sigma(Y - \overline{Y})^2 = 1,250$$

To calculate the error sum of squares, we need to obtain the predicted sentence length ($\hat{Y}$) for each defendant, subtract it from the actual sentence length (Y), square the difference, and then sum:

X	Y	$\hat{Y} = a + bX$	$e = Y - \hat{Y}$	e^2
0	12	14	−2	4
3	13	23	−10	100
1	15	17	−2	4
0	19	14	5	25
6	26	32	−6	36
5	27	29	−2	4
3	29	23	6	36
4	31	26	5	25
10	40	44	−4	16
8	48	38	10	100
			$\Sigma e^2 =$	350

Thus,

$$SS_{error} = \Sigma e^2 = 350$$

and so

$$SS_{reg} = SS_{total} - SS_{error} = 1,250 - 350 = 900$$

The ability of a regression line to make predictions can be expressed in what is known as the *proportionate reduction in error* (PRE), that is, the proportion of the prediction error that can be reduced by knowing the independent variable. The proportionate reduction in error (PRE) due to X is

$$
\begin{aligned}
PRE &= \frac{SS_{total} - SS_{error}}{SS_{total}} \\
&= \frac{SS_{reg}}{SS_{total}} \\
&= \frac{900}{1,250} \\
&= .72
\end{aligned}
$$

Thus, .72, or 72%, of the error in predicting sentence length is reduced by taking number of priors into account. Put differently, 72% of the variance in sentence length is explained by the number of priors the defendant has on his record. This is precisely the information that we sought from the beginning.

Regression and Pearson's Correlation

We approached the problem of sentencing disparity by questioning why some defendants receive longer sentences than others. Using regression, we were able to determine that 72% of the variance in sentence length can be explained by the number of prior convictions. Obtaining the regression equation—the intercept and the slope—was fairly straightforward. It came directly from quantities (SP, SS_X, and SS_Y) derived in calculating the correlation (Pearson's r). However, the steps for computing SS_{error} and SS_{reg} were quite laborious, because they involved making a prediction for each person in the sample. Given the usefulness of determining the proportionate reduction in error (PRE), it would be helpful to employ a far simpler method.

If X and Y are uncorrelated (that is, if Pearson's $r = 0$), SS_{total} and SS_{error} will be the same, because X will not help predict Y. The larger the value of r, the smaller the value of SS_{error} relative to SS_{total}. More precisely, the proportionate reduction in error (PRE) is the square of Pearson's r:

$$r^2 = \frac{SS_{total} - SS_{error}}{SS_{total}}$$

The squared correlation (r^2) is called the *coefficient of determination*. That is, r^2 is the proportion of variance in Y determined or explained by X. The range of possible values for r^2 is from 0 to 1; r^2 is always positive, because even a negative correlation becomes positive when it is squared.

The complementary quantity $1 - r^2$ is called the *coefficient of nondetermination*. That is, the proportion of variance in Y that is not explained by X is $1 - r^2$:

$$1 - r^2 = \frac{SS_{error}}{SS_{total}}$$

For the sentencing data, using calculations from Table 11.2,

$$r = \frac{SP}{\sqrt{SS_X SS_Y}}$$

$$= \frac{300}{\sqrt{(100)(1,250)}}$$

$$= \frac{300}{\sqrt{125,000}}$$

$$= \frac{300}{353.55}$$

$$= .85$$

The coefficient of determination is then

$$r^2 = (.85)^2 = .72$$

Thus, 72% of the variance in sentence length is explained by the number of priors. This agrees with the results of the long method from the last section.

The coefficient of nondetermination is

$$1 - r^2 = 1 - (.85)^2 = 1 - .72 = .28$$

Thus, 28% of the variance in sentence length is not explained by priors. This 28% residual could be a result of other factors concerning the defendant or his crime. Some portion of the 28% could even be random error—that is, error that cannot be attributed to any factor. We'll discuss this in more detail later in the chapter.

Regression and Analysis of Variance

The focus on explained and unexplained variance may remind you of analysis of variance in Chapter 8. In that chapter we decomposed the total sum of squares (SS_{total}) into sum of squares between groups ($SS_{between}$) and sum of squares within groups (SS_{within}). In regression, we decompose the total sum of squares into regression sum of squares (SS_{reg}) and error sum of squares (SS_{error}). In fact, there are very strong similarities between analysis of variance and regression analysis. In both we attempt to account for one variable in terms of another. In analysis of variance the independent variable is categorical or in groups (such as social class or religion), whereas in regression the independent variable is at the interval level (such as number of prior convictions or height).

Fortunately, it is not necessary to calculate a predicted value for every respondent, as we did earlier, to decompose the total variation in the dependent variable into portions explained and not explained by the independent variable. Using the coefficients of determination and nondetermination as the proportions of explained and unexplained variation, we can quickly decompose the total sum of squares (SS_{total} or SS_Y) using these formulas:

$$SS_{reg} = r^2 SS_{total}$$
$$SS_{error} = (1 - r^2) SS_{total}$$

Just as in Chapter 8, an analysis of variance summary table is a convenient way of presenting the results of regression analysis. In Table 11.3, for example, we display under

TABLE 11.3 *Analysis of Variance Summary Table for Sentencing Data*

Source of Variation	SS	df	MS	F
Regression	900	1	900.00	20.57
Error	350	8	43.75	
Total	1,250	9		

the heading "Source of Variation" the regression, error, and total sum of squares for our regression of sentence length on the number of priors. Next the regression and error sums of squares are associated with degrees of freedom. The regression sum of squares has only 1 degree of freedom:

$$df_{reg} = 1$$

For the error sum of squares,

$$df_{error} = N - 2$$

where N is the sample size.

As in Chapter 8, we can next calculate the *mean square regression* (MS_{reg}) and *mean square error* (MS_{error}) by dividing the sums of squares by their respective degrees of freedom:

$$MS_{reg} = \frac{SS_{reg}}{df_{reg}}$$

$$MS_{error} = \frac{SS_{error}}{df_{error}}$$

Finally, by dividing the mean square regression by the mean square error, we obtain an F ratio for testing the significance of the regression—that is, whether the regression explains a significant amount of variation:

$$F = \frac{MS_{reg}}{MS_{error}}$$

To determine if the calculated F ratio is significant, it must exceed the critical value in Appendix C, Table D for 1 and $N - 2$ degrees of freedom.

For our results in Table 11.3 for the sentencing data,

$$F = \frac{900.00}{43.75} = 20.57$$

In Table D, we find that for $\mathbf{a} = .05$ with 1 and 8 degrees of freedom, the critical F is 5.32. Thus, the number of prior convictions explains a significant portion of the variance in sentence length.[1]

BOX 11.1 • *Step-by-Step Illustration: Regression Analysis*

To review the steps of regression analysis, let's reconsider the height–weight data of Chapter 10. In this illustration, we will use computational raw-score formulas for sums of squares to simplify the calculations. Height *(X)*, the independent variable, and weight *(Y)*, the dependent variable, for the eight children are given in Table 11.4, along with the squares and products.

Step 1   Calculate the mean of X and the mean of Y.

$$\bar{X} = \frac{\Sigma X}{N} = \frac{432}{8} = 54$$

$$\bar{Y} = \frac{\Sigma Y}{N} = \frac{720}{8} = 90$$

Step 2   Calculate SS_X, SS_Y, and SP.

$$\begin{aligned} SS_X &= \Sigma X^2 - N\bar{X}^2 \\ &= 23{,}460 - (8)(54)^2 \\ &= 23{,}460 - 23{,}328 \\ &= 132 \end{aligned}$$

$$\begin{aligned} SS_Y &= \Sigma Y^2 - N\bar{Y}^2 \\ &= 65{,}002 - (8)(90)^2 \\ &= 65{,}002 - 64{,}800 \\ &= 202 \end{aligned}$$

$$\begin{aligned} SP &= \Sigma XY - N\bar{X}\bar{Y} \\ &= 38{,}980 - (8)(54)(90) \\ &= 38{,}980 - 38{,}880 \\ &= 100 \end{aligned}$$

---

[1]This F test of the explained variance is equivalent to the test for the significance of the Pearson's correlation presented in Chapter 10. In fact, with one independent variable as we have here, $F = t^2$.

TABLE 11.4 *Calculations for Height–Weight Data*

X	Y	X^2	Y^2	XY	
49	81	2,401	6,561	3,969	$N = 8$
50	88	2,500	7,744	4,400	$\Sigma X = 432$
53	87	2,809	7,569	4,611	$\Sigma Y = 720$
55	99	3,025	9,801	5,445	$\overline{X} = 54$
60	91	3,600	8,281	5,460	$\overline{Y} = 90$
55	89	3,025	7,921	4,895	$\Sigma X^2 = 23,460$
60	95	3,600	9,025	5,700	$\Sigma Y^2 = 65,002$
50	90	2,500	8,100	4,500	$\Sigma XY = 38,980$
432	720	23,460	65,002	38,980	

Step 3 Determine the regression line.

$$b = \frac{SP}{SS_X}$$
$$= \frac{100}{132}$$
$$= .76$$

$$a = \overline{Y} - b\overline{X}$$
$$= 90 - (.76)(54)$$
$$= 48.96$$

$$\hat{Y} = 48.96 + .76X$$

Step 4 Determine correlation and coefficients of determination and nondetermination.

$$r = \frac{SP}{\sqrt{SS_X SS_Y}}$$
$$= \frac{100}{\sqrt{(132)(202)}}$$
$$= \frac{100}{163.29}$$
$$= .6124$$

Thus,

$$r^2 = (.6124)^2$$
$$= .3750$$
$$= 38\%$$

(continued)

BOX 11.1   Continued

$$1 - r^2 = 1 - (.6124)^2$$
$$= 1 - .3750$$
$$= .6250$$
$$= 62\%$$

Step 5   Calculate SS_{total}, SS_{reg}, and SS_{error}.

$$SS_{total} = SS_Y = 202$$

$$SS_{reg} = r^2 SS_{total} = (.3750)(202) = 75.75$$

$$SS_{error} = (1 - r^2) SS_{total} = (.6250)(202) = 126.25$$

Step 6   Calculate regression mean square and error mean square.

$$df_{reg} = 1$$

$$MS_{reg} = \frac{SS_{reg}}{df_{reg}}$$

$$= \frac{75.75}{1} = 75.75$$

$$df_{error} = N - 2 = 8 - 2 = 6$$

$$MS_{error} = \frac{SS_{error}}{df_{error}}$$

$$= \frac{126.25}{6} = 21.04$$

Step 7   Calculate F and compare with the critical value from Table D.

$$F = \frac{MS_{reg}}{MS_{error}}$$

$$= \frac{75.75}{21.04}$$

$$= 3.60$$

$$df = 1 \text{ and } 6$$

$$\alpha = .05$$

$$F = 5.99 \quad \text{(critical value from Table D)}$$

Because the calculated F is smaller than the critical value, height does not explain a significant amount of variance in weight. Although the correlation was fairly strong ($r = +.61$), the sample was too small to obtain significant results.

Step 8   Construct an analysis of variance summary table.

Source of Variation	SS	df	MS	F
Regression	75.75	1	75.75	3.60
Error	126.25	6	21.04	
Total	202.00	7		

Multiple Regression

We have attempted to account for variation in sentence length on the basis of the criminal history of the defendants—that is, to predict sentence length from the number of prior convictions. Overall, 72% of the variance was explained and 28% was not explained by criminal history. This approach is called *simple regression*—one dependent variable and one independent variable—just as Pearson's correlation is often called simple correlation.

Naturally, prior convictions is not the only factor relevant in analyzing sentencing data. It would be foolish to suggest that sentence length depends on just one factor, even though as much as 72% of the variance was explained. We might want to test sentencing models that include such variables as the age of the defendant and whether the defendant had pled guilty or not guilty to the charge. Perhaps in doing so, we can account for even more of the 28% of variance that sentence length could not.

Let's consider the case of two predictors: prior convictions and age. We add to our set of data another independent variable, the age of the defendant *(Z)*.

Priors (X)	Age (Z)	Sentence (Y)
0	18	12
3	22	13
1	27	15
0	28	19
6	30	26
5	35	27
3	36	29
4	29	31
10	34	40
8	31	48

The means, variances, and standard deviations calculated for all three variables as well as their inter-correlations, are as follows:

	Mean	Variance	Standard Deviation	Correlation Matrix Priors (X)	Correlation Matrix Age (Z)	Correlation Matrix Sentence (Y)
Priors	4	10	3.16	1.00	.59	.85
Age	29	29	5.39	.59	1.00	.69
Sentence	26	125	11.18	.85	.69	1.00

One might be tempted also to use the simple regression approach for predicting sentence based on age, for example, and then combine the results with those from regressing sentence on priors performed earlier. Unfortunately, this would produce erroneous results. By squaring the correlations, priors explains $(.85)^2 = .72 = 72\%$ of the variance in sentence length, and age explains $(.69)^2 = .48 = 48\%$ of the variance in sentence length. If we tried to add together these percentages, it would exceed 100%, which is impossible.

The problem is that, to some extent, age and priors overlap in their abilities to explain sentence: Older defendants have accumulated more priors during their lifetimes. Therefore, because age and priors are themselves correlated ($r = .59$), part of the percentage of variance in sentences explained by priors is also explained by age, and vice versa. By adding the proportions that they each explain, a certain portion is double counted.

The situation is illustrated in Figure 11.5, using overlapping circles to represent the three variables and their shared variance. Notice that priors covers 72% of the sentence circle, reflecting the explained variance. Age covers 48% of the sentence circle, reflecting its explained variance. For these to be possible, however, priors and age must, to some extent, overlap. The objective in multiple regression is to determine what percentage of variance in sentence is explained by priors and age together—that is, what percentage of the sentence circle is eclipsed by priors, age, or both.

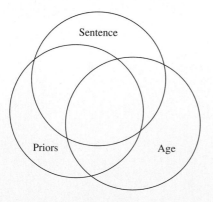

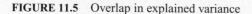

FIGURE 11.5   Overlap in explained variance

Multiple linear regression (or just *multiple regression*) is a generalization of simple regression when one uses two or more predictors. The regression model introduced at the start of this chapter for predicting Y from X extends to predicting Y from a linear combination of X and Z:

$$Y = b_0 + b_1 X + b_2 Z + e$$

where b_0 (rather than a) is used for the Y-intercept, b_1 and b_2 for the slopes of X and Z, respectively, and e for the residual or error term. Similar to the case of one predictor, the predicted values of the dependent variable are

$$\hat{Y} = b_0 + b_1 X + b_2 Z$$

The Y-intercept or constant term, b_0, is the expected value of Y when both X and Z are 0. The slope b_1 reflects the expected change in Y for every unit increase in X, holding constant Z. Similarly, the slope b_2 reflects the expected change in Y for every unit increase in Z, holding constant X.

The calculations for estimating the multiple regression coefficients are complex, especially when the number of predictors is more than just a few, and so computers are almost always needed. Moreover, it entails a number of issues and complications that are beyond the scope of this book. But just to whet your appetite for your next course in statistics, the case of two predictors can be worked out on a calculator, yet includes all the concepts of more elaborate multiple regression analyses for which computers are essential.

The formulas for calculating regression coefficients with two predictor variables (such as priors and age) are a bit complex. The regression coefficient (or slope) for each predictor must be adjusted for whatever explanatory overlap exists with the other predictor. The adjustment is rather similar to the partial correlation calculation introduced in Chapter 10.

For the case of two predictor variables, the regression coefficients, including the Y-intercept, are given by:

$$b_1 = \frac{s_Y}{s_X}\left(\frac{r_{YX} - r_{YZ}r_{XZ}}{1 - r_{XZ}^2}\right)$$

$$b_2 = \frac{s_Y}{s_Z}\left(\frac{r_{YZ} - r_{XY}r_{XZ}}{1 - r_{XZ}^2}\right)$$

$$b_0 = \bar{Y} - b_1\bar{X} - b_2\bar{Z}$$

Based on the means, standard deviations and correlations among sentence, priors, and age, we can calculate

$$b_1 = \frac{11.18}{3.16}\left(\frac{.85 - (.69)(.59)}{1 - (.59)^2}\right)$$

$$= 3.54\left(\frac{.4439}{.6519}\right)$$

$$= 3.54(.6794)$$
$$= 2.395$$

$$b_2 = \frac{11.18}{5.39}\left(\frac{.69 - .85(.59)}{1 - (.59)^2}\right)$$

$$= 2.07\left(\frac{.1885}{.6519}\right)$$

$$= .605$$

$$b_0 = 26 - 2.395(4) - .605(29)$$
$$= -1.132$$

With these results, the regression equation predicting sentence *(Y)* on the basis of priors *(X)* and age *(Z)* is as follows:

$$\hat{Y} = -1.132 + 2.395X + .605Z$$

Notice that each of the two predictors has its own regression coefficient or slope. Again, these slopes indicate the expected change in the dependent variable associated with a unit increase in a given predictor, holding constant the other predictor.

Based on these results, we can say that each additional prior *(X)* tends to carry a 2.395-month increase in sentence length *(Y)*, holding constant defendant age *(Z)*. For example, given two defendants of the same age, but where one has two more priors than the other, the first defendant can be expected to receive a sentence that is 2(2.395) or about 4.8 months longer than the other defendant.

Also, based on these results, we can say that each additional year of age *(Z)* tends to be associated with a .605-month increase in sentence length *(Y)*, holding constant the number of priors *(X)*. If two defendants have similar criminal records, but one is 5 years older, this defendant can be expected to receive a sentence length that is 5(.605) or about 3 months longer than the other defendant.

Finally, the *Y*-intercept represents the base line, or expected sentence length, when both predictors are equal to zero. It shouldn't bother us in the least that the *Y*-intercept here is negative (−1.132 months); after all, how often do we encounter a newborn defendant with no priors?

An overall measure of fit of the actual *Y* values to those predicted by the regression line is given by the *multiple coefficient of determination* (also known as the *squared multiple correlation*). Symbolized by R^2 (in contrast to r^2 as in Pearson's coefficient for bivariate relationships), the multiple coefficient of determination is the proportion of variance in a dependent variable explained by a set of independent variables in combination. For two independent variables (*X* and *Z*),

$$R^2 = \frac{r_{YX}^2 + r_{YZ}^2 - 2r_{YX}r_{YZ}r_{XZ}}{1 - r_{XZ}^2}$$

Based on the correlations among sentence, priors, and age,

$$R^2 = \frac{(.85)^2 + (.69)^2 - 2(.85)(.69)(.59)}{1 - (.59)^2}$$

$$= \frac{.7225 + .4761 - .6921}{1 - .3481}$$

$$= \frac{.5065}{.6519}$$

$$= .78$$

Thus, age and number of prior convictions together explain 78% of the variance in sentence length. Apparently, the inclusion of age adds only a modest amount to the overall prediction of sentence length. Because the number of prior convictions by itself explains 72%, the inclusion of age as an additional variable increases the percentage of explained variance by only 6%.

Dummy Variables

All three variables used in the regression analysis above are measured at the interval level, an assumption in regression analysis. As with other parametric techniques covered in this book, it is generally safe to use ordinal variables so long as the points along the continuum are approximately evenly spaced, as in a 7-point scale from strongly agree to strongly disagree. Nominal variables cannot be used as predictors, unless they are converted into a special kind of variable, called a *dummy variable,* having only two values and generally coded as 0 and 1.

To illustrate, suppose we wanted to include another variable, trial, coded as 1 if the defendant was convicted following a not-guilty plea and 0 if the defendant was sentenced after a guilty plea and no trial. If coded in this fashion, the coefficient of the dummy variable trial will represent the average difference in sentence for defendants convicted after a not guilty plea and those pleading guilty, holding constant the other predictor variables.

Priors (X)	Age (Z)	Sentence (Y)	Trial (D)
0	18	12	1
3	22	13	0
1	27	15	0
0	29	19	1
6	30	26	0
5	35	27	1
3	36	29	1
4	29	31	0
10	34	40	0
8	31	48	1

With more than two predictors, the calculations for multiple regression are so elaborate that they are virtually impossible to accomplish without a computer and special statistical software (like SPSS). But the interpretation of the results is a direct extension of the two-predictor case.

Let's examine the most important portions of the output produced by SPSS for regressing (or predicting) sentence on priors, age, and trial. Table 11.5 reports regression coefficients, their respective standard errors, *t*-values, and significance levels.

Shown in the first column, each of the regression coefficients (*b*) represents the expected change in *Y* corresponding to a unit increase in that predictor while holding constant all other predictors. The coefficient for priors, after controlling for age and trial, is 2.866. Thus, each additional prior offense tends to result in a 2.866 months increase in sentence length, controlling for age and whether the defendant went to trial. Suppose, for example, that hypothetical defendants A and B were the same age, and both had entered the same plea; but A had one more prior than B. Then one would expect A to receive a 2.866 month longer sentence than B.

The other two coefficients would be interpreted in a similar fashion. Every year of age is expected to increase the sentence by .379 months, controlling for both priors and trial status. Finally, the coefficient for the dummy variable trial indicates that the sentence expected from a trial (if convicted) tends to be 6.061 months longer than that from a guilty plea, assuming the other two variables are held constant.

The *t* ratios are calculated by dividing the coefficient by its corresponding standard error. The *t* ratio (with $N - k - 1$ degrees of freedom, where *k* is the number of predictors) allows us to test if the regression coefficient is significantly different from 0 (the null hypothesis of no effect). We could compare each of the *t* values against the critical value of 2.447 found in Table B, in Appendix C for 6 degrees of freedom (here $N = 10$ and $k = 3$ predictors) and a .05 level of significance. In this case, only priors has a significant effect on sentence length. Alternatively, most regression software programs calculate the exact *p* value for each coefficient (denoted by "Sig."), representing the exact probability of obtaining a regression coefficient of at least this size if the null hypothesis of no effect were true. For priors, the exact probability $(p = .010)$ is quite low and certainly below the .05 significance level. For the other two predictors, however, we fail to reject the null hypothesis of no effect.

It may seem a bit odd that the coefficient for priors (2.866) is significant while a much larger coefficient for trial (6.130) is not. This reflects the fact that an increase of one unit in trial (from guilty plea to conviction from a trial) is a major difference in criminal

TABLE 11.5 *Multiple Regression of Sentence Data*

	Unstandardized Coefficients		Standardized Coefficients		
	b	*Std. Error*	*Beta*	*t*	*Sig.*
Constant	.469	10.986		.043	.967
Priors	2.866	.781	.811	3.670	.010
Age	.379	.446	.183	.850	.428
Trial	6.130	4.023	.274	1.524	.178

procedure, while an increase of one prior is not so dramatic. The reason why it is difficult to compare the numerical values of the two regression coefficients is that the underlying variables have rather different standard deviations (or scales).

The column headed by standardized coefficient provides a measure of the impact of each predictor when placed on the same (standardized) scale. This makes substantive interpretation less meaningful. Each standardized coefficient represents the expected change in the dependent variable in terms of standard deviation units corresponding to a one-standard-deviation increase in the predictor while holding constant the other predictors. Although not particularly illuminating, at least the relative sizes of the standardized coefficients indicate their relative strengths in predicting the dependent variable. The .811 standardized coefficient for priors is much larger than those for the other two variables, clearly indicating the greater importance of this variable in predicting sentence lengths.

As in the case of regression with one predictor, we can divide the overall or total sum of squares in the dependent variable into two parts (the portions explained and unexplained by the group of predictor variables), and summarize and test the results using an analysis of variance table, as shown in Table 11.6.

Previously, when using only priors as a predictor, we were able to explain 900 of the total 1,250 sum of squares. With three predictors, this improves to 1047.843, leaving 202.157 as error (or variation in sentence length unaccounted for by the three variables). As before, we obtain the mean square regression and mean square error by dividing the sums of squares by their corresponding degrees of freedom. Finally, the F ratio, found by dividing the mean square regression by the mean square error (or residual), can be compared to the critical value from Table D in Appendix C (4.76 for numerator df = 3, denominator df = 6, and a .05 level of significance). Alternatively, we can simply determine if the exact significance value ($p = .009$) provided by the software is less than our .05 significance level. Either way, $F = 10.367$ is statistically significant, allowing us to reject the null hypothesis of no predictability or effect within the entire regression equation.

Finally, the sums of squares can, as before, be used to calculate the coefficient of determination or squared multiple correlation:

$$R^2 = \frac{SS_{total} - SS_{error}}{SS_{total}} = \frac{SS_{reg}}{SS_{total}}$$

$$= \frac{1047.843}{1250}$$

$$= .838$$

TABLE 11.6 *Analysis of Variance Table for Regression of Sentence Data*

	Sum of Squares	df	Mean Square	F	Sig.
Regression	1047.843	3	349.281	10.367	.009
Error	202.157	6	33.693		
Total	1250.000	9			

Thus, while priors alone explained 72% of the variance in sentence length, the group of three variables can explain 84%. And based on the F ratio from the analysis of variance table ($F = 20.367$), we can conclude that the proportion of variance explained by the regression model, this $R^2 = .838$, is significant.

Interaction Terms

The regression coefficients for priors, age, and trial reflect the independent or separate effect of each variable holding constant the other two. These impacts are sometimes known as *main effects*. It is also possible for variables to have interaction effects—that is, for the size of the effect of one independent variable on the dependent variable to depend on the value of another independent variable.

To illustrate the distinction between main and interaction effects, consider a study that examines the benefits of diet and exercise on weight loss. Suppose that dieting alone produces an average weight loss of 2 pounds for the week and that an hour of daily exercise alone produces an average weight loss of 3 pounds for the week. What results would be expected if combining diet with exercise? Would it be the sum of the diet and exercise effects, that is $2 + 3 = 5$ pounds? That would be true if there is no interaction. An interaction would exist if diet and exercise produced, say, an average weight loss of 6 pounds. In other words, the effect of dieting increases with exercise and the effect of exercise increases with diet.

Multiple regression can easily incorporate and test interaction effects by including the product of variables in the regression model. For example, we could create a new variable, priors $\times$ trial, by multiplying priors times trial. The revised regression results are as shown in Table 11.7. The interpretation would be that each prior offense increases the expected sentence by 2.490 years if there is no trial (when trial $= 0$, the interaction variable is zero and drops out of the equation). For those defendants who go to trial, trial $= 1$ and, thus, priors $\times$ trial $=$ priors), and the effect of each prior is $2.490 + 0.801 = 3.291$. Thus, although the interaction effect is not statistically significant, particularly with such a small sample, going to trial increases the effect of each prior offense by .801 of a year, or about 9 months.

TABLE 11.7   *Multiple Regression of Sentence Data with Interaction*

	Unstandardized Coefficients		Standardized Coefficients		
	b	*Std. Error*	*Beta*	*t*	*Sig.*
Intercept	2.591	12.028		.215	.838
Priors	2.490	1.007	.704	2.472	.056
Age	.368	.463	.177	.795	.463
Trial	2.905	6.536	.130	.445	.675
Priors $\times$ Trial	.801	1.262	.193	.635	.554

Logistic Regression

It should be fairly clear that multiple regression is an extremely useful tool for predicting a wide variety of characteristics, such as sentence lengths for convicted criminals, test scores for undergraduate majors, weight loss for dieters and exercisers—virtually any variable that is measured at the interval level. But what is the social researcher to do when predicting categorical variables—especially dichotomous ones—such as whether someone supports or opposes the legalization of gay marriage, which of two major party candidates will carry various states in a presidential election, or whether survey respondents report owning a handgun. While these types of variables can be used as dummy variable predictors, if that were the research objective, there would be problems in using them as dependent variables in linear regression.

Consider the following data for a randomly selected sample of applicants to a particular college, data that include sex (gender) coded as 0 for females and 1 for males, high school grade average (HSavg) measured on a typical 100-point scale, and admissions outcome (decision) coded as 0 for rejected and 1 for admitted.

Gender	HSavg	Decision
1	76	0
0	78	0
1	81	0
1	83	0
1	84	0
1	85	0
0	86	0
0	87	0
0	86	1
1	87	1
1	88	0
1	89	1
0	89	0
1	90	0
1	90	1
0	91	1
1	92	0
1	93	0
0	92	1
0	94	0
0	96	1
0	97	1
0	98	1
1	98	1
0	99	1

Let's focus first on the impact of HSavg on decision and the scatter plot in Figure 11.6. The scatter of points looks rather unusual—two horizontal strings of dots, the upper

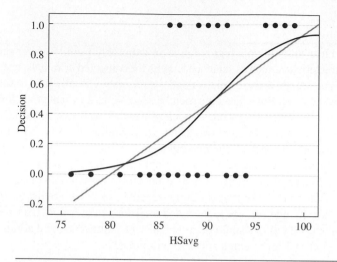

FIGURE 11.6   Linear and logistic regression

for admitted applicants and the lower for rejected applicants. It certainly appears that HSavg can, to some extent, predict decision. The HSavg scores for the admitted applicants (the upper string) tend to be further to the right (mean HSavg of 93.0) than those for the rejected applicants (mean HSavg of 86.1). Were we to apply traditional regression methods, the gray regression line (shown in Figure 11.6) would yield some curious results.

The slope of the line, $b = 0.047$, would suggest that every point increase in HSavg would increase decision by 0.047. Because decision has only two values (0 for reject and 1 for accept), we could interpret the regression coefficient in terms of improving the probability of being admitted. Specifically, the chance of being admitted increases by 0.047 for every unit increase in HSavg. Continuing with this logic, there would be some applicants whose predicted probability of admission would fall below 0, as in fact happens for applicants with HSavg values less than 80. While they may have rather limited chances of being admitted, it certainly would be improper to say that they are less than none. At the other extreme, an applicant with a perfect HSavg of 100 would almost assuredly be admitted, but with straight-line regression, nothing prevents the predicted probability from exceeding 1.

A bit more subtle, it would also seem implausible that the benefit of increasing HSavg is the same at all levels. We might expect, for example, that the difference between 85 and 88 would boost the probability of admission more than the difference between 95 and 98. The extent of positive impact of HSavg on the likelihood of admission would likely diminish a certain point.

The trouble is that linear regression is not suited for dependent variables constrained to be either 0 or 1. A more reasonable model would be one where below some HSavg level the chance of being admitted is virtually 0, then the chances increase for some mid-range levels of HSavg, and then approach a maximum of virtual certainty for especially high HSavg. The S-shaped curve shown in Figure 11.6, known as a *logistic curve*, captures this relationship between HSavg and decision. Logistic regression is the preferred approach for predicting dichotomous variables like decision.

Without getting bogged down in the mathematical details, some basic understanding of the concept of *odds* is needed to interpret logistic regression results. The odds of some event is defined as the probability that event will occur divided by the probability it will not occur:

$$Odds = \frac{P}{1 - P}$$

where P is the usual probability of occurrence. Odds are typically expressed against a benchmark of 1, as in "something-to-one." Thus, if an event has a $P = .75$ chance of occurring, its odds are 3-to-1 $(.75/.25 = 3)$. Alternatively, an event with only a $P = .25$ probability, has odds of .33-to-1 $(.25/.75 = .33)$. A 50/50 proposition, often called "even odds," has odds of 1-to-1 $(.50/.50 = 1)$.

Unlike probability, such as the probability of admission, which varies from 0 to 1, the odds of an event has no ceiling or maximum. Odds can range from 0 (impossible) to infinity (a virtual certainty). Odds are typically used, moreover, to make comparisons between groups. For example, we might contrast the odds of being admitted between males and females based on the cross-tabulation in Table 11.8 of decision by gender. Of the 12 female applicants, $7/12 = 58.33\%$ were admitted, and $5/12 = 41.67\%$ were rejected, so that the odds of admission were $58.33/41.67 = 1.4$, or far more directly from dividing the number admitted by the number rejected, $7/5 = 1.4$. For the 13 male applicants, $4/13 = 30.77\%$ were admitted and $9/13 = 69.33\%$ were rejected. The odds for males are $30.77/69.33$ (or $4/9$) $= 0.44$.

The *odds ratio* is defined as the odds for one group divided by the odds for another group. Here, the odds of admission for females divided by the odds of admission for males is $1.4/0.44 = 3.15$. In other words, females have 3.15 *times* the odds of being admitted than do males. This, of course, begs the question of why females enjoy more than three times the odds of being admitted? Are admission decisions at this college unfair and discriminatory? Or perhaps does the greater odds for the female applicant reflect differences in their academic profiles and not just their gender.

This odds–ratio approach works fine for comparing one group to another, such as the odds of admission for females to that of males. However, when examining the impact of several variables simultaneously (including interval-level predictors) on the odds of some event (for example, how gender and high school grade average impact the odds of being admitted), a regression approach is preferred.

TABLE 11.8 *Cross-Tabulation of Decision by Gender*

		Gender		
		0 = Female	*1 = Male*	**Total**
Decision	**0 = Rejected**	5	9	14
	1 = Admitted	7	4	11
	Total	12	13	25

Logistic regression accomplishes the same objectives with dichotomous dependent variables as multiple linear regression does for interval-level outcomes. The big difference is that logistic regression models the natural logarithm of the odds, thus the *log* in "logistic." The dependent variable is transformed into what is known as the log-odds, or *logit*, and is symbolized by *L*,

$$L = \log\left(\frac{P}{1 - P}\right)$$

The logit *L* has no bounds whatsoever. It ranges from minus infinity (for a virtual impossibility) to positive infinity (a virtual certainty). Moreover, negative values of *L* reflect less than even odds, an *L* of 0 reflects even odds, and positive values of *L* represent greater than even odds.

Fortunately, it is not necessary to work with logarithms in order to use this technique, except to know that logistic regression involves predicting the log-odds of a dependent variable, like decision, based on a linear combination of independent variables. The logistic regression equation for two predictor variables, for example, would be

$$L = b_0 + b_1 X_1 + b_2 X_2 + e$$

where b_0 is the constant (or *Y*-intercept in terms of the log-odds), b_1 and b_2 are the coefficients for predictor variables X_1 and X_2, and *e* is the residual or error term. Predicted log-odds would look the same but without the error term:

$$\hat{L} = b_0 + b_1 X_1 + b_2 X_2$$

Finally, the anti-log applied to each side gives us a prediction of the odds:

$$\text{Predicted } Odds = e^{b_0 + b_1 X_1 + b_2 X_2} = e^{b_0} e^{b_1 X_1} e^{b_2 X_2}$$

where $e = 2.718$ is the special constant known as Euler's number. Furthermore, the probability of admission can be predicted for any applicant given HSavg and gender by

$$\hat{P} = \frac{e^{b_0 + b_1 X_1 + b_2 X_2}}{1 + e^{b_0 + b_1 X_1 + b_2 X_2}}$$

This rather formidable-looking equation actually produces the S-shaped curve used to predict admission decisions.

It is also not at all critical that one understand these algebraic manipulations involving logs and anti-logs. Moreover, the process of estimating the logistic regression coefficients b_0, b_1, and b_2 is quite complex, as is their interpretation. All that we really need to determine is whether or not they are statistically significant. However, the exponentiation of these coefficients (e^b) in the predicted odds equation provides a fairly straight-forward approach to interpreting the effect of each predictor.

TABLE 11.9   *Logistic Regression Results for Predicting Admission*

Independent Variables	*b*	S.E.	Wald	df	Sig.	Exp(b)
Constant	−24.105	10.911	4.881	1	.027	.000
HSavg	.269	.120	5.019	1	.025	1.309
Gender	−.562	1.011	.309	1	.578	.570

Table 11.9 provides the logistic regression estimates from predicting admission based on HSavg and gender. The coefficients *(b)* and their standard errors (S.E.) are used to determine the statistical significance of the respective independent variables. The Wald statistic *(W)*, used to determine statistical significance, is obtained from the square of a logistic regression coefficient divided by its standard error:

$$W = \left(\frac{b}{SE_b}\right)^2$$

W is treated like a chi-square value (with 1 degree of freedom) to assess when the null hypothesis of no effect in the population can be rejected. The significance value shown in the table is then the probability of obtaining a logistic regression coefficient at least as large as that estimated from the sample data by chance alone if the true effect in the population is zero. We see from the table that HSavg is a significant predictor at the .05 level ($p = .025$). However, the coefficient for gender is not statistically significant ($p = .578$) since the likelihood of obtaining a *b* value as large in magnitude as −.562 purely as a result of sampling is quite high ($p = .578$). Thus, what initially appeared to be a decided advantage for females over males largely disappears once HSavg is entered into the logistic regression.

Significant or not, we still cannot easily interpret the estimated *b* coefficients because they represent changes in log-odds. However, the far-right column of the table gives the exponentiated coefficients which can be interpreted in terms of a variable's effect on the odds of the dependent variable, like we did previously with the impact of gender alone on decision. Specifically, the e^b values indicate the multiplier by which each unit increase in the independent variable changes the predicted odds of the dependent variable, holding constant the other variables.

If $b > 0$, then $e^b > 1$, thereby magnifying the odds (multiplying by a factor greater than 1). On the other hand, if $b < 0$, then $e^b < 1$, thereby diminishing the odds (multiplying by a factor less than 1). Finally, if $b = 0$, then $e^b = 1$, thereby leaving the odds unchanged (multiplying by 1).

Consider the results contained in Table 11.9. For HSavg, $e^b = 1.309$, which indicates that every one-point increase in HSavg tends to magnify the odds of being admitted by a factor of 1.309, that is, about a 30% increase in the odds. Also, a two-point increase would predict a change of multiplying the odds twice by 1.301. A five-point increase, say from 85 to 90, would be associated with multiplying the odds by 1.301 five times, or by 3.73, thereby almost quadrupling the odds.

For gender, here are only two possible values, of course, 0 for female and 1 for male. The odds-multiplier, $e^b = .570$, suggests that the odds of admission for males is less than that for females, because the odds are reduced when multiplying by a number less than one. Specifically, a male applicant would tend to have 43% lower odds of being admitted as a female applicant with an identical grade average (multiplying a number by .57 reduces it by 43%). Though this finding could be disturbing for many potential applicants, remember that the estimated coefficient for gender was not statistically significant, leaving wide open the possibility that the apparent gender difference was a mere consequence of sampling error. Finally, this impact of gender holding constant HSavg, though in the same direction, is far less than we previously calculated for gender alone.

Finally, with mixed results in terms of statistical significance of the predictors, how well does the model as a whole predict decisions on admission? There is, unfortunately, no exact R^2 statistic, as in linear regression, to indicate what percentage of the variance in the dependent variable is explained by the independent variables in combination. However, logistic regression does have several options for assessing the overall fit, the most widely used of which is Nagelkerke's R^2, which varies from 0 for no fit to 1 for perfect fit. In our case, the SPSS results indicate that Nagelkerke's $R^2 = .442$, reflecting a fairly good fit, even with only two variables. Were we to include other predictors, such as SAT or ACT scores, the predictive fit would most likely improve.

Summary

In this chapter, we extended the concept of correlation to situations in which a social researcher is concerned with the effect of one variable (the independent variable) on another (the dependent variable). Whereas Pearson's r measures the strength and direction of correlation, the regression line, comprised of a slope and Y-intercept, specifies the exact nature of the impact of the independent variable on the dependent variable. In regression analysis, the value of the dependent variable (denoted Y) is predicted on the basis of the independent variable (denoted X). The proportion of variance in Y explained by X (known as the coefficient of determination) is separated from the proportion of variance in Y not explained by X (known as the coefficient of nondetermination). The procedure known as multiple regression goes beyond the simple regression technique introduced in this chapter to incorporate not just one but two or more predictors simultaneously. In this case, the regression coefficients (slopes) indicate the effect of one independent variable on the dependent variable, holding constant the other predictor variables in the model. In addition, the multiple coefficient of determination assesses the proportion of variance in the dependent variable accounted for by the set of independent variables in combination. Although multiple regression is generally used with interval-level variables, dichotomous predictors—known as dummy variables—can be incorporated as well. In addition, interaction effects can be included by multiplying predictors. Finally, logistic regression provides a very useful extension to the case of predicting dichotomous dependent variables. Logistic regression focuses on the odds of an event, and the extent to which various predictor variables magnify or reduce the odds of one outcome over another.

Terms to Remember

Regression analysis
Y-intercept
Slope (regression coefficient)
Error term
Regression line
Error sum of squares
Regression sum of squares
Coefficient of determination

Coefficient of nondetermination
Mean square regression
Mean square error
Multiple regression
Dummy variable
Odds and odds ratio
Logit
Logistic regression

Questions and Problems

1. In regression analysis, we predict the value of *Y* from *X* based on a
   a. significant correlation coefficient.
   b. a mathematical equation.
   c. a weak positive correlation.
   d. a normal curve.

2. Which of the following denotes the slope or the regression coefficient for *X*?
   a. The *Y*-intercept
   b. The term *a*
   c. The error term
   d. The term *b*

3. The increase or decrease in *Y* expected with each unit change in *X* is known as
   a. the *Y*-intercept, *a*
   b. the error term, *e*
   c. the regression coefficient, *b*.
   d. PRE.

4. A measure of the proportion of the variance in *Y* explained by *X* is known as
   a. the coefficient of nondetermination.
   b. the coefficient of determination.
   c. multiple regression.
   d. the slope.

5. Using two or more predictors, we conduct a
   a. simple regression.
   b. multiple regression.
   c. partial correlation.
   d. Pearson's correlation.

6. Suppose that a researcher collected the following set of data on years of education *(X)* and number of children *(Y)* for a sample of 10 married adults:

X	Y	X	Y
12	2	9	3
14	1	12	4
17	0	14	2
10	3	18	0
8	5	16	2

a. Draw a scatter plot of the data.
b. Calculate the regression slope and *Y*-intercept.
c. Draw the regression line on the scatter plot.
d. Predict the number of children for an adult with 11 years of education.
e. Find the coefficients of determination and nondetermination. What do they mean?
f. Construct an analysis of variance table and perform an *F* test of the significance of the regression.

7. This same researcher then selected at random one child from each of the eight families in which there was at least one child. Interested in the effects of number of siblings *(X)* on happiness of the child *(Y)* on a scale, assumed to be interval, from 1 (very unhappy) to 10 (very happy), she had these data:

X	Y
1	5
0	3
2	4
4	7
2	5
3	9
1	8
1	4

a. Draw a scatter plot of the data.
b. Calculate the regression slope and *Y*-intercept.
c. Draw the regression line on the scatter plot.
d. Predict the happiness of an only child and of a child with two siblings.
e. Find the coefficients of determination and nondetermination. What do they mean?
f. Construct an analysis of variance table and perform an *F* test of the significance of the regression.

8. A legal researcher wanted to measure the effect of length of a criminal trial on the length of jury deliberation. He observed in a sample of 10 randomly selected courtroom trials the following data on length of trial (in days) and length of jury deliberation (in hours)

X (days)	Y (hours)
2	4
7	12
4	6
1	4
1	1
3	4
2	7
5	2
2	4
3	6

a. Draw a scatter plot of the data.
b. Calculate the regression slope and Y-intercept.
c. Draw the regression line on the scatter plot.
d. Predict the length of jury deliberation for a recently completed trial that lasted 6 days.
e. Find the coefficients of determination and nondetermination. What do they mean?
f. Construct an analysis of variance table and perform an F test of the significance of the regression.

9. A personnel specialist with a large accounting firm is interested in determining the effect of seniority (the number of years with the company) on hourly wages for secretaries. She selects at random 10 secretaries and compares their years with the company (X) and hourly wages (Y):

X	Y
0	12
2	13
3	14
6	16
5	15
3	14
4	13
1	12
1	15
2	15

a. Draw a scatter plot of the data.
b. Calculate the regression slope and Y-intercept.
c. Draw the regression line on the scatter plot.
d. Predict the hourly wage of a randomly selected secretary who has been with the company for 4 years.
e. Find the coefficients of determination and nondetermination. What do they mean?
f. Construct an analysis of variance table and perform an F test of the significance of the regression.
g. Based on these results, what is the typical starting wage per hour, and what is the typical increase in wage for each additional year on the job?

10. A communications researcher wanted to measure the effect of television viewing on aggressive behavior. He questioned a random sample of 14 children as to how many hours of television they watch daily (X) and then, as a measure of aggression, observed the number of schoolmates they physically attacked (shoved, pushed, or hit) on the playground during a 15-minute recess (Y). The following results were obtained:

X	Y
0	0
6	3
2	2
4	3
4	4

(continued)

X	Y
1	1
1	0
2	3
5	3
5	2
4	3
0	1
2	3
6	4

a. Draw a scatter plot of the data.
b. Calculate the regression slope and Y-intercept.
c. Draw the regression line on the scatter plot.
d. Predict the number of schoolmates attacked by a child who watches television 3 hours daily.
e. Find the coefficients of determination and nondetermination. What do they mean?
f. Construct an analysis of variance table and perform an F test of the significance of the regression.

11. Crime-related television shows and movies often paint a picture in which police must go "above the law" to catch criminals, implying that to do their jobs the police must circumvent the court's strict rules protecting offenders' civil liberties. Do people who watch these programs come to view the government as being too protective of criminals' rights? To find out, a researcher selected a random sample of people and asked them *(X)* how many hours of crime-related television they watch per week and *(Y)* the extent to which they agree with a statement that the government is too protective of criminals' rights (measured on a scale from 1 to 9, with 1 being "strongly disagree" and 9 being "strongly agree"). The following data were collected:

X	Y
2.0	3
5.5	4
9.5	6
12.0	8
0.0	1
10.5	9
1.5	2
3.0	3
2.0	3

a. Draw a scatter plot for the data.
b. Calculate the regression slope and Y-intercept.
c. Draw the regression line on the scatter plot.
d. Predict the attitude toward the government of someone who watches 7.5 hours of crime-related television per week.
e. Find the coefficients of determination and nondetermination. What do they mean?
f. Construct an analysis of variance table and perform an F test on the significance of the regression.

12. A researcher is interested in the association between *(X)* people's attitudes toward the war in Iraq and *(Y)* how much they support President George W. Bush. Just before the 2004 presidential election, she takes a random sample of voters and collects the following data (both variables being measured on a scale from 1 to 9, with 9 representing strongest support):

Person	X	Y
A	2	3
B	1	1
C	8	7
D	9	8
E	6	4
F	5	6
G	8	9
H	7	6

    a. Draw a scatter plot for the data.
    b. Calculate the regression slope and *Y*-intercept.
    c. Draw the regression line on the scatter plot.
    d. Predict the attitude toward President Bush of someone who scores a 3 on their attitude toward the war.
    e. Find the coefficients of determination and nondetermination. What do they mean?
    f. Construct an analysis of variance table and perform an *F* test on the significance of the regression.

13. Psychologist Erik Erikson suggested that individuals go through eight developmental stages throughout their lives. He believed that during stage 8 (age 65 to death), older adults will either fear or accept death depending on how much pleasure and satisfaction they feel they have experienced when reflecting back on their lives: a person who is satisfied with his or her life will accept death, whereas a person who looks back on life and sees failure and disappointment will fear death. To test this theory, a psychiatrist interviewed a random sample of people aged 65 and older and scored them on *(X)* how fulfilling they feel their lives have been and *(Y)* how much they fear death (both variables scored on a scale from 1 to 10, with a higher score indicating more fulfillment and more fear):

X	Y
7	4
5	6
2	8
9	2
2	7
10	3
6	3

    a. Draw a scatter plot of the data.
    b. Calculate the regression slope and *Y*-intercept.

c. Draw the regression line on the scatter plot.

d. Predict how fearful of death a person would be if she scored a 3 on life fulfillment.

e. Find the coefficients of determination and nondetermination. What do they mean?

f. Construct an analysis of variance table and perform an F test of the significance of the regression.

14. A psychiatrist wondered how much a night's sleep could affect a person's alertness and reaction time. She collected a random sample of people and asked them (X) how many hours of sleep they got the night before and then used a simple test to determine (Y) their reaction time in seconds. The following results were obtained:

X	Y
5.0	13
10.5	8
4.0	15
6.0	12
7.5	11
6.5	13
2.5	21
6.0	10
4.5	17

a. Draw a scatter plot of the data.

b. Calculate the regression slope and Y-intercept.

c. Draw the regression line on the scatter plot.

d. Predict the number of seconds it would take for someone who got 8.5 hours of sleep to react.

e. Find the coefficients of determination and nondetermination. What do they mean?

f. Construct an analysis of variance table and perform an F test on the significance of the regression.

15. Does the number of years that a woman is on birth control (X) have an effect on how long it takes for her to become pregnant (Y) once she ceases to use it and tries to have children? A curious researcher collected the following data for a sample of pregnant women who had previously used birth control:

Years on Birth Control	Time Until Pregnant (in days)
7.5	25
10.0	31
3.5	42
4.0	36
7.0	13
2.5	22
5.0	56

a. Draw a scatter plot of the data.

b. Calculate the regression slope and Y-intercept.

c. Draw the regression line on the scatter plot.

d. Can you predict how long it will take for a woman who was on birth control for 8 years to get pregnant once she stops using it?

e. Find the coefficients of determination and nondetermination. What do they mean?

f. Construct an analysis of variance table and perform an F test on the significance of the regression.

16. An educational researcher was interested in the effect of academic performance in high school on academic performance in college. She consulted the school records of 12 college graduates, all of whom had attended the same high school, to determine their high school cumulative grade average *(X)* and their cumulative grade average in college *(Y)*. The following results were obtained:

X	Y
3.3	2.7
2.9	2.5
2.5	1.9
4.0	3.3
2.8	2.7
2.5	2.2
3.7	3.1
3.8	4.0
3.5	2.9
2.7	2.0
2.6	3.1
4.0	3.2

a. Draw a scatter plot of the data.

b. Calculate the regression slope and *Y*-intercept.

c. Draw the regression line on the scatter plot.

d. Predict the college grade average of a student who attains a 3.0 grade average in high school.

e. Find the coefficients of determination and nondetermination. What do they mean?

f. Construct an analysis of variance table and perform an F test of the significance of the regression.

17. For the variables in problem 21 in Chapter 10, find the multiple coefficient of determination (with *Y* the dependent variable).

18. For the correlation matrix in problem 22 in Chapter 10, find the proportion of variance in the weekly grocery bill *(Y)* determined by family size *(X)* alone, by income *(Z)* alone, and by family size and income together.

19. A researcher wishes to predict the graduation rates in a sample of 240 colleges and universities based on the percentage of freshmen who were in the top tier (10%) of their high school class and the percentage of classes that have fewer than 20 students. The correlation matrix of graduation rate (GRADRATE), percentage of freshmen in the top tier of their high school graduating class (TOPFRESH), and the percentage of classes at the college with fewer than 20 students (SMALLCLS) is shown as follows:

	GRADRATE	TOPFRESH	SMALLCLS
GRADRATE	1.000	.613	.544
TOPFRESH	.613	1.000	.404
SMALLCLS	.544	.404	1.000

a. What percentage of the variance in GRADRATE does each of the predictor variables explain individually?

b. What percentage of the variance in GRADRATE do the two predictors explain in combination?

c. Why doesn't the answer in (b) equal the sum of the two answers in (a)?

20. Following is a portion of an analysis of variance table from a multiple regression predicting students' quantitative SAT scores (QSAT) based on their high school grade average (HSAVG) and their gender (GENDER):

Source	SS	df	MS	F
Regression	127,510	2	?	?
Error	82,990	53	?	
Total	?			

a. Fill out the portions indicated with a question mark.

b. What is the sample size (*N*)?

c. What percent of the variance in QSAT is explained by the two predictors combined?

d. Comment on the overall significance of the results.

21. Suppose that an emergency room doctor is interested in the relationship between gender, age, and sensitivity to pain among adult patients. She collects from a sample of 57 hospital files information on patients' self-reported pain scores, ranging from 0 for no discomfort to 10 for the worst pain imaginable, along with the patients' ages in years and gender (1 for male and 0 for female). The results of her regression analysis predicting pain based on age and gender are shown in the following table:

Variable	*b*	SE	*t*	Sig.
Constant	1.321	0.239	5.527	0.000
Age	0.065	0.041	1.585	0.119
Gender	0.857	0.665	1.289	0.203

a. Interpret the constant (*Y*-intercept)

b. Interpret the regression coefficients for age and gender.

c. Indicate which, if either, of these two variables is a significant predictor of pain score.

22. The following table contains responses to a survey concerning support for same-sex marriage, comparing positions on the issue with political party affiliation of respondents:

Party Affiliation	Position toward Same-Sex Marriage		Total
	Support	*Oppose*	**Total**
Democrat	45	64	109
Republican	21	72	99
Independent	24	60	84
Total	80	212	292

a. Calculate the odds of supporting same-sex marriage for each category of party affiliation.

b. Calculate the odds ratio for each pair of categories of party affiliation.

23. Referring back to problem 21, the same doctor then used the three variables—pain level (pain), age, and gender—to predict whether or not patients were admitted to the hospital (admit = 1) by the ER or sent home (admit = 0). She performs a logistic regression which produces the following results:

Variable	b	SE	Wald	Sig.	e^b
Constant	−11.236	0.239	2210.2	0.000	0.000
Age	0.054	0.022	6.1342	0.013	1.056
Gender	−0.020	0.018	1.2597	0.262	0.980
Pain	0.263	0.043	37.447	0.000	1.301

a. Why did she use logistic regression rather than multiple regression?

b. Interpret the results for each of the three variables.

SPSS Exercises

1. Is marijuana a gateway to other drugs? Using SPSS to analyze the Monitoring the Future Study, test the null hypothesis that there is no relationship between the likelihood of a high school student having smoked marijuana or hashish in their lifetime (V115) and cocaine use in their lifetime (V124). Hint: ANALYZE, REGRESSION, LINEAR and then specify variables. Cocaine use is the dependent variable.

   a. Find the ANOVA test of regression. Is it significant?

   b. What is the regression slope and intercept?

   c. Predict the amount of cocaine use for a student who has smoked marijuana or hashish once or twice.

   d. Find the coefficient of determination. What does it mean?

2. Binge drinking in the United States is defined as consumption of five or more drinks in a row for men and four or more drinks in a row for women. Using SPSS to analyze the Monitoring the Future Study, test the null hypothesis that there is no relationship between binge drinking (V108) and skipping of classes during the school day (V178). Binge drinking is the dependent variable.

    a. Find the ANOVA test of regression. Is it significant?

    b. What is the regression slope and intercept?

    c. Predict the number of times that a high school student will have had five or more drinks in the past two weeks if they have skipped 12 classes in the past two weeks. Hint: 12 skipped classes is not a 12. Run a frequency distribution on both variables to see the values on each scale.

    d. Find the coefficient of determination. What does it mean?

3. Using SPSS, perform a regression analysis to predict binge drinking (V108) on the basis of both truancy (V178) and number of dollars earned per week from a job (V192).

    a. Test the null hypothesis that the number of dollars per week from a job is not related to binge drinking, holding truancy constant.

    b. What are the regression slopes and the intercept?

    c. Which variables are significant at the .05 level?

    d. Find the multiple coefficient of determination (R^2). What does it mean? Did it improve over the amount explained in the two variable regression model?

    e. What is the predicted value of binge drinking for a person who has missed five classes and earns $50 per week from a job? (Remember to look for the values on each of the scales before doing your calculation.)

4. Using SPSS with the Best Places Study, conduct a regression analysis to determine what predicts best and worst cities for stress in America. The stress index ranges from 0 to 100 where 0 is the worst and 100 is the best. Add unemployment, cloudy days, commuting time, and property crime into the regression analysis as independent variables.

    a. Find the ANOVA test of regression. Is it significant?

    b. What are the regression slopes and intercept?

    c. Which variables are significant at the .05 level?

    d. Find the multiple coefficient of determination (R^2). What does it mean?

5. Using SPSS to analyze the General Social Survey, conduct a logistic regression analysis to test the null hypothesis that males and females do not differ in their fear of crime (sex and fear). Hint: ANALYZE, REGRESSION, BINARY LOGISTIC, select variables, and specify whether independent variables are categorical.

    a. What is the logistic regression slope and intercept?

    b. Find the numerical value of e^b for sex. What does it mean?

    c. Find the Nagelkerke R^2. What does this statistic mean?

12

Nonparametric Measures of Correlation

We saw previously that nonparametric alternatives to t and F were necessary for testing group differences with nominal and ordinal data. Similarly, we need nonparametric measures of correlation to use when the requirements of Pearson's r cannot be met. Specifically, these measures are applied if we have nominal or ordinal data or if we cannot assume normality in the population. This chapter introduces some of the best-known nonparametric measures of correlation: Spearman's rank-order correlation coefficient, Goodman's and Kruskal's gamma, phi coefficient, contingency coefficient, and Cramér's V.

Spearman's Rank-Order Correlation Coefficient

We turn now to the problem of finding the degree of association for ordinal data; such data have been ranked or ordered with respect to the presence of a given characteristic.

To take an example from social research, consider the relationship between socio-economic status and amount of time spent using the Internet per month among PC owners.

Although Internet time clearly could be measured at the interval level, socioeconomic status is considered ordinal, and thus a correlation coefficient for ordinal or ranked data is required. Imagine that a sample of eight respondents could be ranked as in Table 12.1.

TABLE 12.1 *Sample Ranked by Socioeconomic Status and Internet Usage*

Respondent	Socioeconomic Status (X) Rank	Time Spent on the Internet (Y) Rank
Max	1	2
Flory	2	1
Tong	3	3
Min	4	5
Juanita	5	4
Linda	6	8
Carol	7	6
Steve	8	7

Highest in socioeconomic status

Most time on the Internet

Max ranked first with respect to socioeconomic status, but second with regard to the amount of time spent on the Internet; Flory's position was second with respect to socioeconomic status, and first in terms of time spent using the Internet, and so on.

To determine the degree of association between socioeconomic status and amount of time on the Internet, we apply *Spearman's rank-order correlation coefficient* (r_s). By formula,

$$r_s = 1 - \frac{6\Sigma D^2}{N(N^2 - 1)}$$

where r_s = rank-order correlation coefficient
D = difference in rank between X and Y variables
N = total number of cases

We set up the present example as shown in Table 12.2. By applying the rank-order correlation coefficient to the data in this table,

$$r_s = 1 - \frac{(6)(10)}{(8)(64 - 1)}$$

$$= 1 - \frac{60}{(8)(63)}$$

$$= 1 - \frac{60}{504}$$

$$= 1 - .12$$

$$= +.88$$

TABLE 12.2 *Relationship between Socioeconomic Status and Interset Usage*

Respondent	Socioeconomic Status (X)	Time Spent on Internet (Y)	D	D²
Max	1	2	−1	1
Flory	2	1	1	1
Tong	3	3	0	0
Min	4	5	−1	1
Juanita	5	4	1	1
Linda	6	8	−2	4
Carol	7	6	1	1
Steve	8	7	1	1
				$\Sigma D^2 = 10$

As with Pearson's r, Spearman's coefficient varies from −1 to +1. Therefore, we find a strong positive correlation ($r_s - +.88$) between socioeconomic status and time spent on the Internet. Respondents having high socioeconomic status tend to use the Internet a good deal; respondents who have low socioeconomic status tend to spend little time on the Internet.

Dealing with Tied Ranks

In actual practice, it is not always possible to rank or order our respondents, avoiding ties at each position. We might find, for instance, that two or more respondents spend exactly the same amount of time on the Internet, that the academic achievement of two or more students is indistinguishable, or that several respondents have the same IQ score.

To illustrate the procedure for obtaining a rank-order correlation coefficient in the case of tied ranks, let us say we are interested in determining the degree of association between position in a graduating class and IQ. Suppose also we are able to rank a sample of 10 graduating seniors with respect to their class position and to obtain their IQ scores as follows:

Respondent	Class Standing (X)	IQ (Y)
Jim	10 ◄—(last)	110
Tracy	9	90
Leroy	8	104
Mike	7	100
Mario	6	110
Kenny	5	110
Mitchell	4	132
Minny	3	115
Cori	2	140
Kumiko	1 ◄—(first)	140

Before following the standard procedure for obtaining a rank-order correlation coefficient, let us first rank the IQ scores of our 10 graduating seniors:

Respondent	IQ	IQ Rank
Jim	110	7
Tracy	90	10
Leroy	104	8
Mike	100	9
Mario	110	6
Kenny	110	5
Mitchell	132	3
Minny	115	4
Cori	140	2
Kumiko	140	1

Positions 5, 6, and 7 are tied

Positions 1 and 2 are tied

The table shows that Cori and Kumiko received the highest IQ scores and are, therefore, tied for the first and second positions. Likewise, Kenny, Mario, and Jim achieved an IQ score of 110, which places them in a three-way tie for the fifth, sixth, and seventh positions.

To determine the exact position in the case of ties, we must *add the tied ranks and divide by the number of ties*. Therefore, the position of a 140 IQ, which has been ranked as 1 and 2, would be the average rank:

$$\frac{1+2}{2} = 1.5$$

In the same way, we find that the position of an IQ score of 110 is

$$\frac{5+6+7}{3} = 6.0$$

Having found the ranked position of each IQ score, we can proceed to set up the problem at hand, as shown in Table 12.3.

We obtain the rank-order correlation coefficient for the problem in Table 12.3 as follows:

$$r_s = 1 - \frac{(6)(24.50)}{(10)(100-1)}$$

$$= 1 - \frac{147}{990}$$

$$= 1 - .15$$

$$= +.85$$

TABLE 12.3 *Relationship between Class Standing and IQ*

Respondent	Class Standing (X)	IQ (Y)	D	D²
Jim	10	6	4.0	16.00
Tracy	9	10	−1.0	1.00
Leroy	8	8	.0	.00
Mike	7	9	−2.0	4.00
Mario	6	6	.0	.00
Kenny	5	6	−1.0	1.00
Mitchell	4	3	1.0	1.00
Minny	3	4	−1.0	1.00
Cori	2	1.5	.5	.25
Kumiko	1	1.5	−.5	.25
				$\Sigma D^2 = 24.50$

The resulting rank-order correlation coefficient indicates a rather strong *positive* correlation between class standing and IQ. That is, students having *high* IQ scores tend to rank *high* in their class; students who have *low* IQ scores tend to rank *low* in their class.

Testing the Significance of the Rank-Order Correlation Coefficient

How do we go about testing the significance of a rank-order correlation coefficient? For example, how can we determine whether the obtained correlation of +.85 between class standing and IQ can be generalized to a larger population? To test the significance of a computed rank-order correlation coefficient, we turn to Table G in Appendix C, where we find the critical values of the rank-order coefficient of correlation for the .05 and .01 significance levels. Notice that we refer directly to the number of pairs of scores *(N)* rather than to a particular number of degrees of freedom. In the present case, $N = 10$ and the rank-order correlation coefficient must exceed .648 to be significant. We therefore reject the null hypothesis that $\rho_s = 0$ and accept the research hypothesis that class standing and IQ are actually related in the population from which our sample was drawn.

Table G only applies to samples of up to 30 cases. What then should we do if N exceeds this number? Consider an example having a somewhat larger number of observations. Table 12.4 provides U.S. Census Bureau rankings of the states according to infant mortality rates and the number of physicians per 100,000 residents for the year 2001, after resolving ties by the averaging approach presented earlier.

Based on these figures,

TABLE 12.4 *State Rankings of Infant Mortality and Physicians per 100,000 Residents, 2001*

State	Infant Mortality	Physicians per 100,000	D	D^2	State	Infant Mortality	Physicians per 100,000	D	D^2
Alabama	4	39	−35	1225	Montana	29	34	−5	25
Alaska	11.5	44	−32.5	1056.25	Nebraska	27.5	28	−0.5	0.25
Arizona	26	41	−15	225	Nevada	42	46	−4	16
Arkansas	10	43	−33	1089	New Hampshire	50	16	34	1156
California	45.5	14	31.5	992.25	New Jersey	30	7	23	529
Colorado	40	25	15	225	New Mexico	31	32	−1	1
Connecticut	34.5	4	30.5	930.25	New York	40	2	38	1444
Delaware	1	20	−19	361	North Carolina	9	23	−14	196
Florida	21.5	24	−2.5	6.25	North Dakota	6	29	−23	529
Georgia	8	36	−28	784	Ohio	14.5	19	−4.5	20.25
Hawaii	32.5	9	23.5	552.25	Oklahoma	21.5	49	−27.5	756.25
Idaho	32.5	50	−17.5	306.25	Oregon	45.5	22	23.5	552.25
Illinois	14.5	11	3.5	12.25	Pennsylvania	23.5	8	15.5	240.25
Indiana	17	37	−20	400	Rhode Island	27.5	6	21.5	462.25
Iowa	43	45	−2	4	South Carolina	5	31	−26	676
Kansas	19	35	−16	256	South Dakota	19	42	−23	529
Kentucky	37	33	4	16	Tennessee	7	18	−11	121
Louisiana	3	12	−9	81	Texas	37	38	−1	1
Maine	34.5	15	19.5	380.25	Utah	49	40	9	81
Maryland	11.5	3	8.5	72.25	Vermont	44	5	39	1521
Massachusetts	48	1	47	2209	Virginia	16	13	3	9
Michigan	13	27	−14	196	Washington	40	17	23	529
Minnesota	47	10	37	1369	West Virginia	23.5	30	−6.5	42.25
Mississippi	2	48	−46	2116	Wisconsin	25	21	4	16
Missouri	19	26	−7	49	Wyoming	37	47	−10	100
									$\Sigma D^2 = 24466$

$$r_s = 1 - \frac{6\Sigma D^2}{N(N^2 - 1)}$$

$$= 1 - \frac{6(24,466)}{50(2,500 - 1)}$$

$$= -0.175$$

The negative correlation suggests that the higher the number of physicians per 100,000 residents, the lower is the infant mortality rate. While this result is reasonable, is the value of Spearman's rank-order correlation large enough to be considered statistically significant? Table G, of course, provides r_s critical values for N only up to 30.

For large samples (for $N > 30$), the following expression tends to have a normal distribution and can be compared with critical values of z:

$$z = r_s\sqrt{N - 1}$$

For our example,

$$z = -0.175\sqrt{50 - 1} = -0.175(7) = -1.225$$

Here z does not nearly reach the 1.96 critical value of z at the .05 level of significance. Therefore, we must retain the null hypothesis of no relationship in the population between infant mortality rates and the availability of physicians.

BOX 12.1 • *Step-by-Step Illustration: Spearman's Rank-Order Correlation Coefficient*

We can summarize the step-by-step procedure for obtaining the rank-order correlation coefficient with reference to the relationship between the degree of participation in voluntary associations and number of close friends. This relationship is indicated in the following sample of six respondents:

Respondent	Voluntary Association Participation (X) Rank	Number of Friends (Y)
A	1 ← Participates	6
B	2 most	4
C	3	6
D	4	2
E	5 Participates	2
F	6 ← least	3

To determine the degree of association between voluntary association participation and number of friends, we carry through the following steps:

Step 1   Rank respondents on the X and Y variables.

As the previous table shows, we rank respondents with respect to X, participation in voluntary associations, assigning the rank of 1 to the respondent who participates most and the rank of 6 to the respondent who participates least.

We must also rank the respondents in terms of Y, the number of their friends. In the present example, we have instances of tied ranks, as shown in the following:

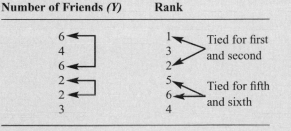

Number of Friends (Y)	Rank

(continued)

BOX 12.1   Continued

To convert tied ranks, we take an average of the tied positions. For first and second positions:

$$\frac{1 + 2}{2} = 1.5$$

For fifth and sixth positions:

$$\frac{5 + 6}{2} = 5.5$$

Therefore, in terms of ranks, where X_R and Y_R denote the ranks on X and Y, respectively,

X_R	Y_R
1	1.5
2	3.0
3	1.5
4	5.5
5	5.5
6	4.0

Step 2   To find ΣD^2, we must find the difference between X and Y ranks *(D)*, square each difference (D^2), and sum these squares (ΣD^2):

X_R	Y_R	D	D^2
1	1.5	−.5	.25
2	3.0	−1.0	1.00
3	1.5	1.5	2.25
4	5.5	−1.5	2.25
5	5.5	−.5	.25
6	4.0	2.0	4.00
		$\Sigma D^2 =$	10.00

Step 3   Plug the result of Step 2 into the formula for the rank-order correlation coefficient.

$$r_s = 1 - \frac{6\Sigma D^2}{N(N^2 - 1)}$$

$$= 1 - \frac{(6)(10)}{(6)(36 - 1)}$$

$$= 1 - \frac{60}{210}$$
$$= 1 - .29$$
$$= +.71$$

Step 4 Compare the obtained rank-order correlation coefficient with the critical value of r_s in Table G.

obtained $r_s = .71$

table $r_s = .886$

$N - 6$

$\alpha = .05$

Because N is less than 30, we can test the significance using Table G in Appendix C. We learn that a correlation coefficient of .886 is necessary to reject the null hypothesis at the .05 level of significance with a sample size of 6. Therefore, although we have uncovered a strong positive correlation between voluntary association participation and number of friends, we must still retain the null hypothesis that population correlation $\rho_s = 0$. Our result cannot be generalized to the population from which our sample was taken.

Requirements for Using the Rank-Order Correlation Coefficient

The rank-order correlation coefficient should be employed when the following conditions can be satisfied:

1. *Ordinal data.* Both X and Y variables must be ranked or ordered.
2. *Random sampling.* Sample members must have been taken at random from a larger population to apply a test of significance.

Goodman's and Kruskal's Gamma

As we saw in connection with the rank-order correlation coefficient, it is not always possible to avoid tied ranks at the ordinal level of measurement. In fact, social researchers frequently work with crude ordinal measures that produce large numbers of tied ranks. Typically, this occurs with ordinal variables that are ranked in categories, such as high, medium, and low. When two such ordinal variables are cross-tabulated, *Goodman's and Kruskal's gamma (G)* is a particularly useful measure of association.

The basic formula for gamma is

$$G = \frac{N_a - N_i}{N_a + N_i}$$

where N_a = number of agreements
        N_i = number of inversions

Agreements and inversions can be understood as expressing the direction of correlation between X and Y variables. Perfect agreement indicates a perfect positive correlation (+1.00): All individuals being studied have been ranked in exactly the same order on both variables.

By contrast, perfect inversion indicates a perfect negative correlation (−1.00), so that the individuals being studied are ranked in exactly reverse order on both variables.

The logic of agreements, inversions, and tied pairs can be illustrated by examining the following simple cross-tabulation in which the frequencies are supplemented by the 10 respondents' names:

		X	
	High	**Medium**	**Low**
High	$f = 2$ Sam Mary	$f = 1$ Ann	$f = 0$
Low	$f = 2$ Alex Jack	$f = 3$ Paul John Lisa	$f = 2$ Sue Bob

Y

Mary and John are in agreement because she is higher than he is on both variables. Similarly, the pair Sam and Sue are in agreement because Sam is at a higher level than Sue on both variables. In all, there are 12 agreements ($N_a = 12$). Can you find all 12?

In contrast, the pair Alex and Ann is an inversion, because Alex exceeds Ann on X but falls below her on Y. Overall, there are two inversions ($N_i = 2$) Can you find both pairs?

Finally, any pair that is at the same level on one or both variables represents a tie. For example, Alex and Jack are tied on both variables (that is, they are in the same cell). Furthermore, although Alex surpasses Sue on X, they are at the same level on Y (that is, they are in the same row), and thus they are counted as a tied pair. Fortunately, you can ignore ties in calculating gamma, because they do not enter into its formula.

BOX 12.2 • *Step-by-Step Illustration: Goodman's and Kruskal's Gamma*

Using a larger example, let us now illustrate the procedure for obtaining a gamma coefficient for cross-tabulated ordinal variables. Suppose that a researcher wanting to examine the relationship between social class *(X)* and faith in the fairness of local police *(Y)* obtained the following data from a questionnaire study of 80 city residents: Among 29 upper-class respondents, 15 were high, 10 were medium, and 4 were low with respect to faith in the police; among 25 respondents who were middle class, 8 were high, 10 were medium, and 7 were low with respect to faith in the police; and among 26 lower-class respondents, 7 were high, 8 were medium, and 11 were low with respect to faith in the fairness of local police. Notice that tied ranks occur at every position. For instance, there were 29 respondents who tied at the rank of upper social class, the highest rank on the X variable.

Step 1 Rearrange the data in the form of a cross-tabulation.

Faith in Fairness of Local Police *(Y)*	Social Class *(X)*		
	Upper	*Middle*	*Lower*
High	15	8	7
Medium	10	10	8
Low	4	7	11
	29	25	26
		$N = 80$	

Notice that the preceding table is a 3×3 cross-tabulation containing nine cells (3 rows $\times$ 3 columns = 9). To ensure that the sign of the gamma coefficient is accurately depicted as either positive or negative, the X variable in the columns must always be arranged in decreasing order from left to right. In the table, for example, social class decreases—upper, middle, lower—from left to right columns. Similarly, the Y variable in the rows must decrease from top to bottom. In the preceding table, faith in the fairness of local police decreases—high, medium, low—from top to bottom rows.

Step 2 Obtain N_a

To find N_a, begin with the cell ($f = 15$) in the upper-left corner. Multiply this number by the sum of all numbers that fall *below and to the right of it*. Reading from left to right, we see that all frequencies below *and* to the right of 15 are 10, 8, 7, and 11. Now repeat this procedure for all cell frequencies that have cells below and to the right of them. By working from left to right in the table,

Upper-class/high faith in police	$(15)(10 + 8 + 7 + 11) = (15)(36) = 540$
Middle-class/high faith in police	$(8)(8 + 11) = (8)(19) = 152$
Upper-class/medium faith in police	$(10)(7 + 11) = (10)(18) = 180$
Middle-class/medium faith in police	$(10)(11) = (110)$

(Note that none of the other cell frequencies in the table—7 in the top row, 8 in the second row, and 4, 7, and 11 in the bottom row—has cells below *and* to the right.) N_a is the sum of the products obtained in the previous table.

$$N_a = 540 + 152 + 180 + 110$$
$$= 982$$

(continued)

BOX 12.2   Continued

Step 3   Obtain N_i.

To obtain N_i, reverse the procedure for finding agreements and begin in the upper-right corner of the table. This time, each number is multiplied by the sum of all numbers that fall *below and to the left of it*. Reading from right to left, we see that frequencies below *and* to the left of 7 are 10, 10, 7, and 4. As before, repeat this procedure for all frequencies having cells below and to the left of them.

Working from right to left,

Lower-class/high faith in police	$(7)(10 + 10 + 7 + 4) = (7)(31) = 217$
Middle-class/high faith in police	$(8)(10 + 4) = (8)(14) = 112$
Lower-class/medium faith in police	$(8)(7 + 4) = (8)(11) = 88$
Middle-class/medium faith in police	$(10)(4) = 40$

(Note that none of the other cell frequencies in the table—15 in the top row, 10 in the middle row, 11, 7, and 4 in the bottom row—has cells below *and* to the left.)

N_i is the sum of the products computed in the previous table. Therefore,

$$N_i = 217 + 112 + 88 + 40$$
$$= 457$$

Step 4   Plug the results of Steps 2 and 3 into the formula for gamma.

$$G = \frac{N_a - N_i}{N_a + N_i}$$
$$= \frac{982 - 457}{982 + 457}$$
$$= \frac{525}{1,439}$$
$$= +.36$$

A gamma coefficient of $+.36$ indicates a moderate positive correlation between social class and faith in local police. Our result suggests a correlation based on a dominance of agreements. (Note that a gamma coefficient of $-.36$ would have indicated instead a moderate *negative* correlation based on a dominance of *inversions.*)

Before continuing, let's take another look at how gamma functions in the context of cross-tabulations. Consider, for example, the center (middle–medium) cell in the cross-tabulation of social class and faith in the fairness of local police. Cases that are below and to the right are agreements because they are lower on both variables.

Cases that are below and to the left are inversions because they are lower on Y but greater on X. Why not count the 15 upper–high cases as agreements too with respect to our middle–medium reference cell? Although these are agreements, they would have already been counted: The 15 middle–medium cases are counted as agreements when upper–high is the reference cell. Counting agreements and inversions only toward the bottom of the table avoids any mistake of double counting.

Testing the Significance of Gamma

To test the null hypothesis that X and Y are not associated in the population—that $\gamma = 0$—we convert our calculated G to a z score by the following formula:

$$z = G\sqrt{\frac{N_a + N_i}{N(1 - G^2)}}$$

where G = calculated gamma coefficient
N_a = number of agreements
N_i = number of inversions
N = sample size

In the foregoing step-by-step illustration, we found that $G = .36$ for the correlation between social class and faith in the fairness of local police. To test the significance of our finding, we substitute in the formula as follows:

$$z = (.36)\sqrt{\frac{982 + 457}{(80)[1 - (.36)^2]}}$$

$$= (.36)\sqrt{\frac{1,439}{(80)(.87)}}$$

$$= (.36)\sqrt{\frac{1,439}{69.60}}$$

$$= (.36)\sqrt{20.68}$$

$$= (.36)(4.55)$$

$$= 1.64$$

Turning to Table A (or the bottom row of Table B or C) in Appendix C, we see that z must exceed 1.96 to reject the null hypothesis at the .05 level of significance. Because our calculated z (1.64) is smaller than the required table value, we must retain the null hypothesis that faith in police $\gamma = 0$ and reject the research hypothesis that $\gamma \neq 0$. Our obtained correlation cannot be generalized to the population from which our sample was drawn.

Requirements for Using Gamma

The following requirements must be taken into account to employ gamma as a measure of association for cross-tabulations.

1. *Ordinal data.* Both X and Y variables must be ranked or ordered.
2. *Random sampling.* To test the null hypothesis ($\gamma = 0$), sample members must have been taken on a random basis from some specified population.

Correlation Coefficient for Nominal Data Arranged in a 2 × 2 Table

In Chapter 9 a test of significance for frequency data known as chi-square was introduced. By a simple extension of the chi-square test, the degree of association between variables at the nominal level of measurement can now be determined. Let us take another look at the null hypothesis that

> *the proportion of smokers among American college students is the same as the proportion of smokers among Canadian college students.*

In Chapter 9 this null hypothesis was tested within a sample of 21 American students and a sample of 15 Canadian students attending the same university. It was determined that 15 out of the 21 American students but only 5 out of the 15 Canadian students were nonsmokers. Thus, we have the 2 × 2 problem shown in Table 12.5.

The relationship between nationality and smoking status among college students was tested in Chapter 9 by applying the 2 × 2 chi-square formula (with Yates's correction because of small expected frequencies) using a summary table as follows:

$$\chi^2 = \Sigma \frac{(|f_0 - f_e| - .5)^2}{f_e}$$

TABLE 12.5 *Smoking among American and Canadian College Students: Data from Table 9.7*

Smoking Status	Nationality		Total
	Americans	*Canadians*	**Total**
Nonsmokers	15	5	20
Smokers	6	10	16
Total	21	15	$N = 36$

f_o	f_e	$\lvert f_o - f_e \rvert$	$\lvert f_o - f_e \rvert - .5$	$(\lvert f_o - f_e \rvert - .5)^2$	$\dfrac{(\lvert f_o - f_e \rvert - .5)^2}{f_e}$
15	11.67	3.33	2.83	8.01	.69
5	8.33	3.33	2.83	8.01	.96
6	9.33	3.33	2.83	8.01	.86
10	6.67	3.33	2.83	8.01	1.20

$$\chi^2 = 3.71$$

Having calculated a chi-square value of 3.71, we can now obtain the *phi coefficient (ϕ)*, which is a measure of the degree of association for 2×2 tables. By formula,

$$\phi = \sqrt{\frac{\chi^2}{N}}$$

where ϕ = phi coefficient
χ^2 = calculated chi-square value
N = total number of cases

By applying the foregoing formula to the problem at hand,

$$\phi = \sqrt{\frac{3.71}{36}}$$
$$= \sqrt{.1031}$$
$$= .32$$

Our obtained phi coefficient of .32 indicates the presence of a moderate correlation between nationality and smoking.

Testing the Significance of Phi

Fortunately, the phi coefficient can be easily tested by means of chi-square, whose value has already been determined, and Table E in Appendix C:

obtained χ^2 = 3.71

table χ^2 = 3.84

df = 1

α = .05

Because our calculated chi-square value of 3.71 is less than the required table value, we retain the null hypothesis of no association and reject the research hypothesis that nationality and smoking are associated in the population. Therefore, although moderate in size, the obtained phi coefficient is not statistically significant.

Requirements for Using the Phi Coefficient

To employ the phi coefficient as a measure of association between X and Y variables, we must consider the following requirements:

1. *Nominal data.* Only frequency data are required.
2. *A 2 × 2 table.* The data must be capable of being cast in the form of a 2 × 2 table (2 rows by 2 columns). It is inappropriate to apply the phi coefficient to tables larger than 2 × 2, in which several groups or categories are being compared.
3. *Random sampling.* To test the significance of the phi coefficient, sample members must have been drawn on a random basis from a larger population.

Correlation Coefficients for Nominal Data in Larger than 2 × 2 Tables

Until this point, we have considered the correlation coefficient for nominal data arranged in a 2 × 2 table. As we saw in Chapter 9, there are times when we have nominal data, but are comparing several groups or categories. To illustrate, let us reconsider the null hypothesis that

> *the relative frequency of permissive, moderate, and authoritarian child-discipline is the same for fathers, stepfathers, and live-in partners.*

In Chapter 9 this hypothesis was tested with the data in the 3 × 3 table in Table 12.6. Differences in child-discipline by type of relationship were tested by applying the chi-square formula using a summary table as follows:

f_o	f_e	$f_o - f_e$	$(f_o - f_e)^2$	$\dfrac{(f_o - f_e)^2}{f_e}$
7	10.79	−3.79	14.36	1.33
9	10.11	−1.11	1.23	.12
14	9.10	4.90	24.01	2.64
10	10.07	−.07	.00	.00
10	9.44	.56	.31	.03
8	8.49	−.49	.24	.03
15	11.15	3.85	14.82	1.33
11	10.45	.55	.30	.03
5	9.40	−4.40	19.36	2.06
				$\chi^2 = 7.58$

In the present context, we seek to determine the correlation or degree of association between type of relationship *(X)* and child-discipline method *(Y)*. In a table larger than 2 × 2, this can be done by a simple extension of the chi-square test, which is referred to as the *contingency coefficient (C)*. The value of C can be found by the formula,

TABLE 12.6 *Cross-Tabulation of Child-Discipline Practices and Type of Relationship*

| | | Type of Relationship | | | |
		Father	Stepfather	Live-in Partner	Total
Discipline Practices	Permissive	7	9	14	30
	Moderate	10	10	8	28
	Authoritarian	15	11	5	31
	Total	32	30	27	89

$$C = \sqrt{\frac{\chi^2}{N + \chi^2}}$$

where C = contingency coefficient
$\quad\quad N$ = total number of cases
$\quad\quad \chi^2$ = calculated chi-square value

In testing the degree of association between religion and child-discipline method,

$$C = \sqrt{\frac{7.58}{89 + 7.58}}$$
$$= \sqrt{\frac{7.58}{96.58}}$$
$$= \sqrt{.0785}$$
$$= .28$$

Our obtained contingency coefficient of .28 indicates that the correlation between type of relationship and child discipline can be regarded as a rather weak one. The variables are related, but many exceptions can be found.

Testing the Significance of the Contingency Coefficient

Just as in the case of the phi coefficient, whether the contingency coefficient is statistically significant can be easily determined from the size of the obtained chi-square value. In the present example, we find that the relationship between the variables is nonsignificant and therefore confined to the members of our samples. This is true because the calculated chi-square of 7.58 is smaller than the required table value:

$$\text{obtained } \chi^2 = 7.58$$
$$\text{table } \chi^2 = 9.49$$
$$df = 4$$
$$\alpha = .05$$

Requirements for Using the Contingency Coefficient

To appropriately apply the contingency coefficient, we must be aware of the following requirements:

1. *Nominal data.* Only frequency data are required. These data may be cast in the form of a 2×2 table or larger.
2. *Random sampling.* To test the significance of the contingency coefficient, all sample members must have been taken at random from a larger population.

An Alternative to the Contingency Coefficient

Despite its great popularity among social researchers, the contingency coefficient has an important disadvantage: The number of rows and columns in a chi-square table will influence the maximum size taken by C. That is, the value of the contingency coefficient will not always vary between 0 and 1.0 (although it will never exceed 1.0). Under certain conditions, the maximum value of C may be .94; at other times, the maximum value of C may be .89; and so on. This situation is particularly troublesome in nonsquare tables—that is, tables that contain different numbers of rows and columns (for example, 2×3, 3×5, and so on).

To avoid this disadvantage of C, we may decide to employ another correlation coefficient, which expresses the degree of association between nominal-level variables in a table larger than 2×2. Known as Cramér's V, this coefficient does not depend on the size of the χ^2 table and has the same requirements as the contingency coefficient. By formula,

$$V = \sqrt{\frac{\chi^2}{N(k-1)}}$$

where V = Cramér's V
  N = total number of cases
  k = number of rows *or* columns, whichever is smaller (if the number of rows equals the number of columns as in a 3×3, 4×4, or 5×5 table, either number can be used for k)

Let us return to the cross-tabulation of child-discipline and type of relationship as shown in Table 12.6 (a 3×3 table):

$$V = \sqrt{\frac{7.58}{(89)(3-1)}}$$
$$= \sqrt{\frac{7.58}{(89)(2)}}$$
$$= \sqrt{\frac{7.58}{178}}$$
$$= \sqrt{.0426}$$
$$= .21$$

As a result, we find a Cramér's V correlation coefficient equal to .21, indicating a weak relationship between the variables. As with phi and the contingency coefficient, the significance of the relationship is determined by the chi-square test.

Summary

As we saw in Chapter 9 in connection with tests of significance, there are nonparametric tests to employ instead of t and F when the assumptions about a normal distribution or the interval-level of measurement are not appropriate. Similarly, when the requirements of Pearson's r cannot be met, nonparametric alternatives become attractive to a social researcher who seeks to measure the degree of association between two variables, particularly when both variables are not at the internal level or when there is substantial non-normality within the data. To determine the correlation between variables at the ordinal level of measurement, a researcher can apply Spearman's rank-order correlation coefficient (r_s). To use this measure appropriately, both X and Y variables must be ranked or ordered. When ordinal data are arranged in a cross-tabulation, Goodman's and Kruskal's gamma coefficient *(G)* becomes a useful correlation coefficient. For variables measured at the nominal level and displayed in a cross-tabulation, several nonparametric measures of association are available and are derived from the chi-square statistic that we encountered in Chapter 9. Specifically, the phi coefficient (ϕ) is used to assess the strength of association in a cross-tabulation with two rows and two columns. For tables larger than 2×2, it is appropriate to use the contingency coefficient C (for square tables) and Cramér's V.

Terms to Remember

Spearman's rank-order correlation coefficient
Goodman's and Kruskal's gamma
Phi coefficient

Contingency coefficient
Cramér's V

Questions and Problems

1. When the requirements of Pearson's r cannot be met, we might still be able to employ
   a. a t ratio.
   b. a parametric measure of correlation.
   c. a nonparametric measure of correlation.
   d. None of the above

2. Why would we decide to test the significance of a rank-order, gamma, or contingency correlation coefficient?
   a. To measure its strength
   b. To measure its direction
   c. To see whether we can generalize it to the population from which our sample was drawn
   d. To see whether we have really met the requirements for using the test

3. The phi and contingency coefficients are extensions of which test?
   a. Pearson's *r*
   b. Spearman's rank order
   c. Chi-square
   d. Gamma

4. The following five students were ranked in terms of when they completed an exam (1 = first to finish, 2 = second to finish, and so on) and were given an exam grade by the instructor. Test the null hypothesis of no relationship between grade *(X)* and length of time to complete the exam *(Y)* (that is, compute a rank-order correlation coefficient and indicate whether it is significant).

X	Y
53	1
91	2
70	3
85	4
91	5

5. In Chapter 10 data were presented on the distance to school *(X)* and number of clubs joined *(Y)* for 10 high school students to which the Pearson's correlation coefficient was applied. One of the students (David), who previously lived 10 miles from school, has now moved farther away. He is now 17 miles from school. Because of the apparent skewness created by this extreme value, Pearson's *r* may give a distorted result. As an alternative, calculate Spearman's rank-order correlation for the following data, and indicate whether it is significant:

	Distance to School (miles)	Number of Clubs Joined
Lee	4	3
Rhonda	2	1
Jess	7	5
Evelyn	1	2
Mohammed	4	1
Steve	6	1
George	9	9
Juan	7	6
Chi	7	5
David	17	8

6. A medical sociologist investigated the relationship between severity of illness *(X)* and length of stay in a hospital *(Y)*. Choosing eight patients at random, she ranked the seriousness of their ailment and determined the number of days they were hospitalized. Her results were as follows:

Patient	X	Y
A	6	12
B	4	19
C	1	18
D	8	3
E	3	21
F	2	21
G	7	5
H	5	10

Compute a rank-order correlation coefficient and indicate whether there is a significant relationship between *X* and *Y*.

7. A demographer was interested in the relationship between population density *(X)* and quality of life *(Y)*. Ranking 10 major cities on both variables, he obtained the following results:

City	X	Y
A	8	2
B	1	7
C	3	8
D	7	1
E	4	5
F	10	3
G	2	10
H	5	6
I	6	9
J	9	4

Compute the rank-order correlation and indicate whether there is a significant relationship between *X* and *Y*.

8. A researcher is interested in the relationship between political participation *(X)* and years of education *(Y)*. After selecting a random sample of people over the age of 18, he ranked them in terms of their amount of political participation and determined their years of education. He obtained the following results:

Person	X	Y
A	5	16
B	9	9
C	1	20
D	4	15
E	6	8
F	2	17
G	8	15
H	3	18
I	7	7

Compute a rank-order correlation coefficient and indicate whether there is a significant relationship between *X* and *Y*.

9. A psychology professor decides to give a quiz at the beginning of every class in hopes that it will encourage her students to do the assigned readings. However, after a few quizzes have been given, she begins to wonder if her quizzes are truly separating those who have done the reading from those who have not, or if they are merely penalizing those students who read the material but did not understand it. To find out, she asks students to write down on their quizzes how much time they spent reading the assignment. She then ranks them in terms of the amount of time spent reading the material *(X)* and their grades on the quiz *(Y)*. The following results were obtained:

Student	X	Y
A	3	6
B	7	11
C	2	1
D	10	8
E	8	12
F	1	4
G	9	10
H	11	7
I	4	5
J	12	9
K	5	2
L	6	3

Compute a rank-order correlation coefficient and indicate whether there is a significant relationship between *X* and *Y*.

10. A social worker notices that some of her most anxious clients often come to see her with coffee mugs in hand. Wondering if their anxiousness *(X)* is related to their amount of caffeine consumption *(Y)*, she randomly samples 10 of her clients and ranks them on both variables. The following results were obtained:

Client	X	Y
A	4	3
B	1	2
C	5	6
D	10	9
E	8	10
F	2	4
G	9	8
H	3	1
I	7	7
J	6	5

Compute a rank-order correlation coefficient and indicate whether there is a significant relationship between X and Y.

11. An educational sociologist studied 12 college graduates—the entire graduating class of a small college—to determine whether their academic achievement as high school students was related to how well they performed in college. As shown, cumulative grade-point average in high school and class rank in college were employed as measures of academic performance:

GPA in High School	College Class Rank
3.3	1
2.9	12
2.5	9
4.0	8
2.8	11
2.5	10
3.7	2
3.8	7
3.5	3
2.7	6
2.6	4
4.0	5

Determine the strength and direction of correlation between GPA in high school and college class rank for the 12 members of this graduating class. Is it necessary to test the significance of the correlation?

12. A sociologist interested in the relationship between social class and amount of reading asked a sample of 105 children from varying socioeconomic backgrounds about the number of books they had read outside of school during the last year. For the following data, compute a gamma coefficient to determine the degree of association between socioeconomic status *(X)* and number of books read *(Y)*, and indicate whether the relationship is significant.

	Socioeconomic Status			
Number of Books Read	*Lower*	*Lower Middle*	*Upper Middle*	*Upper*
None	15	12	6	5
One	12	8	10	8
More than one	4	6	9	10

13. A researcher was interested in whether job satisfaction is related to commuting time to work. He interviewed 185 employees of a particular company concerning their commuting

time *(X)* and their job satisfaction *(Y)*. For the following data, compute a gamma coefficient to determine the degree of association between commuting time and job satisfaction, and indicate whether there is a significant relation between X and Y.

Job Satisfaction	Commuting Time (minutes)			
	60+	*30–59*	*15–29*	*Under 15*
Very satisfied	8	12	25	22
Somewhat satisfied	9	20	23	11
Dissatisfied	13	18	17	7
		$N = 185$		

14. The following 96 students were ranked from high to low with respect to their consumption of alcoholic beverages *(X)* and their daily use of marijuana *(Y)*. For these data, compute a gamma coefficient to determine the degree of association between consumption of alcohol and use of marijuana, and indicate whether there is a significant relation between X and Y.

Use of Marijuana	Consumption of Alcohol		
	High	*Medium*	*Low*
High	5	7	20
Medium	10	8	15
Low	15	6	10
		$N = 96$	

15. A researcher is interested in determining the degree of association between the political party affiliation of husbands and wives. For the following cross-tabulation, calculate chi-square and determine the phi coefficient.

Husband's Political Party Affiliation	Wife's Political Party Affiliation	
	Democrat	*Republican*
Democrat	24	6
Republican	8	22

16. A curious graduate student wanted to determine the degree of association between successful campaign fund raising and political success in elections. The following cross-tabulation displays the data she collected about the amount of fund raising and overall political success:

Political Success

Fund Raising	High	Medium	Low
High	12	4	2
Medium	6	6	4
Low	1	2	9

Calculate chi-square and determine the contingency coefficient.

17. A researcher is interested in determining the degree of association between the political leanings of husbands and wives. For the following cross-tabulation, calculate chi-square and determine the contingency coefficient.

Husband's Political Leaning	Wife's Political Leaning		
	Liberal	Moderate	Conservative
Liberal	10	6	4
Moderate	7	11	5
Conservative	3	5	9

18. A medical sociologist internist is curious as to whether there is any connection between various behaviors that could be viewed as risk taking. She collects data from her records on 162 patients concerning their usage of seat belts and of cigarettes. For the following cross-tabulation, calculate chi-square and Cramér's *V*.

	Use of Cigarettes	
Use of Seat Belts	Smoker	Nonsmoker
Always	10	32
Usually	30	26
Seldom	46	18

19. To collect evidence pertaining to gender discrimination, a sample of college professors was classified by academic rank and gender. For the following cross-tabulation, calculate chi-square and Cramér's *V*.

	Gender	
Rank	Male	Female
Full professor	8	2
Associate professor	10	6
Assistant professor	25	20

20. Does birth order influence how much a child obeys his or her parents? For the following cross-tabulation, calculate chi-square and Cramér's *V*.

	Obedience	
Birth Order	*Obedient*	*Disobedient*
First child	11	14
Second child	15	12
Third child	17	20

21. A researcher collected data from 198 people about their political leaning and whether they would favor reinstating the assault weapons ban. For the following cross-tabulation, calculate chi-square and Cramér's *V*.

	Reinstate Assault Weapons Ban?	
Political Leaning	*Yes*	*No*
Liberal	56	7
Moderate	33	28
Conservative	11	63

22. To find out if wealthier women tend to have Cesarean sections (C-sections) more often than poorer women, a sample of recent mothers was classified by socioeconomic status and whether they delivered their baby via C-section. For the following cross-tabulation, calculate the chi-square and Cramér's *V*.

	C-section Delivery?	
Socioeconomic Status	*Yes*	*No*
High	14	8
Medium	9	12
Low	3	23

SPSS Exercises

1. Using SPSS, analyze the General Social Survey to find out if respondents' attitudes toward assistance with healthcare costs for the sick (HELPSICK) are related to their attitudes about the assistance with healthcare costs for the poor (HELPPOOR). Test the null hypothesis using Spearman's rank order correlation coefficient. Hint: To obtain a Spearman's rank order correlation, click on ANALYZE, CORRELATE, BIVARIATE, and check the box for Spearman (to obtain less output, unclick Pearson).

2. A researcher wonders whether people will give similar answers to questions about abortion depending upon the circumstances. In particular, she wants to know whether people will differ (or not) when the issue is asking about abortion after rape (ABRAPE) or abortion under any circumstances (ABANY). Using the General Social Survey, find the contingency coefficient to test the null hypothesis that there is no relationship in how people assess questions about abortion comparing abortion after rape and under any circumstances. Hint: The contingency coefficient is available as an optional statistic in the Crosstabs procedure.

3. A researcher suspects that there may be a relationship between marijuana use and speeding. Using SPSS, analyze the Monitoring the Future Survey to find out whether she is right. Apply gamma to test the null hypothesis that there is no relationship between lifetime use of marijuana or hashish (V115) and getting a traffic ticket or warning for a moving violation (V197). Hint: Gamma is available as an optional statistic in the Crosstabs procedure.

4. The General Social Survey includes a variety of questions on activities that Americans may do at night. Apply gamma to test the null hypothesis that there is no relationship between each of the different nighttime activities (SOCBAR, SOCCOMMUN, SOCFREND, SOCREL).

5. Are high school boys more likely than girls to have drunk alcohol? Use SPSS to find Phi for testing the null hypothesis that there is no relationship between lifetime consumption of alcohol (V103) and sex among high school seniors (V115). Hint: Phi is available as an optional statistic in the Crosstabs procedure.

6. A researcher is curious as to whether there is any connection between where you live and drinking of alcoholic beverages. Using the Monitoring the Future Data Set, find Cramer's *V* to test the null hypothesis that regional location (V13) is not related to lifetime consumption of alcohol (V103). Hint: Phi and Cramér's *V* are found in the same place in SPSS. Using the Best Places Study, conduct a similar analysis to find Cramér's *V* to test the null hypothesis that regional location (REGION) is not related to average consumption of alcohol (ALCOHOL). Comment on the similarities or differences in the results between the two studies.

Looking at the Larger Picture: Measuring Association

In the last part of the text, we considered tests for differences between groups. Using the *z* test of proportions, for example, we tested differences by sex in the percentage of students who smoke. Using the *t* test and then analysis of variance we assessed differences between sexes and then among races in terms of their mean levels of smoking and drinking. Using chi-square, we also tested for differences between groups (race and parents as smokers/nonsmokers) in terms of the percentage of respondents who were smokers. In this final portion of the text, we focused more on establishing the degree of association between variables, as opposed to focusing on differences between groups.

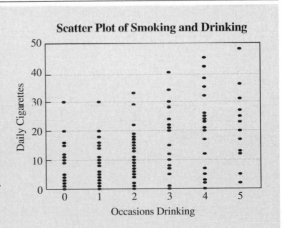

Scatter Plot of Smoking and Drinking

Continued

Let's look at the correlation between levels of smoking and drinking. Are both these forms of consumption connected in any way? We see in the scatterplot that there is a tendency for the two variables to correlate. For those students who seldom drank, the dots tend to cluster at the low end of smoking; for those who drank more often, the dots spread over a wider range of smoking levels, especially the highest. The Pearson's correlation coefficient, $r = +.34$, indicates a fairly moderate and positive association. With $N = 155$ (only smokers included), the correlation is statistically significant ($t = 4.47$, 153 df). For the population, therefore, we can conclude that students who drink a lot also tend to smoke a lot.

Next, we consider the relationship between age and drinking. By law, none of these students is supposed to drink, but clearly many of them report doing so. Is there a tendency for those who are closer to the legal drinking age to drink more often than younger students? As indicated by the correlation coefficient, $r = +.62$, older students do tend to drink more often. Since this correla-

tion is significant, ($t = 12.44$, 248 df), we can conclude that age and smoking are correlated in the population of students.

Cross-Tabulation of Extent of Participation in Sports/Exercise and Smoking

	Smoke	Not Smoke
Very frequent	8	35
Often	48	39
Seldom	65	16
Never	34	5

$\chi^2 = 58.0$, 3 df (significant), Cramér's $V = .48$

Finally, we are interested in the association between athletic involvement and whether a student smokes. We can display these data in the form of a cross-tabulation and calculate Cramer's V, a measure of association designed for non-square tables.

Part V

Applying Statistics

Chapter 13
Choosing Statistical Procedures for Research Problems

13

Choosing Statistical Procedures for Research Problems

It surely has been a long and sometimes difficult journey through the chapters of this book, from the early stages of statistical description through a set of some of the most commonly used techniques for statistically analyzing social science data. In Chapter 1 we attempted to provide a roadmap of the major themes that you would encounter along this journey into statistics. At this final stop, it may be helpful to look back at the highlights of the trip and what you have learned on route.

Chapter 1 introduced some basic concepts concerning variables, the raw ingredients used in any statistical recipe. Most important, variables come in three types, depending on the precision and complexity of measurement. Nominal-level variables (such as whether last Saturday night's date was to "a movie," "a club," or "a party") can be categorized and counted. Ordinal-level variables (such as whether your Saturday night date was "great," "pretty good," or "never again") can likewise be categorized and counted, but they also can be ranked from one extreme to the other (say from best to worst). Finally, interval-level variables (such as how long the Saturday night date lasted) can be categorized, counted, ranked, and scored on a fixed numerical scale (say length in hours).

Chapter 2 then presented methods for tabulating and graphing variables as a first step in summarizing and describing their distributions. Frequency distributions and pie or bar graphs help us to gain a clearer idea about our data. We might summarize, for example, the Saturday-night date experiences of a group of 40 classmates—what percentage went to the movies, what percentage were pleased with their date. Grouped frequency distributions were also used to help to summarize interval-level variables like length of date into some manageable set of categories (say, "Under 2 hours," "2 hours up to 4 hours," "4 hours up to 6 hours," and "6 hours until dawn"). Frequency polygons were also used to explore the shape of a distribution, for example, whether date length tends to be symmetric or skewed among the 40 date-length scores.

Equipped with a basic summary of the distribution of a variable, Chapters 3 and 4 next introduced measures for describing in precise terms certain characteristics of a distribution.

Specifically, three measures of central tendency provide different perspectives on "the average," depending on the research objective and level of measurement. For the date experiences of the 40 classmates, we might calculate the modal (most common) date type as "to a party." We might also determine the median (or middlemost) date quality as "pretty good," and the mean date length as 4.3 hours.

There is more to a distribution of scores than the mode, median, and mean, of course. Measures of variability (range, variance, and standard deviation) were used to characterize how the data are dispersed around the center of a distribution. To say that the 40 classmates had a mean date length of 4.3 hours is only half the story. Learning that their dates ranged from 1.5 to 12.0 hours raises all sorts of interesting questions about how they got along and what they did while together.

Chapter 5 focused on the normal curve, a particular distributional form (symmetric and bell-shaped) having most cases falling near the middle and a few falling in either the high or low extremes or tails. Despite its familiarity to anyone who has been graded "on a curve," there are actually rather few variables (measures like IQ, SAT scores, height, and weight) that possess this kind of normal distribution. Still, the normal curve plays a very critical role in helping us to make inferences from samples to populations.

Chapter 6 provided the theoretical foundation for understanding how we can generalize from modest size samples to large populations, for example, how we might generalize the average date length of a sample of 40 classmates to the population of the entire student body, if not to all college-age students across the country. It is perhaps the most important statistical "fact of life" that the sample mean (if based on a reasonably large number of cases) is normally distributed, no matter what the distribution of the raw scores on which the mean is based.

The sample of 40 classmates may have yielded a mean date length of 4.3 hours, yet another similarly drawn sample of 40 students might have produced a shorter average (say 4.1 hours) if there were no "all-nighters" in the group. Or another sample could perhaps have produced a mean of 4.8 hours. The actual sample mean obtained depends to some extent on the luck of the draw.

In practice, we only draw one sample of cases (for example, $N = 40$ cases), yet use the theory about the sampling distribution of all possible sample means of the date length of 40 students (which is normal in shape) to generalize to what the population mean date length would likely be. In practice, we use the sample mean and its standard error to construct an interval representing the range spanning 95% of all possible sample means. That is, we can conclude with 95% confidence that the true population mean is within a certain margin of error from our sample mean date length, for example, from 4.15 hours to 4.45 hours. Chapter 6 also presented a method for estimating the population proportion based on a sample, for example, the percentage of all Saturday night dates in the population that involved a movie.

Chapter 7 introduced the logic of hypothesis testing in the context of making comparisons between two groups. For example, do residential students or commuters stay out longer on a Saturday night date? We might test the null hypothesis of no difference in mean date length for the populations of residential and commuter students at the .05 level of significance. The *t* ratio offers a method for determining whether a difference in mean date length between independently drawn samples of 65 dormitory residents and 48 commuters is large enough to allow us to reject the null hypothesis. We may obtain a mean of 4.39 for residential students and 4.13 for commuters; the *t* test helps us to assess whether the difference between sample means (4.39 − 4.13 = .26 hour) could just be the result of sampling error.

Chapter 7 also presented an approach for testing differences between means for the same sample measured twice. For example, we might compare the mean date length of the first and second dates for a sample of 20 students to see whether the dates tend to grow longer the second time around. Chapter 7 also presented a method for testing differences between groups in terms of proportions. For example, we might compare samples of residential and commuter students in terms of the percentage of dates involving clubbing to test the null hypothesis of no population difference between residential students and commuters in their likelihood of going to a club on Saturday night dates.

Chapter 8 extended the comparison of means approach to several groups, that is, samples from three or more populations. Through the analysis of variance, we can test whether the population mean date lengths for business majors, science majors, humanities majors, and social science majors are equal. We might obtain sample means of 4.4, 4.1, 4.3, and 4.8, respectively. We then examine whether the variability among these four group means is larger than what we could expect by chance if the population means were equal.

Chapter 9 introduced nonparametric techniques. Whereas the *t* test and analysis of variance make strong assumptions about the nature of the data, specifically interval-level scores (as with date length in hours), nonparametric approaches focus on counts or frequencies as well as percentages—basic characteristics of all types of variables. Thus, nonparametric tests are more widely applicable, but they also are less powerful, less able to reject the null hypothesis when it is false.

The chi-square test was presented for analyzing cross-tabulated data. In essence, the focus of chi-square is on whether the frequency distribution of some variable differs significantly between groups defined by some other variable. For example, we might examine whether type of date (movie, club, or party) differs by type of major. The chi-square test assesses whether the frequencies observed in a cross-tabulation diverge from the frequencies one would expect under the null hypothesis of no difference between groups (type of major) in terms of the distribution of date type.

Chapter 10 moved the focus to measuring the correlation between variables. Whereas Chapters 7 through 9 had focused on differences in the mean or proportion of some variable between groups defined by another variable, the measures of association presented in Chapters 10 through 12 consider the degree to which two variables are related to one another.

Introduced in Chapter 10, Pearson's correlation coefficient measures the linear correlation between two interval-level variables. Pearson's *r* examines the extent to which two variables move in the same or opposite direction, whether changes in one variable tend to be linked statistically to changes in the other. We might correlate student age with date length, for example, to determine if older or younger students stay out longer on a Saturday night date. Furthermore, regression, described in Chapter 11, allows us to measure the impact of age (the independent variable) on date length (the dependent variable). A slope of .40 would suggest that among college-age students dates tend to increase by .4 hours (or 24 minutes) with each year of age.

Finally, Chapter 12 extended the correlational approach to variables that are measured at less than the interval level. Spearman's correlation is specifically designed to compare ranks on two ordinal-level variables. For example, you might rank your classmates in terms of both popularity and date length and use Spearman's correlation to test their association. Do more popular students stay out longer on dates than their less popular classmates? In addition, Gamma (for ordinal measures) and phi, contingency coefficient, and Cramér's V (for nominal measures) can be used to assess the strength of association for data presented in the form of a cross-tabulation.

As noted throughout the text, each statistical procedure has a set of assumptions for its appropriate application. In selecting among procedures, any researcher must therefore consider a number of factors, such as the following:

1. Whether the researcher seeks to test for statistically significant differences, degree of association, or both
2. Whether the researcher has achieved the nominal, ordinal, or interval level of measurement of the variables being studied
3. Whether the variables being studied are normally distributed in the population from which they were drawn (and, if not, whether the sample size is large enough to relax this requirement)

This chapter provides a number of hypothetical research situations in which the foregoing criteria are specified. The reader is asked to choose the most appropriate statistical procedure for each research situation from among the following tests that have been covered in Parts III and IV of the text:

1. t ratio
2. Analysis of variance
3. Chi-square test
4. Median test
5. Pearson's r correlation coefficient and regression
6. Spearman's rank-order correlation coefficient
7. Goodman's and Kruskal's gamma
8. Phi coefficient
9. Contingency coefficient
10. Cramér's V

Table 13.1 locates each statistical procedure with respect to some of the important assumptions that must be considered for its appropriate application. Looking at the columns of the table, we face the first major decision related to the selection of a statistical procedure: Do we wish to determine whether a relationship exists? The tests of significance discussed in Chapter 7 to 9 are designed to determine whether an obtained sample difference reflects a true population difference. Or do we seek instead to establish the strength of the relationship between two variables? This is a question of correlation that can be addressed by means of the statistical procedures introduced in Chapters 10 to 12.

The rows of Table 13.1 direct our attention to the level at which our variables are measured. If we have achieved the interval level of measurement, we may well consider employing a parametric procedure such as t, F, or r. If, however, we have achieved either

TABLE 13.1 *Choosing an Appropriate Statistical Technique*

Level of Measurement	Test of Difference (Chapters 7–9)	Measure of Correlation (Chapters 10–12)
Nominal	Chi-square test (nonparametric test for comparing two or more samples)	Phi coefficient (nonparametric measure for 2×2 tables) Contingency coefficient and Cramér's V (nonparametric measures for larger than 2×2 tables)
Ordinal	Median test (nonparametric test for comparing two samples)	Spearman's rank-order correlation coefficient (nonparametric measure for ranked data) Goodman's and Kruskal's gamma (nonparametric measure for cross tabulations)
Interval	t ratio (parametric test for comparing two samples or the same sample measured twice) Analysis of variance (parametric test for comparing three or more samples)	Pearson's r correlation coefficient and regression (parametric measurement for scores)

These are the minimal levels of measurement required to apply a particular technique.

the nominal or ordinal level of measurement, the choice is limited to several nonparametric alternatives. Finally, for correlation problems involving variables of different levels of measurement, a correlation measure appropriate for the lower level of the two is used. For example, Spearman's rank-order coefficient can be used to measure the correlation between an interval and ordinal variable; Cramér's V can be used to measure the association between an ordinal and nominal variable.

The solutions to the following research situations can be found at the end of the chapter.

Research Situations

Research Situation 1

A researcher conducted an experiment to determine the effect of a lecturer's age on student preferences to hear him lecture. In a regular classroom situation, 20 students were told that the administration wished to know their preferences regarding a forthcoming visiting lecturer series. In particular, they were asked to evaluate a professor who "might be visiting the campus." The professor was described to all students in the same way with one exception: One-half of the students were told the professor was 65 years old; one-half were told the professor was 25 years old. All students were then asked to indicate their willingness to attend the professor's lecture (higher scores indicate greater willingness). The following results were obtained:

X_1 (Scores of Students Told Professor Was 25 Years Old)	X_2 (Scores of Students Told Professor Was 65 Years Old)
65	78
38	42
52	77
71	50
69	65
72	70
55	55
78	51
56	33
80	59

Which statistical procedure would you apply to determine whether there is a significant difference between these groups of students with respect to their willingness to attend the lecture?

Research Situation 2

A researcher conducted an experiment to determine the effect of a lecturer's age on student preferences to hear her lecture. In a regular classroom situation, 30 students were told that the administration wished to know their preferences regarding a forthcoming visiting lecturer series. In particular, they were asked to evaluate a professor who "might be visiting the campus." The professor was described to all students in the same way with one exception: One-third of the students were told that the professor was 75 years old; one-third were told the professor was 50 years old; and one-third were told the professor was 25 years old. All students were then asked to indicate their willingness to attend the professor's lecture (higher scores indicate greater willingness). The following results were obtained:

X_1 (Scores of Students Told Professor Was 25 Years Old)	X_2 (Scores of Students Told Professor Was 50 Years Old)	X_3 (Scores of Students Told Professor Was 75 Years Old)
65	63	67
38	42	42
52	60	77

71	55	32
69	43	52
72	36	34
55	69	45
78	57	38
56	67	39
80	79	46

Which statistical procedure would you apply to determine whether there is a significant difference between these groups of students with respect to their willingness to attend the lecture?

Research Situation 3

To investigate the relationship between spelling and reading ability, a researcher gave spelling and reading examinations to a group of 20 students who had been selected at random from a large population of undergraduates. The following results were obtained (higher scores indicate greater ability):

Student	X (Spelling Score)	Y (Reading Score)
A	52	56
B	90	81
C	63	75
D	81	72
E	93	50
F	51	45
G	48	39
H	99	87
I	85	59
J	57	56
K	60	69
L	77	78
M	96	69
N	62	57
O	28	35
P	43	47
Q	88	73
R	72	76
S	75	63
T	69	79

Which statistical procedure would you apply to determine the degree of association between spelling and reading ability?

Research Situation 4

To investigate the validity of a particular reading test, researchers gave the reading test to a sample of 20 students whose ability to read had been previously ranked by their teacher. The test score and teacher's rank for each student are listed in the following table:

Student	X (Reading Score)	Y (Teacher's Rank)
A	28	18
B	50	17
C	92	1
D	85	6
E	76	5
F	69	10
G	42	11
H	53	12
I	80	3
J	91	2
K	73	4
L	74	9
M	14	20
N	29	19
O	86	7
P	73	8
Q	39	16
R	80	13
S	91	15
T	72	14

Which statistical procedure would you apply to determine the degree of association between reading scores and teacher's ranking?

Research Situation 5

To investigate regional differences in helpfulness toward strangers, a researcher dropped 400 keys (all of which had been stamped and tagged with a return address) around mailboxes in the northeastern, southern, midwestern, and western regions of the United States. The number of keys returned by region (as an indicator of helpfulness) is indicated in the following table:

	Region			
	Northeast	*South*	*Midwest*	*West*
Returned	55	69	82	61
Not returned	45	31	18	39
	100	100	100	100

Which statistical procedure would you apply to determine whether these regional differences are statistically significant?

Research Situation 6

To examine the relationship between authoritarianism and prejudice, a researcher administered measures of authoritarianism (the *F* scale) and prejudice (a checklist of negative adjectives to be assigned to African Americans) to a national sample of 950 adult Americans. The following results were obtained: Among 500 authoritarian respondents, 350 were prejudiced and 150 were tolerant. Among 450 nonauthoritarian respondents, 125 were prejudiced and 325 were tolerant.

Which statistical procedure would you apply to study the degree of association between authoritarianism and prejudice?

Research Situation 7

To investigate the relationship between year in school and grade-point average, researchers examined the academic records of 186 college students who were selected on a random basis from the undergraduate population of a certain university. The researchers obtained the following results:

Grade-Point	Year in School			
Average	*1st*	*2nd*	*3rd*	*4th*
A– or better	6	5	7	10
B– to B+	10	16	19	18
C– to C+	23	20	15	7
D+ or worse	15	7	6	2
	54	48	47	37

Which statistical procedure would you apply to determine the degree of association between grade-point average and year in school?

Research Situation 8

To investigate the influence of frustration on prejudice, 10 subjects were asked to assign negative adjectives—such as *lazy, dirty*, and *immoral*—to describe the members of a minority group (a measure of prejudice). All subjects described the minority group both before and after they had taken a series of lengthy and difficult examinations (the frustrating situation). The following results were obtained (higher scores represent greater prejudice):

Subject	X_1 (Prejudice Scores before Taking the Frustrating Examinations)	X_2 (Prejudice Scores after Taking the Frustrating Examinations)
A	22	26
B	39	45
C	25	24

(continued)

Subject	X_1 (Prejudice Scores before Taking the Frustrating Examinations)	X_2 (Prejudice Scores after Taking the Frustrating Examinations)
D	40	43
E	36	36
F	27	29
G	44	47
H	31	30
I	52	52
J	48	59

Which statistical procedure would you apply to determine whether there is a statistically significant difference in prejudice before and after the administration of the frustrating examinations?

Research Situation 9

To investigate the relationship between a respondent's actual occupational status and his or her subjective social class (that is, a respondent's own social class identification), 677 individuals were asked to indicate their occupation and the social class in which they belonged. Among 190 respondents with upper-status occupations (professional–technical–managerial), 56 identified themselves as upper class, 122 as middle class, and 12 as lower class; among 221 respondents with middle-status occupations (sales–clerical–skilled labor), 42 identified themselves as upper class, 163 as middle class, and 16 as lower class; among 266 with lower-status occupations (semiskilled and unskilled labor), 15 identified themselves as upper class, 202 as middle class, and 49 as lower class.

Which statistical procedure would you apply to determine the degree of association between occupational status and subjective social class?

Research Situation 10

To investigate the influence of college major on the starting salary of college graduates, researchers interviewed recent college graduates on their first jobs who had majored in engineering, liberal arts, or business administration. The results obtained for these 21 respondents are the following:

Starting Salaries

Engineering	Liberal Arts	Business
$38,500	$25,000	$30,500
32,300	21,000	28,000
34,000	25,000	23,000
29,500	22,000	24,300

29,000	20,500	38,500
28,500	19,500	40,000
27,500	28,000	21,000

Which statistical procedure would you apply to determine whether there is a significant difference between these groups of respondents with respect to their starting salaries?

Research Situation 11

To investigate the influence of college major on the starting salary of college graduates, researchers interviewed recent college graduates on their first jobs who had majored in either liberal arts or business. The results obtained for these 16 respondents are the following:

Starting Salaries

Liberal Arts	Business
$25,000	$30,500
21,500	28,000
25,000	23,000
22,500	24,300
20,500	38,500
19,000	40,000
28,000	21,000

Which statistical procedure would you apply to determine whether there is a significant difference between liberal arts majors and business majors with respect to starting salaries?

Research Situation 12

A researcher conducted an experiment to determine the effect of a lecturer's age on student willingness to hear him lecture. In a regular classroom situation, 130 students were told that the administration wished to know their preferences regarding a forthcoming visiting lecturer series. In particular, they were asked to evaluate a professor who "might be visiting the campus." The professor was described to all students in the same way with one exception: One-half of the students were told the professor was 65 years old; one-half were told the professor was 25 years old. All students were then asked to indicate their willingness to attend the professor's lecture with the following results: Among those students told that the professor was 65, 22 expressed their willingness to attend his lecture and 43 expressed their unwillingness; among the students told that the professor was 25, 38 expressed their willingness to attend his lecture and 27 expressed their unwillingness.

Which statistical procedure would you apply to determine whether there is a significant difference between these groups of students with respect to their willingness to attend the professor's lecture?

Research Situation 13

To study the influence of teacher expectancy on student performance, achievement tests were given to a class of 15 third-grade students both before and six weeks after their teacher was informed that all of them were gifted. The following results were obtained (higher scores represent greater achievement):

Student	X_1 (Achievement Scores before Teacher Expectancy)	X_2 (Achievement Scores after Teacher Expectancy)
A	98	100
B	112	113
C	101	101
D	124	125
E	92	91
F	143	145
G	103	105
H	110	115
I	115	119
J	98	99
K	117	119
L	93	99
M	108	105
N	102	103
O	136	140

Which statistical procedure would you apply to determine whether there is a statistically significant difference in achievement before and after the introduction of the teacher expectancy?

Research Situation 14

To investigate the relationship between neighborliness and voluntary association membership, a researcher questioned 500 public housing tenants, chosen at random, concerning the amount of time they spend with other tenants (high, medium, or low neighborliness) and the number of clubs and organizations in which they participate (high, medium, or low voluntary association membership). Questionnaire results were as follows: Among 150 tenants who were high in regard to neighborliness, 60 were high, 50 were medium, and 40 were low with respect to voluntary association membership; among 180 tenants who were medium in regard to neighborliness, 45 were high, 85 were medium, and 50 were low with respect to voluntary association membership; and among 170 tenants who were low in regard to neighborliness, 40 were high, 50 were medium, and 80 were low with respect to voluntary association membership.

Which statistical procedure would you apply to determine the degree of association between neighborliness and voluntary association membership?

Research Situation 15

To determine the effect of special needs labels on teachers' judgments of their students' abilities, a researcher asked a sample of 40 elementary school teachers to evaluate the academic potential of an 11-year-old girl who was entering the sixth grade. All teachers were given the report of a "school psychologist" who described the girl in exactly the same way—with one exception: One-fourth were told that the child was emotionally disturbed, one-fourth were told she was mentally retarded, one-fourth were told she was learning disabled, and one-fourth were given no special needs label. Immediately after reading about the girl, all teachers were asked to indicate on the numeric rating scale how successful they felt her academic progress would be (higher scores indicate greater optimism). The following scores were obtained:

X_1 (Scores for Emotionally Disturbed Label)	X_2 (Scores for Mentally Retarded Label)	X_3 (Scores for Learning Disabled Label)	X_4 (Scores for No Label)
23	29	25	42
29	42	46	51
41	32	53	31
36	25	56	37
37	37	44	40
56	48	41	43
45	42	38	55
39	28	32	52
28	30	50	42
32	32	43	24

Which statistical procedure would you apply to determine whether there is a significant difference between these groups of teachers with respect to their evaluations?

Research Situation 16

As an indirect way to study the stress associated with moving to a new city, data were collected on what percentage of the population had moved to the given locale in the past year and the amount of antacid sold in that locale. The percentage of the population that had moved to that particular locale in the past year and rank of that locale in antacid sales are presented in the following table:

Metropolitan Area	X (Percentage of the Population Who Moved There in the Past Year)	Y (Rank with Respect to Antacid Sales)
A	10	1
B	4	6
C	2	3

(continued)

Metropolitan Area	X (Percentage of the Population Who Moved There in the Past Year)	Y (Rank with Respect to Antacid Sales)
D	1	7
E	6	4
F	5	2
G	8	5

Which statistical procedure would you apply to determine the degree of association between the percentage of population who had recently moved and antacid sales?

Research Situation 17

To study the gossip of male and female college students, a researcher unobtrusively sat in the campus center and listened to students' conversations. The researcher was able to categorize the tone of students' gossip as negative (unfavorable statements about a third person), positive (favorable statements about a third person), or neutral. The following results were obtained: For 125 instances of gossip in the conversations of female students, 40 were positive, 36 were negative, and 49 were neutral. For 110 instances of gossip in the conversations of male students, 36 were positive, 32 were negative, and 42 were neutral.

Which statistical procedure would you apply to determine whether there is a significant difference between male and female college students with respect to the tone of their gossip?

Research Situation 18

To investigate the relationship between musical taste and food preference, a random sample of 331 people were asked what kind of music (classical, jazz, rock) and what kind of ethnic food (Chinese, French, Italian) they most preferred. Among the 80 people who preferred classical music, 8 preferred Chinese food, 52 preferred French food, and 20 preferred Italian. Among the 112 people who preferred jazz, 55 preferred Chinese food, 18 preferred French food, and 39 preferred Italian. Among the 139 rock lovers, 47 liked Chinese, only 8 liked French, and 84 were gourmands of Italian cuisine.

Which statistical procedure would you apply to determine the degree of association between the type of music and the type of food the respondents prefer?

Research Situation 19

A researcher interested in the relationship between the percentage of single men and the rape rate collected the following data regarding rapes per 100,000 and the percentage of single men in the population for eight cities over a 1-year period:

City	Percent Single Men	Rapes per 100,000
A	29	55
B	26	40

C	19	37
D	27	34
E	30	48
F	31	42
G	16	31
H	24	51

Which statistical procedure would you apply to determine the degree of association between rape rate and the percentage of single men in the population of these eight cities?

Research Situation 20

To study the effect of classroom size on students' course evaluations, a researcher randomly assigned 20 college students to one of two sections of the same course taught by the same instructor: a large class (50+) and a small class (under 25). The following course evaluations—on a four-point scale from 1 (poor) to 4 (outstanding)—were obtained at the end of the semester:

Large Class	Small Class
3	4
3	2
1	3
4	3
2	4
1	4
2	3
3	3
4	3
3	4

Which statistical procedure would you apply to determine whether a significant difference exists between large and small classes with respect to course evaluations?

Research Situation 21

A researcher was interested in the effects of anomie—normlessness or breakdown of rules in a social setting. She obtained the following suicide rates (the number of suicides per 100,000 population) for five high-anomie, five moderate-anomie, and five low-anomie cities:

	Anomie	
High	*Moderate*	*Low*
19.2	15.6	8.2
17.7	20.1	10.9
22.6	11.5	11.8
18.3	13.4	7.7
25.2	14.9	8.3

Which statistical procedure would you apply to determine whether there is a significant difference by level of anomie in suicide rates?

Research Situation 22

A researcher conducted a study to examine the effect of chronological age on poverty status—in particular, she wanted to compare older people (65+), middle-aged people, young adults, and children in terms of number living below the poverty level. Her results were as follows: Among 200 older people, 36 lived below the poverty level; among 350 middle-aged people, 50 lived below the poverty level; among 240 young adults, 40 lived below the poverty level; and among 250 children, 51 lived below the poverty level.

Which statistical procedure would you apply to determine whether there is a significant difference between age categories with respect to level of poverty?

Research Situation 23

In a study of the relationship between mental and physical health, a random sample of 250 patients was questioned regarding their symptoms of depression (for example, insomnia, lack of concentration, suicidal thoughts) and given a physical examination to uncover any symptoms of physical illness (for example, high blood pressure, erratic EKG, high cholesterol). Among the 100 patients categorized as being in "excellent physical health," only 5 exhibited symptoms of depression; among the 110 patients categorized in "good health," 14 exhibited symptoms of depression; and among the 40 patients categorized in "poor health," 20 exhibited symptoms of depression.

Which statistical procedure would you apply to determine the degree of association between depression and physical illness?

Research Situation 24

In a study of the relationship between gender and physical health, a random sample of 200 patients—100 men and 100 women—was given a physical examination to uncover any symptoms of physical illness (for example, high blood pressure, erratic EKG, high cholesterol). Among the 100 men, 37 were categorized as being in "excellent physical health," 43 were categorized in "good health," and 20 were categorized in "poor health." Among the 100 women, 52 were categorized as being in "excellent physical health," 35 were categorized in "good health," and 13 were categorized in "poor health."

Which statistical procedure would you apply to determine the degree of association between gender and physical illness?

Research Situation 25

Shortly after the robbery trial of O. J. Simpson ended, black and white Americans were asked by pollsters whether they agreed with the jury's guilty verdict. In one of many such studies, a researcher questioned 150 respondents—75 blacks and 75 whites—as to Simpson's guilt or innocence. He determined that 51 white Americans but only 25 black Americans agreed with the jury that O. J. Simpson was guilty.

Which statistical procedure would you apply to determine whether black and white Americans differed significantly in regard to their agreement with the jury's verdict?

Research Situation 26

In a study of the relationship between mental and physical health, a random sample of 15 subjects was questioned regarding their symptoms of depression (for example, insomnia, lack of concentration, suicidal thoughts) and given a physical examination to uncover any symptoms of physical illness (for example, high blood pressure, erratic EKG, high choles terol). All subjects were then scored on a checklist of depression from 0 to 5, depending on how many symptoms of depression they exhibited, and scored on a checklist of physical illness from 1 to 10, indicating how they had fared on the physical exam (higher scores indicate worse health). The following results were obtained:

Subject	Depression	Physical Illness
A	2	0
B	0	2
C	5	7
D	1	1
E	3	8
F	0	2
G	1	3
H	2	5
I	2	3
J	0	2
K	4	5
L	1	5
M	3	6
N	1	3
O	2	3

Which statistical procedure would you apply to determine the degree of association between depression and physical illness?

Research Situation 27

To investigate whether race has an influence on religiosity, a group of researchers surveyed a random sample of people on a range of questions, including how often they attend religious services and how often they pray. After collecting the data, the researchers used the responses to create a religiosity scale ranging from 1 to 7, with 7 being the most religious. After using this scale to rate respondents in terms of their religiosity, the following results were obtained:

Black	White	Hispanic
3	5	7
5	2	3
1	6	2

(continued)

Black	White	Hispanic
7	4	6
	4	5
	7	
	2	

Which statistical procedure would you apply to determine whether there are significant differences among these groups of respondents with respect to their religiosity?

Research Situation 28

A political science graduate student wondered how much the first presidential debate could influence people's choice of candidates in the 2008 election. She selected a random group of voters, and before the first debate she asked them to rate on a scale from 1 to 9 who they intended to vote for in the election (1 = definitely McCain, 5 = undecided, and 9 = definitely Obama). The sampled voters were then asked the same question again after they had viewed the debate. The following results were obtained:

Voter	X_1 (Before First Debate)	X_2 (After First Debate)
A	2	4
B	5	8
C	1	1
D	7	9
E	2	1
F	5	5
G	9	7
H	3	8
I	5	2
J	9	9

Which statistical procedure would you apply to determine whether there is a statistically significant difference in voters' choice of candidates before and after the first presidential debate?

Research Situation 29

A group of 250 male athletes undergo a series of medical and psychological tests prior to a large competition, and 78 of them test positive for steroid use. Out of these 78 athletes, 46 are classified as having "high aggression," 21 are classified as having "medium aggression," and 42 are classified as having "low aggression." Of the 172 athletes who did not test positive for steroids, 34 are classified as having "high aggression," 96 are classified as having "medium aggression," and 42 are classified as having "low aggression."

Which statistical procedure would you apply to determine the degree of association between steroid use and aggression?

Research Situation 30

In a study of the relationship between seasons and depression, a Canadian researcher collects data about the number of suicides (both completed and attempted) that have occurred in his city throughout the past year, as well as the average number of daylight hours during the weeks in which the suicides occurred. The following results were obtained:

Hours of Daylight	Number of Suicides
18	6
17	5
16	9
15	11
14	13
13	12
12	15
11	18
10	24
9	35

Which statistical procedure would you apply to determine the degree of association between hours of daylight and number of suicides?

Research Situation 31

For years, women's rights groups have urged for the passage of the Equal Rights Amendment (ERA), which would officially give women the same rights as men under the U.S. Constitution. However, this amendment has yet to be ratified by all 50 states. To find out if there is a relationship between a state legislator's gender and whether he or she supports the passage of the ERA, a random sample of 84 state legislators was surveyed and the following results were obtained: Among 35 female legislators, 27 supported the ERA and 8 did not; and among 50 male legislators, 17 supported the ERA and 33 did not.

Which statistical procedure would you apply to study the degree of association between legislator's gender and support for the ERA?

Research Situation 32

Following the terrorist attacks of September 11, 2001, the U.S. government took several steps to help prevent such attacks from occurring in the future. This included the passage of the Patriot Act, which gave the government more power in its fight against terrorism. Do Americans from both ends of the political spectrum support this act equally? To find out, a researcher surveyed 20 random people aged 18 and over and ranked them in terms of their political leaning (with 1 being the most conservative and 20 being the most liberal). The respondents were also scored in terms of their support of the Patriot Act (with higher scores indicating more support). The results were as follows:

Subject	X (Political Leaning)	Y (Support of Patriot Act)
A	15	11
B	9	55
C	3	83
D	17	24
E	19	36
F	10	48
G	1	97
H	8	78
I	13	50
J	5	88
K	2	91
L	16	15
M	4	77
N	18	13
O	20	10
P	12	65
Q	7	74
R	6	81
S	11	42
T	14	33

Which statistical procedure would you apply to determine the degree of association between political leaning and support of the Patriot Act?

Research Situation 33

To investigate whether hospitalized patients recover more quickly when they receive frequent visits from friends and family members, a group of researchers selected a random sample of 160 patients and collected data about their speed of recovery and frequency of visits. The following results were obtained:

Speed of Recovery	Frequency of Visits			
	Often	*Sometimes*	*Rarely*	*Never*
Fast	33	8	6	5
Medium	12	25	11	3
Slow	6	16	26	9
	51	49	43	17

Which statistical procedure would you apply to determine the degree of association between frequency of hospital visits and speed of recovery?

Research Situation 34

A researcher wonders if a 4-week public speaking course can really help people overcome their anxiety about talking in front of an audience. Before the first class, she selects a random group of people who have enrolled in the class due to anxiety problems and asks them to rank on a scale from 1 to 10 how anxious they feel about public speaking. The respondents are then asked the same question again after the last class. The following results are obtained (with higher scores indicating more anxiety):

Subject	X_1 (Anxiety before Class)	X_2 (Anxiety after Class)
A	10	6
B	9	5
C	10	8
D	7	3
E	9	5
F	8	6
G	10	6
H	9	4
I	10	7

Which statistical procedure would you apply to determine whether there is a statistically significant difference in anxiety levels before and after the public speaking course?

Research Situation 35

A marriage counselor wonders if couples who seek her services due to an infidelity (either on the part of the husband, the wife, or both) end up getting divorced more often than do couples who seek her services for reasons other than infidelity. She consults her files and collects the following data about a random sample of 45 couples:

	Reason for Marriage Counseling	
	Infidelity	*Not Infidelity*
Divorced	19	13
Not Divorced	6	7

Which statistical procedure would you apply to determine whether there is a statistically significant difference between the divorce rates of these two groups of couples?

Research Solutions

Solution to Research Situation 1 (t Ratio)

This situation represents a comparison between the scores of two independent samples of students. The *t* ratio (Chapter 7) is employed to make comparisons between two means when interval data have been obtained. The median test (Chapter 9) is a nonparametric alternative that can be applied when we suspect that the scores are not normally distributed in the population or that the interval level of measurement has not been achieved.

Solution to Research Situation 2 (Analysis of Variance)

This situation represents a comparison of the scores of three independent samples of students. The *F* ratio (analysis of variance, Chapter 8) is employed to make comparisons between three or more independent means when interval data have been obtained. The median test (Chapter 9) is a nonparametric alternative that can be applied when we have reason to suspect that the scores are not normally distributed in the population or when the interval level of measurement has not been achieved.

Solution to Research Situation 3 (Pearson's r Correlation Coefficient)

This situation is a correlation problem, because it asks for the degree of association between *X* (spelling ability) and *Y* (reading ability). Pearson's *r* (Chapter 10) can be employed to detect a straight-line correlation between *X* and *Y* variables when both of these variables have been measured at the interval level. If *X* (spelling ability) and *Y* (reading ability) are not normally distributed in the population, we could consider applying a nonparametric alternative such as Spearman's rank-order correlation coefficient (Chapter 12).

Solution to Research Situation 4 (Spearman's Rank-Order Correlation Coefficient)

This situation is a correlation problem, asking for the degree of association between *X* (reading scores) and *Y* (teacher's rankings of reading ability). Spearman's rank-order correlation coefficient (Chapter 12) can be employed to detect a relationship between *X* and *Y* variables when both of these variables have been ordered or ranked. Pearson's *r* cannot be employed, because it requires interval-level measurement of *X* and *Y*. In the present case, reading scores *(X)* must be ranked from 1 to 20 before rank-order is applied.

Solution to Research Situation 5 (Chi-Square Test)

This situation is a comparison between the frequencies (returned versus not returned) found in four groups (northeast, south, midwest, and west). The chi-square test of significance (Chapter 9) is used to make comparisons between two or more samples. Only nominal data are required. Present results can be cast in the form of a 2×4 table, representing 2 rows and

4 columns. Notice that the degree of association between return rate *(X)* and region *(Y)* can be measured by means of Cramér's *V* (Chapter 12). Note that the contingency coefficient is not a preferred measure here because the contingency table is not square.

Solution to Research Situation 6 (Phi Coefficient)

This situation is a correlation problem that asks for the degree of association between X (authoritarianism) and Y (prejudice). The phi coefficient (Chapter 12) is a measure of association that can be employed when frequency or nominal data can be cast in the form of a 2×2 table (2 rows by 2 columns). In the present problem, such a table would take the following form:

Level of Prejudice	Level of Authoritarianism	
	Authoritarian	*Nonauthoritarian*
Prejudiced	350	125
Tolerant	150	325
	$N = 950$	

Solution to Research Situation 7 (Goodman's and Kruskal's Gamma)

This situation is a correlation problem that asks for the degree of association in a cross-tabulation of X (grade-point average) and Y (year in school). Goodman's and Kruskal's gamma coefficient (Chapter 12) is employed to detect a relationship between X and Y, when both variables are ordinal and have been cast in the form of a cross-tabulation. In the present problem, grade-point average has been ranked from A to D or worse and year in school has been ranked from first to fourth. The contingency coefficient *(C)* or Cramér's *V* (Chapter 12) represents an alternative to gamma that assumes only nominal-level data. However, because these variables are ordinal, gamma is preferable.

Solution to Research Situation 8 (t Ratio)

This situation represents a before–after comparison of a single sample measured at two different points in time. The *t* ratio (Chapter 7) can be employed to compare two means from a single sample arranged in a before–after panel design.

Solution to Research Situation 9 (Goodman's and Kruskal's Gamma)

This situation is a correlation problem that asks for the degree of association between X (occupational status) and Y (subjective social class). Gamma (Chapter 12) is especially well suited to the problem of detecting a relationship between X and Y, when both variables are ordinal and can be arranged in the form of a cross-tabulation. In the present situation, occupational status and subjective social class have been ordered from upper to middle to lower, generating a very large number of tied ranks (for example, 221 respondents had middle-status occupations). To obtain the gamma coefficient, the data must be cast in the form of a cross-tabulation as follows:

Subjective Social Class *(Y)*	Occupational Status *(X)*		
	Upper	*Middle*	*Lower*
Upper	56	42	15
Middle	122	163	202
Lower	12	16	49
	190	221	266

The contingency coefficient *(C)* and Cramér's *V* are alternatives to gamma that assume only nominal data. Because these variables are ordinal, gamma would be preferable.

Solution to Research Situation 10 (Analysis of Variance)

This situation is a comparison of the scores of three independent samples of respondents. The *F* ratio (Chapter 8) is used to make comparisons between three or more independent means when interval data have been obtained. The median test (Chapter 9) is a nonparametric alternative that can be employed when we suspect that the scores may not be normally distributed in the population or when the interval level of measurement has not been achieved.

Solution to Research Situation 11 (t Ratio)

This situation is a comparison between the scores of two independent samples of respondents. The *t* ratio (Chapter 7) is employed to compare two means when interval data have been obtained. The median test (Chapter 9) is a nonparametric alternative that can be applied when we cannot assume that the scores are normally distributed in the population or when the interval level of measurement has not been achieved.

Solution to Research Situation 12 (Chi-Square Test)

This situation represents a comparison of the frequencies (willingness versus unwillingness) in two groups of students (those told the professor was 65 versus those told the professor was 25). The chi-square test of significance (Chapter 9) is used to make comparisons between two or more samples when either nominal or frequency data have been obtained. Present results can be cast in the form of the following 2×2 table, representing two rows and two columns:

Willingness to Attend	Experimental Condition	
	Students Told Professor Was 65	*Students Told Professor Was 25*
Willing	22	38
Unwilling	43	27

Solution to Research Situation 13 (t Ratio)

This situation is a before–after comparison of a single sample that is measured at two different points in time. The t ratio (Chapter 7) can be used to compare two means drawn from a single sample arranged in a before–after design.

Solution to Research Situation 14 (Goodman's and Kruskal's Gamma)

This situation is a correlation problem that asks for the degree of association between X (neighborliness) and Y (voluntary association membership). Goodman's and Kruskal's gamma coefficient (Chapter 12) is applied to detect a relationship between X and Y, when both variables are ordinal and can be arranged in a cross-tabulation. In the present situation, both neighborliness and voluntary association participation have been ranked from high to low. The contingency coefficient *(C)* or Cramér's V (Chapter 12) represents an alternative to gamma that assumes only nominal-level data.

Voluntary Association Participation *(Y)*	Neighborliness *(X)*		
	Low	*Medium*	*High*
Low	80	50	40
Medium	50	85	50
High	40	45	60
	170	180	150

Solution to Research Situation 15 (Analysis of Variance)

This situation represents a comparison of the scores of four independent samples of respondents. The F ratio (Chapter 8) is employed to make comparisons between three or more independent means when interval data have been achieved. The median test (Chapter 9) is a nonparametric alternative to be employed when we suspect that the scores may not be normally distributed in the population or when the interval level of measurement has not been achieved.

Solution to Research Situation 16 (Spearman's Rank-Order Correlation Coefficient)

This situation is a correlation problem asking for the degree of association between X (percentage of new population) and Y (amount of antacid sales). Spearman's rank-order correlation coefficient (Chapter 12) can be applied to detect a relationship between X and Y variables when both have been ranked or ordered. Pearson's r cannot be employed because it requires interval data on X and Y. In the present case, percentage of new population must be ranked from 1 to 7 before rank-order correlation is applied.

Solution to Research Situation 17 (Chi-Square Test)

This situation is a comparison of the frequencies (negative, positive, and neutral tone) in female and male college students. The chi-square test of significance (Chapter 9) is employed to make comparisons between two or more samples when nominal data have been obtained. Present results can be cast in the form of the following 2×3 table, representing two columns and three rows:

	Sex of Students	
Tone of Gossip	*Female Conversations*	*Male Conversations*
Negative	36	32
Positive	40	36
Neutral	49	42

Solution to Research Situation 18 (Contingency Coefficient)

This situation is a nonparametric correlation problem (Chapter 12) that asks for the degree of association between two variables measured at the nominal level: preference for type of music and preference for type of food. The contingency coefficient and Cramér's V are measures of association for comparing several groups or categories at the nominal level. To obtain a contingency coefficient or Cramér's V, the data must be arranged in the form of a frequency table as follows:

	Food Preference *(X)*		
Musical Taste *(Y)*	*Chinese*	*French*	*Italian*
Classical	8	52	20
Jazz	55	18	39
Rock	47	8	84

Solution to Research Situation 19 (Pearson's r Correlation Coefficient)

This situation is a correlation problem, because it asks for the degree of association between the percentage of single men in the population and the rape rate. Pearson's r (Chapter 10) is employed to detect a straight-line correlation between X and Y variables when both characteristics have been measured at the interval level. If the rape rate and percentage of single men are not normally distributed across the populations of cities, we might consider applying a nonparametric alternative such as Spearman's rank-order correlation coefficient (Chapter 12).

Solution to Research Situation 20 (t Ratio)

This situation is a comparison between the course evaluation scores of two independent samples of students. The t ratio (Chapter 7) is used to compare two means when interval data have been obtained. The median test (Chapter 9) is a nonparametric alternative that can be employed when we believe that the scores are not normally distributed in the population or that the interval level has not been attained.

Solution to Research Situation 21 (Analysis of Variance)

This situation represents a comparison of the suicide rates in cities representing three anomie levels—low, moderate, and high. The F ratio (Chapter 8) is employed to make comparisons between three or more independent means when interval data have been achieved. The median test (Chapter 9) is a nonparametric alternative to be employed when we believe that the scores may not be normally distributed in the population or when the interval level of measurement has not been achieved.

Solution to Research Situation 22 (Chi-Square Test)

This situation represents a comparison of the frequencies (those below the poverty level versus those not below the poverty level) in four age groups (elders versus middle-aged adults versus young adults versus children). The chi-square test of significance (Chapter 9) is employed to compare two or more samples when nominal or frequency data have been obtained. Present results can be cast in the form of the following 3×2 table, representing three columns and two rows:

	Age of Respondent			
Poverty Status	*Older*	*Middle Age*	*Young*	*Children*
Below poverty level	36	50	40	51
Not below poverty level	164	300	200	199

Solution to Research Situation 23 (Goodman's and Kruskal's Gamma)

This situation is a correlation problem that asks for the degree of association between two variables, X and Y, measured at the ordinal level: physical health and depression (note that an underlying order exists, even though only two levels of depression are indicated). Gamma (Chapter 12) is especially applicable when both variables are ordinal and can be arranged in the form of a cross-tabulation as follows:

	Physical Health		
Depression	*Excellent*	*Good*	*Poor*
Depressed	5	14	20
Not depressed	95	96	20

The contingency coefficient *(C)* and Cramér's *V* are alternatives to gamma that can be used to compare several groups or categories at the nominal level. Because these variables are ordinal, gamma would be preferable.

Solution to Research Situation 24 (Cramér's V)

This situation is a nonparametric correlation problem (Chapter 12) that asks for the degree of association between two variables, one measured at the nominal level (gender) and the other measured at the ordinal level (physical health). Cramér's *V* is a measure of association for comparing several groups or categories at the nominal level when a table has different numbers of rows and columns. When variables are measured at two different levels, statistical tests for the lower level are usually appropriate to apply. For example, Spearman's rank-order correlation coefficient is employed when *X* is an interval measure and *Y* is an ordinal measure. In the same way, Cramér's *V* is appropriate when *X* is nominal and *Y* is ordinal. To calculate Cramér's *V,* the data must be arranged in the form of a frequency table as follows:

	Physical Health		
Gender	*Excellent*	*Good*	*Poor*
Men	37	43	20
Women	52	35	13

Solution to Research Situation 25 (Chi-Square Test)

This situation is a comparison of the frequencies (guilty versus not guilty) in two groups of people—black and white Americans. The chi-square test of significance (Chapter 9) is used to make comparisons between two or more samples when nominal or frequency data have been obtained. Present results can be cast in the form of the following 2×2 table, representing two rows and two columns:

	Race of Respondent	
O. J. Simpson Was	*Black*	*White*
Guilty	25	51
Not guilty	50	24

Solution to Research Situation 26 (Pearson's r Correlation Coefficient)

This situation is a correlation problem, because it asks for the degree of association between *X* (depression) and *Y* (physical illness). Pearson's *r* (Chapter 10) is employed to detect a straight-line correlation between *X* and *Y* variables when both characteristics have been measured at the interval level. If the symptoms of depression and physical health are not normally distributed, we might consider applying a nonparametric alternative such as Spearman's rank-order correlation coefficient (Chapter 12).

Solution to Research Situation 27 (Analysis of Variance)

This situation compares the scores of three independent samples of respondents. The F ratio (Chapter 8) is used in making comparisons between three or more independent means when interval data have been achieved. The median test (Chapter 9) is a nonparametric alternative to be applied when the scores may not be normally distributed or when we have only ordinal data.

Solution to Research Situation 28 (t Ratio)

This situation is a before–after comparison of a single sample that has been measured at two different points in time. The t ratio (Chapter 7) can be used to compare two means taken from a single sample arranged in a before–after design.

Solution to Research Situation 29 (Contingency Coefficient)

This situation is a nonparametric correlation problem (Chapter 12) that asks for the degree of association between two variables, both of which contain three categories. The contingency coefficient is a measure of association for comparing several groups or categories when a researcher is working with frequency data. The contingency coefficient is especially useful when the data can be cast in the form of a table that has the same number of rows and columns, for example, 2×2, 3×3, and so on.

Solution to Research Situation 30 (Pearson's r Correlation Coefficient)

This situation is a correlation problem that asks for the degree of association between X (hours of daylight) and Y (number of suicides). Pearson's r (Chapter 10) is used to detect a straight-line correlation between X and Y variables when both of them have been measured at the interval level. As a nonparametric alternative, Spearman's rank-order correlation coefficient (Chapter 12) should also be considered.

Solution to Research Situation 31 (Phi Coefficient)

This situation is a correlation problem that asks for the degree of association between X (legislator's gender) and Y (support for the Equal Rights Amendment). The phi coefficient (Chapter 12) is a measure of association that can be employed when frequency or nominal data are capable of being cast in the form of a 2×2 table (2 rows and 2 columns).

Solution to Research Situation 32 (Spearman's Rank-Order Correlation Coefficient)

This situation is a correlation problem, asking for the degree of association between X (political leaning ranked) and Y (scores of support for the Patriot Act). Spearman's rank-order correlation coefficient (Chapter 12) is useful for detecting a relationship between X and Y variables when at least one of them has been ranked or ordered. Pearson's r cannot be employed, because it requires interval-level measurement of both X and Y. In the present case, political leaning (X) has been ranked from 1 to 20.

Solution to Research Situation 33 (Goodman's and Kruskal's Gamma)

This situation is a correlation problem that asks for the degree of association between frequency of visits by family and friends and speed of recovery in a sample of 160 hospital patients. Gamma (Chapter 12) is especially appropriate for detecting a relationship between X and Y when both variables are ordinal and can be arranged in the form of a cross-tabulation, generating a large number of tied ranks. In the present case, frequency of visits has been ordered as often, sometimes, rarely, and never, and speed of recovery has been ordered as fast, medium, and slow. The contingency coefficient (C) and Cramér's V are alternatives to gamma that assume only nominal data.

Solution to Research Situation 34 (t Ratio)

This situation represents a before–after comparison of a single sample that has been measured at two different points in time. The t ratio (Chapter 7) can be employed to compare two means drawn from a single sample arranged in a before–after design.

Solution to Research Situation 35 (Chi-Square Test)

This situation is a comparison of the frequencies (infidelity versus no infidelity) in two groups of couples, those divorced and those not divorced. The chi-square test of significance (Chapter 9) is employed to make comparisons between two or more samples when either nominal or frequency data have been obtained.

Appendixes

Appendix A

Using ABCalc and SPSS

Several of the questions at the end of each chapter of this textbook assume the use of James Alan Fox's statistics calculator, ABCalc, or a conventional calculator. Others require use of the Statistical Program for the Social Sciences (SPSS). To assist in using these tools, this appendix includes:

- Instructions for downloading James Alan Fox's statistics calculator (ABCalc);
- Instructions for obtaining a copy of the instructional SPSS data sets from the companion website; and
- Basic computer commands for the Statistical Package for the Social Sciences (SPSS 16.0 or higher).

Additionally, these features require:

- Access to SPSS software. SPSS software is available as (1) full version, (2) graduate package, (3) SPSS career starter, and (4) student version. Many universities and colleges provide access to site-licensed use of SPSS (full version) in their computer labs. Other students wishing to work on their own computers may purchase the SPSS student version, which allows users to work with smaller data sets (50 variables and 1,500 cases). The SPSS student packages have a 4-year license for student home use.
- Microsoft Excel 2003 or Excel 2007 in order to use Fox's ABCalc.

Website for Data Sets

The instructional SPSS data sets are located on the Allyn & Bacon website as a supplement to Levin, Fox, and Forde's textbook. You will need to have Internet access using a browser such as Internet Explorer. Type the link address into the address line in the browser as http://ablongman.com/levinfox/

Additional instructions are provided on the website, but the basics are:

- Get a flash drive (portable drive/memory stick or a diskette).
- Go to the website and click on the links to save each file to the device where you wish to store the data.

The ABCalc program files (with separate versions for Microsoft Office 2003 and Office 2007) include the following:

- The ABCalc program in a compressed format (zipped). This requires access to a program such as Winzip to decompress it.
- The ABCalc program (without compression).
- ABCalc program documentation in pdf format (Adobe Acrobat Reader).

There are three SPSS data files to download. Each SPSS file has a codebook prepared in pdf format (requiring Adobe Acrobat Reader):

- BP 2004 for 11E.sav—A subset of variables from the Best Places Study of the most and least stressful cities to live in America.
- GSS 2006 for 11E.por—A subset of variables from the 2006 General Social Survey.
- MTF 2006 for 11E.sav—A subset of variables from the 2006 Monitoring the Future Study.

James Alan Fox's Statistics Calculator

One of the authors (James Alan Fox) has developed the Allyn & Bacon Calculator—ABCalc—for use with this textbook. You must have Microsoft Excel 2003 (or higher) to use this calculator, and have the security level set at medium or low. Once you have the file and you are working on a computer that has Excel, click on the file to start the calculator. You must enable macros for it to work properly.

The main menu is shown as follows:

This statistics calculator is very easy to use. It calculates statistics as you enter data into a spreadsheet. A short manual for ABCalc is also available from the website.

Introduction to SPSS

The Statistical Program for the Social Sciences (SPSS) is a commonly used statistical program because it is relatively user-friendly. The program enables calculation of a wide variety of elementary and complex statistics. This appendix shows you a few basic steps in SPSS so that you may use SPSS drop-down menus to:

- Open an existing data set
- Enter your own data
- Open Portable SPSS files from ICPSR
- Recode or collapse categories
- Analyze data
- Select cases
- Weight cases
- Make a graph or chart
- Save computer output

Opening an Existing Data File

There are many different ways to retrieve an SPSS data file. Let's assume that you've started the SPSS program on a computer with a Microsoft Windows platform. A dialog box, shown as follows, comes up when SPSS is started. By default, it asks you whether you wish to "Run the tutorial." You may want to do so at another time, but right now we shall "Open an existing data source." Click on this bullet and it will be shaded. Also, notice that you may see a list of names of files that can be accessed by clicking on the name of the file.

To open an existing data file, click on OK at the bottom of the screen. It will bring up the next dialog box. In the Open Data box, change the "Look in" location (at the top of the box) to the location where you have downloaded and stored the data set. In this case, the file is stored on a portable jump drive inside a folder called "Levin."

A list of SPSS data files inside the folder will show up on this screen. Notice on the bottom of the screen that you can change the type of file. By default, SPSS data files ending with ".sav" are shown on the screen. Double click on the file that you want to open and it will enter this information into the "File name" box and open the file. If you do not see the SPSS data file listed in this window, check that your flash drive is inserted into the port, or click on "My Computer" and navigate to the correct port. You must specify the correct location of the file. After you have successfully launched the SPSS data file, an output screen appears and the data window is available for analysis.

One of your instructional data sets has been prepared as a "portable" file. The GSS 2006 for 11E.por file can be opened by changing the "Files of type" to "SPSS Portable." Double click on the GSS portable file to open it. Portable files are useful because they enable different computers (for example, Mac and PC) and different software (SPSS and SAS) to work with the same data set. Data libraries sometimes include portable files in their archives.

Entering Your Own Data

A few exercises in this textbook require you to enter your own data into SPSS. If a data set was very large, this would be quite a tedious task. Nevertheless, it is important for you to understand that every data set you use had to be created by someone. Let's make up some hypothetical data for an example with six people answering a survey about their gender and fear of crime. We'll record whether the person is male (1), female (2), or unknown (0). We'll use a question on fear of crime that has been included on several National Crime Victim Surveys asking people, "How safe do you (or would you) feel when walking alone at night in your own neighborhood?" Possible responses are very safe (4), safe (3), unsafe (2), very unsafe (1), and no response (0).

The following are the six people for this exercise:

Male and safe	1, 3
Female and unsafe	2, 2
Female and safe	2, 3
Male and very safe	1, 4
Female and very unsafe	2, 1
Male and no response	1, 0

SPSS version 16 is used in this example. Other recent versions have very similar windows. The screens in SPSS 16 may look somewhat different, but the features are the same. In Windows, click on the Start button, move up to Programs, and down to the SPSS for Windows 16.0 icon. A successful start will take you to the default pop-up window asking you, "What would you like to do?" Click in the circle for "Type in data" and then click on "OK." A new or untitled SPSS data file will be created. The variable names and case numbers are blank.

In SPSS 16, there are "Data View" and "Variable View" tabs at the bottom of the data editor screen. We're going to create two variables and enter some data for each. We will need to define the variables, designate a format, and then label everything! To begin, click on the variable view tab and type "gender" into the top left cell of the "Variable View" dialog box.

Notice that we may need to set the options on all aspects of variables. Right now, gender is set to the default with no labels and a numeric size of 8 digits wide with 2 decimal places and no labels, no missing values, and so on. Let's change each option to something appropriate.

In Chapter 1 we looked at levels of measurement. Note that levels of measurement as shown in the variable view include scale (interval/ratio), ordinal, and nominal. A user may designate for SPSS procedures appropriate for the level of measurement of a variable. If you do this, SPSS will attempt to assist you in selecting appropriate statistics. Next, there are several very important settings in this dialog box: type, labels, and missing values.

There are many types of formats of variables. The most common format is numeric. The numeric form is defined using a code of *x.y*, with *x* being the number of digits and *y* being the number of decimal places. The default format for numeric data in SPSS is 8.2. Let's look at our example. How many possibilities are there for gender? That is, how wide can the variable be? The numbers 0–9 have a width of 1, 10–99 are 2 digits, 100–999 are 3, and so on. There is only one digit (and no decimal places) needed for gender. We will simply be coding male as 1 and female as 2. Therefore, we'll want to change the TYPE to 1 digit wide with 0 decimal places. Throughout this appendix, SPSS commands are shown in uppercase to emphasize the command or word that you should look for in the drop-down boxes. Click on TYPE and change it to 1.0 now.

Labeling of variables and their attributes is an important part of data entry. Labels should be *descriptive* and there should be a label for *every attribute*, including missing values. The only exception to this rule is that on scales we may want to just label the minimum and maximum values (for example, very satisfied [1] to very dissatisfied [7] with intermediate values blank [2–6]). Try to avoid the use of abbreviations and acronyms because you may forget what they stand for and these make reading your work difficult for other people. You will see many acronyms in the Monitoring the Future data set that is used with this textbook. These acronyms were selected long ago when the computer software limited the names to eight characters in length. In some cases, their use of acronyms is clear and in others it is very difficult to interpret the computer output without going to their codebook. The major goal in using labels is to make the computer output readable for you and for others.

There are two kinds of labels to add to a variable: variable labels and value labels. Variable labels should provide a description of the basic content of the variable. Click in the LABEL box for gender and type "Gender of the respondent."

Next we need to add labels for each of the values/categories. Click on VALUES and three dots will appear on the right side of the box. Double click on these dots to bring up the VALUE LABELS dialog box. These labels are entered one at a time by entering the value, the label, and then clicking on ADD. Every possible value including no response (0) should be labeled. Once everything has been entered you will click on CONTINUE. These procedures are repeated for every variable.

The third major feature in defining data is that values of variables may be set as non-valid or *missing*. Statistical analyses will count the number of *valid cases* for a variable. For example, people may refuse to answer a question, or they may not know, or a question may be not applicable. The researcher will want to know the average for people who answered the question, and to know the number of missing cases in his or her statistical analysis. The statistical software must be instructed about the responses that are to be treated as *missing values*. This is done by clicking on the MISSING tab in the variable view. Missing values may be assigned as specific numbers or as a range of numbers. In this example, we want to set zero as a discrete missing value for gender. By setting a missing value, the software will distinguish valid cases (where information is known) from missing cases (where information is unknown). In this example, we want to exclude cases when the gender of the respondent is unknown (coded as 0).

Missing values are set by the researcher. You must look to see whether particular values are excluded or should be excluded from statistical analysis of data. Do not assume that these are set properly when you use a data set that has been put together by someone else. In fact, most researchers will turn all missing values off when they archive their data because different statistical packages may handle missing values in different ways. It is up to you as a researcher to ensure that these are set properly. Enter these data into SPSS and your screen should look like the following box:

The real test to see how well you've entered the data and its associated information comes when you produce a frequency distribution. There should be *labels on everything* and *missing values should be set*. The basic steps to generate a frequency distribution are to click ANALYZE, DESCRIPTIVE STATISTICS, and FREQUENCIES from the pull-down menus, and then select the variables. We'll go over the steps for analyzing data in much more detail shortly. The computer output is shown as follows:

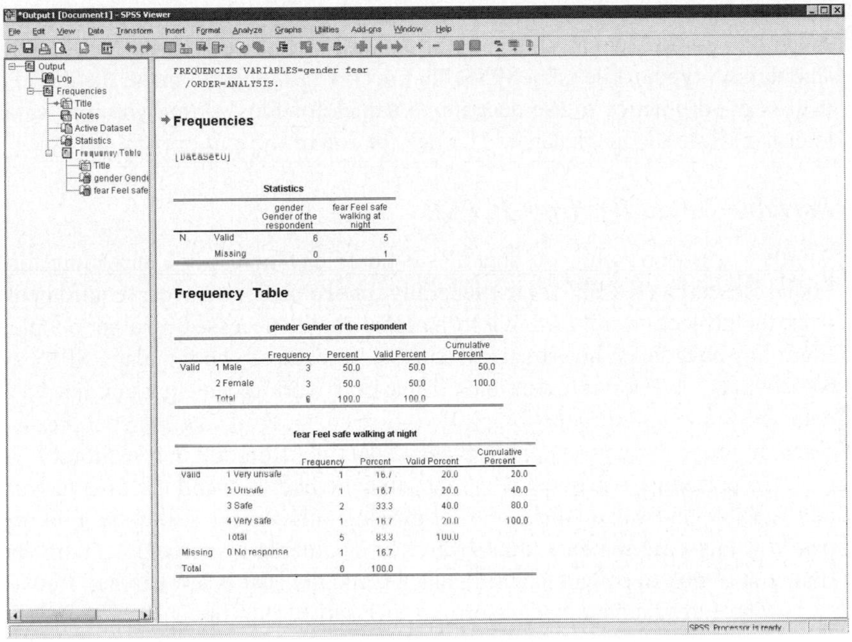

These frequency distributions are "clean" with labels on each of the variables and values. It is a normal part of data entry to examine computer output, and to fix problems as they arise. Putting together a data set can be a lot of work, thus, you may want to save your data file. This is done from the data editor window by clicking on FILE, and then SAVE.

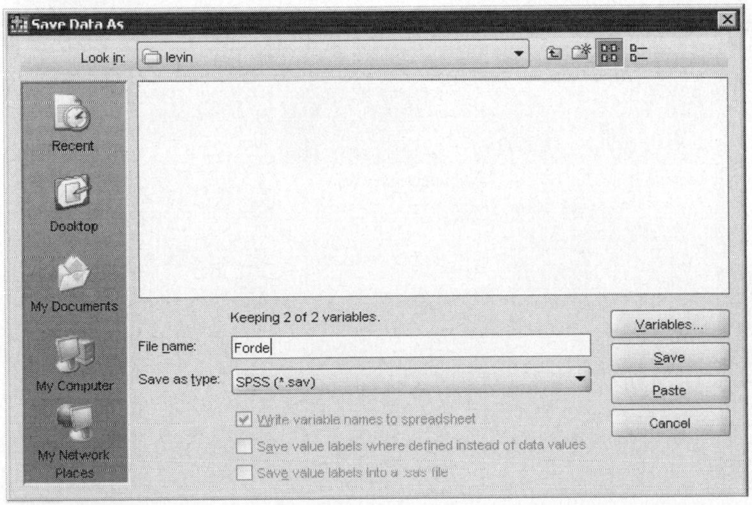

The default folder for SPSS is a folder called "SPSS." In most college computing labs the default folder is the desktop of the computer. However, you may want to keep your file. As one of many possible solutions, we accomplish this by moving the location of the file to a folder on a jump drive that we created and called "levin." The file may also be saved on a jump drive by changing the SAVE IN folder from SPSS to the jump drive (E:). The default type of file is an SPSS file (.sav). You'll need to name the file in order to save it. It is good practice to use descriptive names for files so that you may catalog them for later use. Here the file is named "Forde" for one of the authors of this book.

Portable Data Files from ICPSR

Another important source of data files is the Inter-University Consortium for Political and Social Research (ICPSR). Many federally funded projects have a requirement that the data from the project be archived in a format that can be accessed by other people. ICPSR data librarians have taken investigators' data sets and saved the raw data, SPSS syntax, and/or SAS syntax, and portable data files. Their hard work means that you have access to many data sets in the ICPSR data library. If you are considering a term paper, thesis, or advanced research project you may want to access a data file from the ICPSR library.

Importantly, you now know that someone had to spend the time to enter these data, add variable and value labels, and set missing values. It is still quite a bit of work to prepare a data set from a data library, but it is a valuable resource potentially saving you the time that it took to collect the original data, and the cost is low (in fact, free to most users).

The Inter-University Consortium of Political and Social Research is located at the University of Michigan in Ann Arbor, Michigan. You can access it on the World Wide Web at http://icpsr.umich.edu. Their front page is shown as follows:

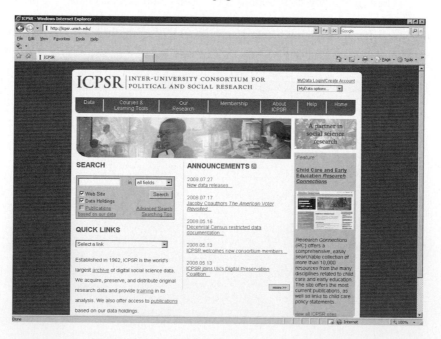

You may search their website for different types of data sets using key terms, investigator's names, and study numbers if you know them. Topical archives cover many different areas, including the census, crime, health, education, voting, and much more. We selected the National Archive of Criminal Justice Data which generated the following screen:

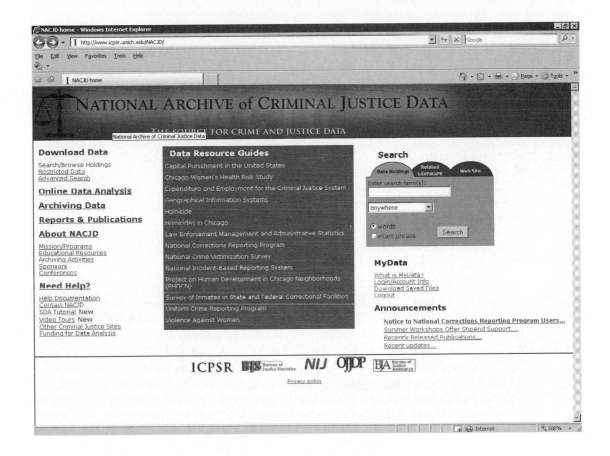

We entered the term "bail" in the search screen on the right side of the window. The search yielded 27 studies sorted by title. The studies may also be sorted by date. You may learn more about each study by clicking on "description." The download link will take you directly to a download screen. The related literature link provides a listing of publications based on the particular data set. When you're starting a library search, you'll most likely want to read more about a study before downloading it. Let's have a look at the description for ICPSR Study 3205 by Eileen Sullivan.

Notice that the description gives you a very large amount of information about each study, including who conducted it, what agency funded it, the methods used, and the access to files and their availability. In particular, we would like to know what the study is about and to find out about the format of the data files, whether there is an on-line codebook, and the structure of the data set(s).

An ICPSR abstract gives a brief description of the purpose of the project. At the end of the abstract references are provided where additional published information about the study can be found. Most data sets also have on-line (machine readable) documentation of codebooks and questionnaire instruments (if applicable). The access and availability section at the end of the description provides documentation on the public availability of the data and how to download and open the file.

The access and availability section is very important. The key elements are descriptions of how the data were archived and the types of statistical software that can open the data sets. The screen tells you that there is an SPSS portable file for this study. As a beginning user, the portable file is easiest to access. Later, you can move to SPSS syntax and raw data files which are more common. Note that the full version of SPSS can easily handle these files.

The availability of data from some studies is restricted to ICPSR member institutions. If your college is a member of ICPSR, your college library should be able to tell you how to obtain permission to access these restricted data files.

Next, we would like to download the data. Go back to the search screen and click on the download tab.

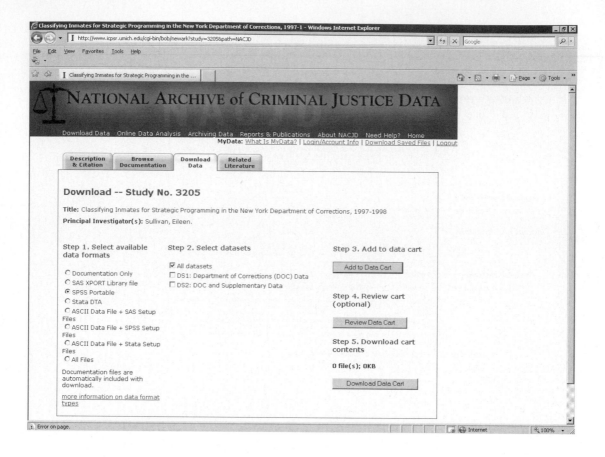

Notice that we selected SPSS Portable as the data format. After doing so, we "Added" the data to the data cart and pressed "Download." When you get to this step, a permissions screen will pop-up asking you to complete a short survey. Many ICPSR data sets are available for free to the general public for research purposes. The permissions procedures require that you provide some information about yourself and a declaration of how you will use the data. ICPSR may ask you to complete a brief questionnaire asking who you are, where you are located, and what you plan to do with the data.

When you are ready to complete an authorized download you will be asked to certify that you agree with the terms and conditions of the download.

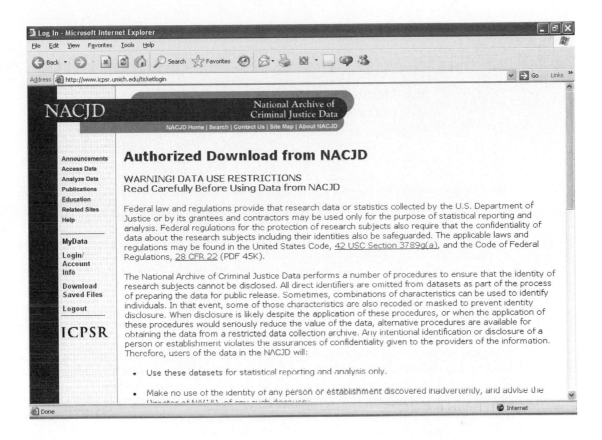

The final steps for pulling together a data set involve selecting the files that you need to download. The main items are documentation (codebook or questionnaires), raw data, and data definition statements (SPSS, SAS, etc.). As shown, we selected all files and all data sets. The third step is "add to data cart." This will put the files in a cart and tell you how large the files are going to be. ICPSR will prepare a compressed file for downloading. The study will be bundled into a compressed file which includes the documentation and data file. We chose to save the file and placed it on the desktop. A program such as Winzip will be needed in order to open this file.

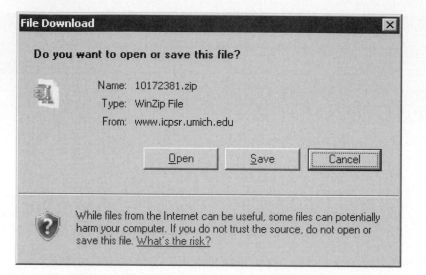

Now you've seen where some of the data came from for this book. You could select your own study, download a codebook, and get started on using SPSS to access the data. Just keep in mind that putting a large and complex data set together may require persistence and patience. Also, many data sets are quite large so that they will not fit on a floppy disk. It may take some planning to download the codebook. Check to see whether the variables that you need are in the data set and, from there, download the SPSS portable data file. It can be a lot of work to get to this point, but research is hard work and many of the data sets from ICPSR are available for you to use in your research.

To open the portable SPSS file, start SPSS, click on "Open an existing data source," and change the file type to "SPSS portable file." It is very similar to opening a regular SPSS data file, but you need to change the type of file from ".sav" to ".por" to complete the process.

Recoding Variables

You may wish to come back to this section on recoding variables when you are more familiar with data analysis. Whether you're working with a data file that someone else has created or with one of your own making, variables in a data file often need to be recoded— or changed—in some way, or values for new variables need to be calculated based on changes in the old variables. Following are some of the main reasons to recode variables:

- You may want to change the order of the categories so that the values go from what is intuitively the lowest to highest.
- You may have two studies that have similar variables but different coding schemes. You might recode to make them as comparable as possible.
- You may want to recode so that you use a different statistic or procedure.
- You may wish to recode to collapse or group a large number of categories into a few categories.
- A recode and computation may allow you to look at combinations across several variables.

Let's look at an example from the Monitoring the Future Study that measured heroin use in the past 30 days as an ordinal variable with the following categories:

V141 062R* :#X "H"/LAST30DA				
	Frequency	Percent	Valid Percent	Cumulative Percent
Valid 1 0 OCCAS:(1)	14177	95.7	99.6	99.6
2 1 2X:(2)	29	.2	.2	99.8
3 3–5X:(3)	10	.1	.1	99.9
4 6–9X:(4)	7	.0	.0	99.9
5 10–19X:(5)	3	.0	.0	100.0
7 40+OCCAS:(7)	5	.0	.0	100.0
Total	14231	96.1	100.0	
Missing −9 MISSING	583	3.9		
Total	14814	100.0		

Given that there are very few high school students who have used heroin in the last 30 days, we might want to recode the variable so that all students who have used it are combined and shown as yes (1) and students who have not used it are shown as no (0). To do this, we would recode 2–7 as 1 and 1 as 0 (it is common practice to code yes–no variables as 1 for yes and 0 for no).

To recode, click on TRANSFORM (located in the pull-down menu at the top of the screen) and then click on RECODE INTO DIFFERENT VARIABLES. It is a good idea to create a new variable rather than writing over the original variable because you may wish to keep the original variable for other analyses and you should verify that the changes were correct.

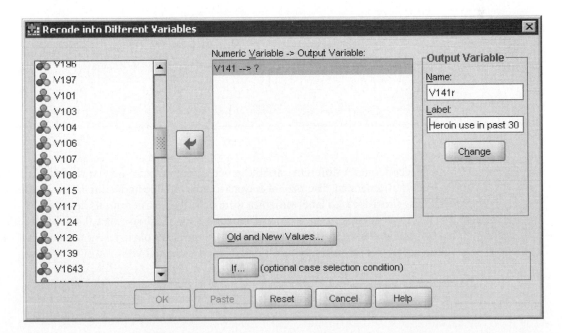

The original variable was v141 and we named the new variable as v141r. We also typed a label for it as "Heroin use in the past 30 days." Click on CHANGE to enter the name of the output variable.

Next, we need to enter the Old and New Values. We've already added 1 to become 0. We're working to recode 2–7 into a new value of 1. Look at the Range (on the left of the screen) with old values between 2 and 7 to become new value of 1. By clicking on ADD this range will be added to the list of values to recode. Finally, click on CONTINUE and the recode is done.

The recode into a different variable process will create a new variable using categories that you've defined. You may also want to add variable and value labels to this new variable. The procedures to label variables were described earlier in this appendix.

The frequency distribution for the recoded variable shows that 0.4% of American high school seniors have tried heroin in the past 30 days. Heroin use is now a nominal variable with categories recoded as 0 and 1 and labeled as no and yes.

V141r Heroin use in past 30 days				
	Frequency	Percent	Valid Percent	Cumulative Percent
Valid 0 No	14177	95.7	99.6	99.6
1 Yes	54	.4	.4	100.0
Total	14231	96.1	100.0	
Missing System	583	3.9		
Total	14814	100.0		

Data Analysis

The usual first step in every statistical analysis in SPSS is to click on ANALYZE at the top of the screen and then to choose your statistical analysis. For example, a frequency distribution in SPSS can be obtained by clicking on ANALYZE, then DESCRIPTIVE STATISTICS, and finally FREQUENCIES. The drop-down box on our computer is shown as follows. We have set the options so that variable names (rather than labels) are shown in the order that they appear in the data file. Other people may prefer labels, or alphabetical listings of acronyms, or the measurement level of variables. The measurement level is a new option in version 16. Most older data sets will not have this feature active in the data set.

You can change these settings (and you may really want to do so) by clicking on EDIT and OPTIONS. You'll need to do this each time you have a session on a lab computer. You'll only need to do it once if it is on your own computer. We have also changed the output labels to "names and labels" and "values and labels." Make these changes and click "Apply" (or close SPSS and restart it).

Changing the options will make it easier for you to read output and to locate variables within a data file. Try it and you'll see that it isn't too complicated. You can also proceed without making these changes.

Move the variables from the list on the left to the right to obtain a frequency distribution. The following box asks for a frequency distribution for lifetime use of marijuana or hashish (V115).

The frequencies procedure is "run" by clicking on OK at the bottom left of the box. If you wish to obtain additional features for this variable you could click on STATISTICS, which would open a box where you may choose from a wide variety of optional statistics. Click on CONTINUE to exit from this box.

There are too many statistical procedures to show all of them in a short appendix. The vast majority of tasks that you are asked to complete for this textbook will involve opening a data set, clicking on ANALYZE, and choosing the statistical procedure and its options. The frequencies procedure was ANALYZE, DESCRIPTIVE STATISTICS, and then FREQUENCIES. The problem sets will provide hints on where a statistical procedure is located in the drop-down menus.

Select Cases

There are often times when a researcher wishes to work with a portion of a data set. For example, Forde's students live in the South and they might want to examine closely high school students located in the South. SPSS can "select cases" for statistical analysis. To do this, click on DATA followed by SELECT CASES to open the following menu:

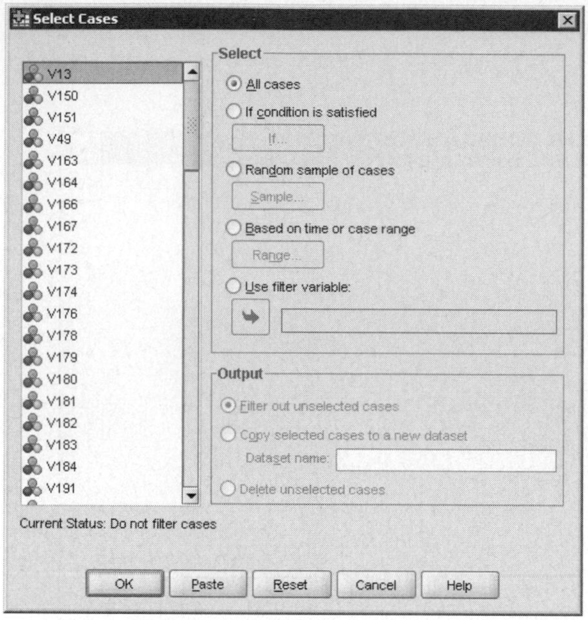

We will use V13 to set the condition that the high school student is from the South (category = 3). To do this, click on the "If" under IF CONDITION IS SATISFIED. We entered V13 = 3 and clicked CONTINUE. By clicking OK this filter is activated. It can be turned off by opening the menu again and clicking RESET.

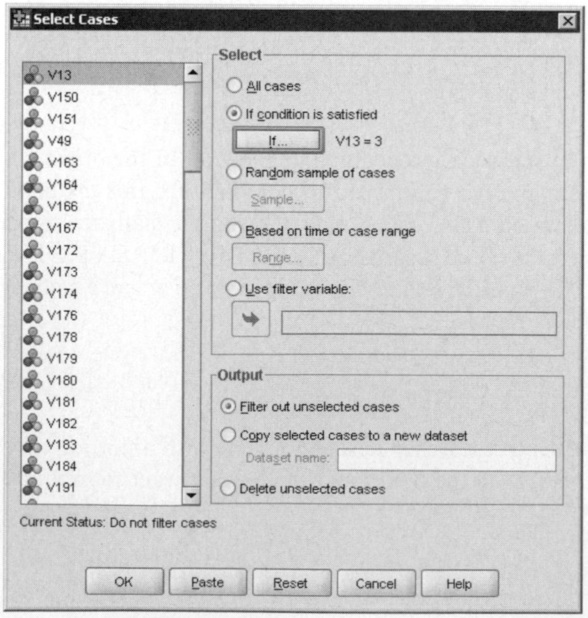

The "Select Cases" feature is very useful if you wish to do repeated analyses on a subset of a larger data set. We could also save the selected cases as output from this window copied to a new data set. We named the data set "South," which generated a new data set of students from the South. You may do this kind of task if you want to limit your analysis to a subset. If you do this kind of task, be careful to keep a copy of your original data set.

Weight Cases

Many complex data sets include a variable to weight cases. There are weighting variables for both the General Social Survey (WTSALL) and the Monitoring the Future Survey (V5). There are many different reasons why data may need to be weighted. For example, the response rates for men may be lower in a survey so that there are fewer men in the sample than there actually are in the population. To correct for this sampling bias, a weighting variable can give more weight to a man's response and less weight to a woman's response so that the totals for the weighted response add up to the known total for the population. The GSS and MTF websites have longer explanations on how their weighting variables are calculated.

If you wish to use weighted cases in statistical analysis, weights in SPSS are activated by clicking on DATA and WEIGHT CASES:

For the Monitoring the Future Study, we selected V5, which was provided with the data set as the sampling variable for weighting. By clicking OK the current status will move to weighted by V5. Note that the bottom right corner of the data editor now shows as "Weight on." Note that weighing the data file does not alter the values of the data, but only affects how much influence each case will have in the analysis.

Graphs and Charts

Graphs provide a method for visually presenting information from a table to highlight an important statistic or trend. SPSS can generate many different types of graphs. Chapter 2 discussed the basics for pie charts, bar graphs, and line charts. As you build graphs in

SPSS, remember that there often are multiple ways to transform tables into graphs and that you need to consider whether your choice is effective in providing appropriate titles, labels, and so on.

The interactive chart builder is a drop-down box accessed by clicking on GRAPHS at the top of the data dialog screen. We'll use it to make a bar chart for V115 to show how often high school seniors say that they have smoked marijuana or hashish in their lifetime. It is quite easy to use, but it does assume that the level of measurement has been set for each of the variables. Move to the variable view in the data window. Notice that every variable in the data set is by default set to nominal. If you wish to make a graph for a variable, you will need to set the level of measurement to the appropriate level. We changed V115 to "ordinal" so that we could produce a bar chart.

To begin, we clicked on "bar chart" on the left side of the screen. Next, we dragged and dropped the simple bar chart into the chart preview area. To select V115, we scrolled down the variable list shown on the left side of the screen, right clicked on V115, and dragged and dropped it into the box for the *x* axis.

Lastly, we changed the *y* axis from "count" to "percentages" using the statistics box for the elements of the chart. This box is shown as an option under statistics on the far left of the screen when the chart builder opens.

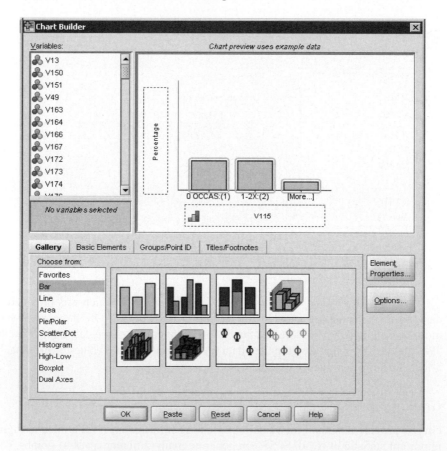

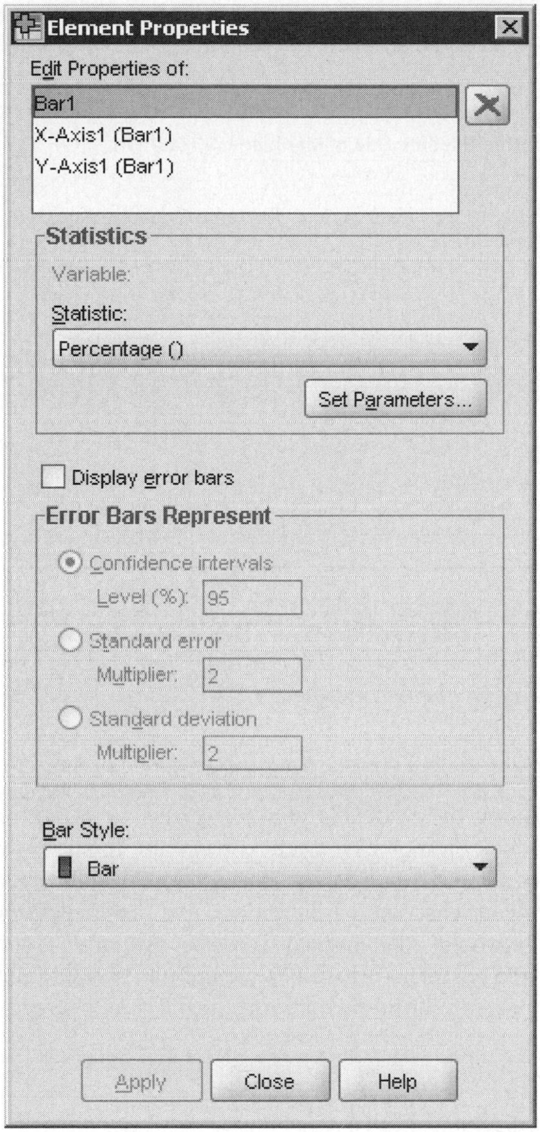
Element Properties
Edit Properties of:
Bar1
X-Axis1 (Bar1)
Y-Axis1 (Bar1)
Statistics
Variable:
Statistic:
Percentage ()
Set Parameters...
Display error bars
Error Bars Represent
Confidence intervals
Level (%): 95
Standard error
Multiplier: 2
Standard deviation
Multiplier: 2
Bar Style:
Bar
Apply
Close
Help

Titles and footnotes can be added to the chart by clicking on their tab in the chart builder. Finally, click on OK and the chart will be generated. The SPSS-generated bar chart for V115 follows:

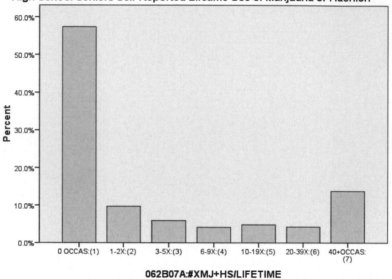

High School Seniors Self-Reported Lifetime Use of Marijuana or Hashish

062B07A:#XMJ+HS/LIFETIME

Source: Author computation based on the Monitoring the Future Study, 2006

To complete the chart using SPSS, double click on it and the "Chart editor" will open. In this editor, you may click on footnote (or click where the footnote should be) to edit it, click on each label to edit and change options, change the title, change fill patterns, and much more. You can save your chart onto a jump drive and make additional changes using SPSS similar to what you would with a regular word processing file when you make revisions. You may also copy the completed chart as a picture into a word processor such as Microsoft Word. You can impress your professor with a neat figure in a report.

Computer Output

SPSS can be used to generate a large amount of computer output. You may save computer output for later use by clicking on FILE, SAVE AS, and then give it a name and location (on your computer or jump drive). You do not have to save computer output. You can delete any output by clicking on it and pressing delete. You can also delete the entire file simply by closing it. If you do this, SPSS will create a new output file the next time you analyze a variable.

It is also possible to copy computer output from SPSS into other computer applications such as Microsoft Word. With some practice you can dazzle your professor with reports including statistical output and graphs.

This appendix has provided a very brief introduction to SPSS. For more information, see www.spss.com. In the meantime, Happy Computing!

Appendix B

A Review of Some Fundamentals of Mathematics

For students of statistics who need to review some of the fundamentals of algebra and arithmetic, this appendix covers the problems of working with decimals, negative numbers, and the summation sign.

Working with Decimals

When adding and subtracting decimals, be sure to place decimal points of the numbers directly below one another. For example, to add 3,210.76, 2.541, and 98.3,

```
3,210.76
    2.541
   98.3
3,311.601
```

To subtract 34.1 from 876.62,

```
 876.62
−34.1
 842.52
```

When multiplying decimals, be sure that your answer contains the same number of decimal places as both multiplicand and multiplier combined. For example,

Multiplicand →	63.41	2.6	.0003	.5
Multiplier →	× .05	× 1.4	× .03	× .5
Product →	3.1705	3.64	.000009	.25

Before dividing, always eliminate decimals from the divisor by moving the decimal point as many places to the right as needed to make the divisor a whole number. Make a corresponding change of the same number of places for decimals in the dividend (that is, if you move the decimal two places in the divisor, then you must move it two places in the dividend). This procedure will indicate the number of decimal places in your answer.

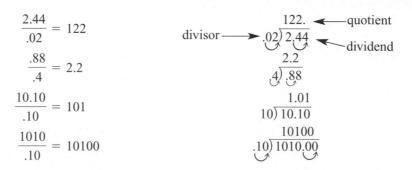

Arithmetic operations frequently yield answers in decimal form; for instance, 2.034, 24.7, and 86.001. The question arises as to how many decimal places we should have in our answers. A simple rule to follow is to carry every intermediate operation as fully as convenient and to round off the final answer to two more decimal places than found in the original set of numbers.

To illustrate, if data are derived from an original set of whole numbers (for instance, 12, 9, 49, or 15), we might carry out operations to three decimal places (to thousandths) but express our final answer to the nearest hundredth. For example,

$$3.889 = 3.89$$
$$1.224 = 1.22$$
$$7.761 = 7.76$$

Rounding to the nearest decimal place is generally carried out as follows: Drop the last digit if it is less than 5 (in the examples that follow, the last digit is the thousandth digit):

less than 5
$$26.234 = 26.23$$
$$14.891 = 14.89$$
$$1.012 = 1.01$$

Add 1 to the preceding digit if the last digit is 5 or more (in the examples that follow, the preceding digit is the hundredth digit):

5 or more
$$26.236 = 26.24$$
$$14.899 = 14.90$$
$$1.015 = 1.02$$

The following have been rounded to the nearest whole number:

3.1 = 3
3.5 = 4
4.5 = 5
4.8 = 5

The following have been rounded to the nearest tenth:

3.11 = 3.1
3.55 = 3.6
4.45 = 4.5
4.17 = 4.2

The following have been rounded to the nearest hundredth:

3.328 = 3.33
4.823 = 4.82
3.065 = 3.07
3.055 = 3.06

Dealing with Negative Numbers

When adding a series of negative numbers, make sure that you give a negative sign to the sum. For example,

$$
\begin{array}{rr}
-20 & -3 \\
-12 & -9 \\
\underline{-6} & \underline{-4} \\
-38 & -16
\end{array}
$$

To add a series containing both negative and positive numbers, first group all negatives and all positives separately; add each group and then subtract their sums (the remainder gets the sign of the larger number). For example,

$$
\begin{array}{rrrr}
-6 & +4 & -6 & +6 \\
+4 & +2 & -1 & -10 \\
+2 & \underline{+6} & \underline{-3} & \underline{-4} \\
-1 & & -10 & \\
-3 & & & \\
\underline{-4} & & &
\end{array}
$$

To subtract a negative number, you must first give it a positive sign and then follow the procedure for addition. The remainder gets the sign of the larger number. For example,

$$\begin{array}{r} 24 \\ -(-6) \\ \hline 30 \end{array}$$ -6 gets a positive sign and is therefore added to 24. Because the larger value is a positive number (24), the remainder (30) is a positive value.

$$\begin{array}{r} -6 \\ -(-24) \\ \hline 18 \end{array}$$ -24 gets a positive sign and is therefore subtracted. Because the larger value is a positive number (remember that you have changed the sign of -24), the remainder (18) is a positive value.

$$\begin{array}{r} -24 \\ -(-6) \\ \hline 18 \end{array}$$ -6 gets a positive sign and is therefore subtracted. Because the larger value is a negative number (-24), the remainder (-18) is a negative value.

When multiplying (or dividing) two numbers that have the same sign, always assign a positive sign to their product (or quotient). For example,

$$(+8) \times (+5) = +40$$

$$+5\overline{)\,+40}^{\,+8} \qquad -5\overline{)\,-40}^{\,+8}$$

$$(-8) \times (-5) = +40$$

In the case of two numbers having different signs, assign a negative sign to their product (or quotient). For example,

$$(-8) \times (+5) = -40 \qquad +5\overline{)\,-40}^{\,-8}$$

The Summation Sign

The greek letter Σ (capital sigma) is used in statistics to symbolize the sum of a set of numbers. Thus, for example, if the variable X has the values

3   6   8   5   6

then

$$\begin{aligned} \Sigma X &= 3 + 6 + 8 + 5 + 6 \\ &= 28 \end{aligned}$$

The summation sign Σ is a very convenient way to represent any kind of sum or total. However, there are a couple of basic rules that will enable you to use it properly in computing various statistics. In evaluating a complex formula having a summation sign, any operation involving an exponent (such as a square), multiplication, or division is performed before the summation, *unless* there are parentheses that dictate otherwise. (In math, parentheses always take precedence, meaning that you always perform the operation within them first.)

Applying these rules, the term ΣX^2 means: Square the X scores and then add. In contrast, the notation $(\Sigma X)^2$ dictates: Add the X scores and then square the total. Let's set the preceding X values and their squares in column form and calculate these two expressions.

X	X^2
3	9
6	36
8	64
5	25
6	36
$\Sigma X = 28$	$\Sigma X^2 = 170$

Thus, whereas $\Sigma X^2 = 170$, $(\Sigma X)^2 = (28)^2 = 784$. It is essential that you keep in mind this distinction through many of the calculations in this book.

To illustrate the same concept with multiplication rather than squares, let's add to our X values another variable Y:

8 2 1 0 3

We now form three columns: one for X, one for Y, and one for their product *(XY)*.

X	Y	XY
3	8	24
6	2	12
8	1	8
5	0	0
6	3	18
$\Sigma X = 28$	$\Sigma Y = 14$	$\Sigma XY = 62$

Thus, similar to what we found with squares, the sum of products is very different from the product of sums. Whereas, $\Sigma XY = 62$, $(\Sigma X)(\Sigma Y) = (28)(14) = 392$.

Appendix C

Tables

TABLE A *Percentage of Area under the Normal Curve*

Column a gives *z*, the distance in standard deviation units from the mean. Column b represents the percentage of area between the mean and a given *z*. Column c represents the percentage at or beyond a given *z*.

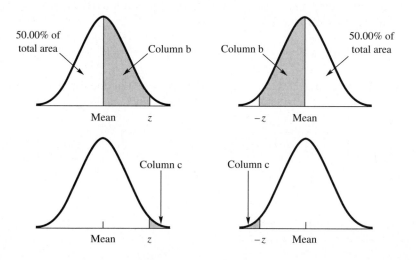

(a) z	(b) Area between Mean and z	(c) Area beyond z	(a) z	(b) Area between Mean and z	(c) Area beyond z
.00	.00	50.00	.42	16.28	33.72
.01	.40	49.60	.43	16.64	33.36
.02	.80	49.20	.44	17.00	33.00
.03	1.20	48.80	.45	17.36	32.64
.04	1.60	48.40	.46	17.72	32.28
.05	1.99	48.01	.47	18.08	31.92
.06	2.39	47.61	.48	18.44	31.56
.07	2.79	47.21	.49	18.79	31.21
.08	3.19	46.81	.50	19.15	30.85
.09	3.59	46.41	.51	19.50	30.50
.10	3.98	46.02	.52	19.85	30.15
.11	4.38	45.62	.53	20.19	29.81
.12	4.78	45.22	.54	20.54	29.46
.13	5.17	44.83	.55	20.88	29.12
.14	5.57	44.43	.56	21.23	28.77
.15	5.96	44.04	.57	21.57	28.43
.16	6.36	43.64	.58	21.90	28.10
.17	6.75	43.25	.59	22.24	27.76
.18	7.14	42.86	.60	22.57	27.43
.19	7.53	42.47	.61	22.91	27.09
.20	7.93	42.07	.62	23.24	26.76
.21	8.32	41.68	.63	23.57	26.43
.22	8.71	41.29	.64	23.89	26.11
.23	9.10	40.90	.65	24.22	25.78
.24	9.48	40.52	.66	24.54	25.46
.25	9.87	40.13	.67	24.86	25.14
.26	10.26	39.74	.68	25.17	24.83
.27	10.64	39.36	.69	25.49	24.51
.28	11.03	38.97	.70	25.80	24.20
.29	11.41	38.59	.71	26.11	23.89
.30	11.79	38.21	.72	26.42	23.58
.31	12.17	37.83	.73	26.73	23.27
.32	12.55	37.45	.74	27.04	22.96
.33	12.93	37.07	.75	27.34	22.66
.34	13.31	36.69	.76	27.64	22.36
.35	13.68	36.32	.77	27.94	22.06
.36	14.06	35.94	.78	28.23	21.77
.37	14.43	35.57	.79	28.52	21.48
.38	14.80	35.20	.80	28.81	21.19
.39	15.17	34.83	.81	29.10	20.90
.40	15.54	34.46	.82	29.39	20.61
.41	15.91	34.09	.83	29.67	20.33

(continued)

TABLE A *Continued*

(a) z	(b) Area between Mean and z	(c) Area beyond z	(a) z	(b) Area between Mean and z	(c) Area beyond z
.84	29.95	20.05	1.27	39.80	10.20
.85	30.23	19.77	1.28	39.97	10.03
.86	30.51	19.49	1.29	40.15	9.85
.87	30.78	19.22	1.30	40.32	9.68
.88	31.06	18.94	1.31	40.49	9.51
.89	31.33	18.67	1.32	40.66	9.34
.90	31.59	18.41	1.33	40.82	9.18
.91	31.86	18.14	1.34	40.99	9.01
.92	32.12	17.88	1.35	41.15	8.85
.93	32.38	17.62	1.36	41.31	8.69
.94	32.64	17.36	1.37	41.47	8.53
.95	32.89	17.11	1.38	41.62	8.38
.96	33.15	16.85	1.39	41.77	8.23
.97	33.40	16.60	1.40	41.92	8.08
.98	33.65	16.35	1.41	42.07	7.93
.99	33.89	16.11	1.42	42.22	7.78
1.00	34.13	15.87	1.43	42.36	7.64
1.01	34.38	15.62	1.44	42.51	7.49
1.02	34.61	15.39	1.45	42.65	7.35
1.03	34.85	15.15	1.46	42.79	7.21
1.04	35.08	14.92	1.47	42.92	7.08
1.05	35.31	14.69	1.48	43.06	6.94
1.06	35.54	14.46	1.49	43.19	6.81
1.07	35.77	14.23	1.50	43.32	6.68
1.08	35.99	14.01	1.51	43.45	6.55
1.09	36.21	13.79	1.52	43.57	6.43
1.10	36.43	13.57	1.53	43.70	6.30
1.11	36.65	13.35	1.54	43.82	6.18
1.12	36.86	13.14	1.55	43.94	6.06
1.13	37.08	12.92	1.56	44.06	5.94
1.14	37.29	12.71	1.57	44.18	5.82
1.15	37.49	12.51	1.58	44.29	5.71
1.16	37.70	12.30	1.59	44.41	5.59
1.17	37.90	12.10	1.60	44.52	5.48
1.18	38.10	11.90	1.61	44.63	5.37
1.19	38.30	11.70	1.62	44.74	5.26
1.20	38.49	11.51	1.63	44.84	5.16
1.21	38.69	11.31	1.64	44.95	5.05
1.22	38.88	11.12	1.65	45.05	4.95
1.23	39.07	10.93	1.66	45.15	4.85
1.24	39.25	10.75	1.67	45.25	4.75
1.25	39.44	10.56	1.68	45.35	4.65
1.26	39.62	10.38	1.69	45.45	4.55

(a) z	(b) Area between Mean and z	(c) Area beyond z	(a) z	(b) Area between Mean and z	(c) Area beyond z
1.70	45.54	4.46	2.14	48.38	1.62
1.71	45.64	4.36	2.15	48.42	1.58
1.72	45.73	4.27	2.16	48.46	1.54
1.73	45.82	4.18	2.17	48.50	1.50
1.74	45.91	4.09	2.18	48.54	1.46
1.75	45.99	4.01	2.19	48.57	1.43
1.76	46.08	3.92	2.20	48.61	1.39
1.77	46.16	3.84	2.21	48.64	1.36
1.78	46.25	3.75	2.22	48.68	1.32
1.79	46.33	3.67	2.23	48.71	1.29
1.80	46.41	3.59	2.24	48.75	1.25
1.81	46.49	3.51	2.25	48.78	1.22
1.82	46.56	3.44	2.26	48.81	1.19
1.83	46.64	3.36	2.27	48.84	1.16
1.84	46.71	3.29	2.28	48.87	1.13
1.85	46.78	3.22	2.29	48.90	1.10
1.86	46.86	3.14	2.30	48.93	1.07
1.87	46.93	3.07	2.31	48.96	1.04
1.88	46.99	3.01	2.32	48.98	1.02
1.89	47.06	2.94	2.33	49.01	.99
1.90	47.13	2.87	2.34	49.04	.96
1.91	47.19	2.81	2.35	49.06	.94
1.92	47.26	2.74	2.36	49.09	.91
1.93	47.32	2.68	2.37	49.11	.89
1.94	47.38	2.62	2.38	49.13	.87
1.95	47.44	2.56	2.39	49.16	.84
1.96	47.50	2.50	2.40	49.18	.82
1.97	47.56	2.44	2.41	49.20	.80
1.98	47.61	2.39	2.42	49.22	.78
1.99	47.67	2.33	2.43	49.25	.75
2.00	47.72	2.28	2.44	49.27	.73
2.01	47.78	2.22	2.45	49.29	.71
2.02	47.83	2.17	2.46	49.31	.69
2.03	47.88	2.12	2.47	49.32	.68
2.04	47.93	2.07	2.48	49.34	.66
2.05	47.98	2.02	2.49	49.36	.64
2.06	48.03	1.97	2.50	49.38	.62
2.07	48.08	1.92	2.51	49.40	.60
2.08	48.12	1.88	2.52	49.41	.59
2.09	48.17	1.83	2.53	49.43	.57
2.10	48.21	1.79	2.54	49.45	.55
2.11	48.26	1.74	2.55	49.46	.54
2.12	48.30	1.70	2.56	49.48	.52
2.13	48.34	1.66	2.57	49.49	.51

(continued)

TABLE A   *Continued*

(a) z	(b) Area between Mean and z	(c) Area beyond z	(a) z	(b) Area between Mean and z	(c) Area beyond z
2.58	49.51	.49	2.97	49.85	.15
2.59	49.52	.48	2.98	49.86	.14
2.60	49.53	.47	2.99	49.86	.14
2.61	49.55	.45	3.00	49.87	.13
2.62	49.56	.44	3.01	49.87	.13
2.63	49.57	.43	3.02	49.87	.13
2.64	49.59	.41	3.03	49.88	.12
2.65	49.60	.40	3.04	49.88	.12
2.66	49.61	.39	3.05	49.89	.11
2.67	49.62	.38	3.06	49.89	.11
2.68	49.63	.37	3.07	49.89	.11
2.69	49.64	.36	3.08	49.90	.10
2.70	49.65	.35	3.09	49.90	.10
2.71	49.66	.34	3.10	49.90	.10
2.72	49.67	.33	3.11	49.91	.09
2.73	49.68	.32	3.12	49.91	.09
2.74	49.69	.31	3.13	49.91	.09
2.75	49.70	.30	3.14	49.92	.08
2.76	49.71	.29	3.15	49.92	.08
2.77	49.72	.28	3.16	49.92	.08
2.78	49.73	.27	3.17	49.92	.08
2.79	49.74	.26	3.18	49.93	.07
2.80	49.74	.26	3.19	49.93	.07
2.81	49.75	.25	3.20	49.93	.07
2.82	49.76	.24	3.21	49.93	.07
2.83	49.77	.23	3.22	49.94	.06
2.84	49.77	.23	3.23	49.94	.06
2.85	49.78	.22	3.24	49.94	.06
2.86	49.79	.21	3.25	49.94	.06
2.87	49.79	.21	3.30	49.95	.05
2.88	49.80	.20	3.35	49.96	.04
2.89	49.81	.19	3.40	49.97	.03
2.90	49.81	.19	3.45	49.97	.03
2.91	49.82	.18	3.50	49.98	.02
2.92	49.82	.18	3.60	49.98	.02
2.93	49.83	.17	3.70	49.99	.01
2.94	49.84	.16	3.80	49.99	.01
2.95	49.84	.16	3.90	49.995	.005
2.96	49.85	.15	4.00	49.997	.003

TABLE B *Random Numbers*

Row	1	2	3	4	5	6	7	8	9	10	11	12	13	14	15	16	17	18	19
										Column Number									
1	9	8	9	6	9	9	0	9	6	3	2	3	3	8	6	8	4	4	2
2	3	5	6	1	7	4	1	3	2	6	8	6	0	4	7	5	2	0	3
3	4	0	6	1	6	9	6	1	5	9	5	4	5	4	8	6	7	4	0
4	6	5	6	3	1	6	8	6	7	2	0	7	2	3	2	1	5	0	9
5	2	4	9	7	9	1	0	3	9	6	7	4	1	5	4	9	6	9	8
6	7	6	1	2	7	5	6	9	4	8	4	2	8	5	2	4	1	8	0
7	8	2	1	3	4	7	4	6	3	0	7	5	0	9	2	9	0	6	1
8	6	9	5	6	5	6	0	9	0	7	7	1	4	1	8	3	1	9	3
9	7	2	1	9	9	8	0	1	6	1	6	2	3	6	9	5	5	8	4
10	2	9	0	7	3	0	8	9	6	3	3	8	5	5	6	5	2	0	9
11	9	3	5	4	5	7	4	0	3	0	1	0	4	3	3	9	5	3	2
12	9	7	5	7	9	4	8	6	8	7	6	1	6	8	2	5	5	5	3
13	4	1	7	8	6	8	1	0	5	8	8	6	1	6	8	2	9	0	4
14	5	0	8	3	3	4	5	4	4	2	5	3	0	4	9	6	1	2	3
15	3	5	0	2	9	4	1	0	0	3	9	0	5	8	6	0	9	9	6
16	0	3	8	2	3	5	1	0	1	0	6	8	5	2	4	8	0	3	8
17	1	7	2	9	1	2	7	8	4	7	0	3	3	1	5	8	2	7	3
18	5	0	5	7	9	5	8	7	8	9	3	5	3	4	4	6	1	1	3
19	7	7	3	3	5	3	6	1	3	2	8	5	4	1	4	8	3	9	0
20	1	0	9	1	3	8	2	5	3	0	3	8	0	9	3	3	0	4	5
21	1	3	8	5	1	8	5	9	4	1	9	3	9	3	6	5	9	8	4
22	8	6	4	7	8	7	5	9	4	1	9	3	9	3	6	5	9	8	4
23	0	6	9	6	5	1	0	3	2	6	7	7	4	9	6	0	3	4	0
24	7	6	7	4	7	0	8	3	8	7	3	2	5	1	2	4	2	9	7
25	3	2	3	8	1	3	1	8	7	4	5	9	0	0	2	4	1	2	1
26	9	2	1	6	4	2	3	8	7	6	2	6	2	6	4	8	1	0	1
27	3	7	4	2	2	8	1	7	8	0	6	0	0	0	3	2	2	9	7
28	0	7	8	0	8	5	1	5	2	6	5	8	7	5	3	0	5	9	6
29	7	4	2	3	3	2	6	0	0	6	5	2	2	3	6	3	9	0	4
30	1	8	2	7	5	9	5	3	6	5	2	9	9	1	1	7	3	4	3
31	4	3	1	8	7	0	6	0	8	6	5	0	1	0	4	0	6	1	5
32	8	5	8	0	6	1	4	1	2	0	4	4	1	4	7	6	3	5	1
33	4	5	8	5	0	4	5	8	3	9	2	8	7	8	9	0	8	4	3
34	5	0	2	5	4	9	2	2	1	1	0	0	5	4	8	7	6	4	0
35	0	8	1	7	0	6	3	3	4	7	6	2	6	8	9	3	4	1	4
36	2	5	9	3	4	6	0	7	5	2	0	0	9	6	0	8	2	2	5
37	2	1	3	1	3	7	8	9	8	4	9	3	8	0	2	2	1	8	1
38	3	8	8	6	8	5	1	3	3	4	6	7	2	6	3	4	8	6	7
39	0	9	9	8	5	9	8	4	4	2	2	1	1	0	1	7	6	1	3
40	2	2	3	5	3	9	7	4	4	2	1	4	0	5	8	2	3	0	8

(continued)

TABLE B   *Continued*

										Column Number											
20	21	22	23	24	25	26	27	28	29	30	31	32	33	34	35	36	37	38	39	40	Row
0	9	7	1	1	9	1	2	7	3	5	1	8	4	0	4	1	0	6	0	3	1
8	3	7	7	9	1	4	9	9	5	9	2	0	1	6	1	2	6	6	7	0	2
2	5	6	3	7	8	3	3	8	4	3	9	3	9	0	0	9	8	3	5	2	3
4	7	0	8	6	6	5	9	6	2	7	3	5	9	0	1	8	0	9	6	9	4
0	9	8	7	3	5	6	8	8	1	2	0	2	3	2	6	4	3	1	9	7	5
5	1	8	8	4	7	0	1	7	6	8	2	1	6	3	2	1	8	1	8	3	6
1	3	7	8	6	9	5	4	1	7	3	8	7	1	5	6	5	6	4	3	6	7
5	9	0	1	5	2	8	6	5	5	7	8	1	8	7	1	2	4	0	4	1	8
2	2	5	5	2	1	8	6	9	8	9	8	0	5	8	9	9	4	1	3	4	9
1	3	4	2	8	5	0	7	9	8	4	3	5	8	0	9	4	6	6	0	5	10
2	6	8	6	6	4	7	1	5	1	6	4	6	7	6	0	8	7	3	5	2	11
8	6	0	1	4	2	9	8	6	8	0	7	6	5	1	9	1	3	7	0	3	12
9	5	7	0	9	8	7	6	9	0	6	5	4	0	3	6	5	6	3	5	0	13
2	2	3	4	7	8	0	2	0	8	0	3	4	9	2	5	7	7	8	6	4	14
2	4	6	1	0	5	0	6	1	4	9	4	7	3	9	1	7	6	4	5	8	15
6	3	4	8	1	6	9	5	6	2	0	4	6	1	6	8	1	9	9	1	1	16
9	0	5	1	3	6	1	9	5	4	1	2	5	4	2	9	5	6	2	4	0	17
3	6	7	0	3	5	3	7	4	1	7	5	4	8	3	7	4	8	5	7	2	18
4	3	6	6	3	6	3	0	0	9	4	2	2	5	1	8	9	5	1	9	7	19
1	0	6	9	0	2	7	3	9	8	4	0	6	9	8	2	3	2	8	0	4	20
9	1	3	5	7	9	6	2	4	3	4	6	4	9	1	3	1	7	5	2	2	21
6	4	2	2	2	1	4	5	2	2	8	3	2	1	2	6	6	0	1	8	9	22
7	2	6	9	0	7	5	3	2	5	6	2	7	6	3	8	1	4	1	5	1	23
8	2	8	2	4	4	4	2	9	1	9	8	3	4	4	1	0	4	6	9	6	24
7	3	1	4	3	0	4	7	1	3	7	4	8	6	7	3	2	6	6	2	0	25
0	6	4	5	8	3	1	4	8	1	8	3	1	6	4	3	0	2	8	7	3	26
4	2	2	8	3	2	1	9	3	0	1	7	5	9	0	9	1	2	5	8	2	27
2	9	8	7	2	0	6	4	0	2	7	1	3	1	6	8	7	0	9	2	5	28
0	8	0	5	6	8	2	4	3	6	1	3	5	2	3	5	9	8	6	2	1	29
0	1	7	6	1	5	7	9	0	3	5	3	4	2	4	8	5	6	4	0	6	30
5	1	9	8	5	2	4	5	1	7	5	3	2	4	6	7	9	9	6	7	2	31
0	3	6	6	3	7	8	6	9	7	2	8	9	0	7	2	9	4	0	8	6	32
5	0	0	0	2	0	8	9	0	1	0	6	2	0	4	6	9	6	5	4	9	33
1	9	4	4	2	6	4	2	4	1	0	2	7	9	6	8	7	5	6	9	3	34
0	0	5	3	8	3	2	7	5	0	4	7	6	4	6	3	0	4	7	5	3	35
6	2	6	2	0	6	0	1	4	8	9	6	5	9	7	3	6	7	6	5	4	36
6	3	9	0	3	5	0	9	1	2	0	5	9	7	3	2	5	9	3	0	2	37
9	7	3	3	5	4	0	6	4	9	4	7	9	1	4	3	9	7	7	1	8	38
1	9	6	2	9	4	2	9	7	0	3	8	9	5	7	0	6	9	7	2	5	39
5	9	4	5	8	6	2	3	0	6	2	9	8	6	3	0	4	1	0	7	6	40

Source: N. M. Downie and R. W. Heath, *Basic Statistical Methods,* 3rd ed., Harper & Row, New York, 1970. Reprinted by permission of Harper & Row.

TABLE C *Critical Values of t*

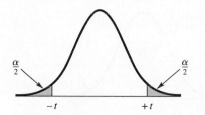

For any given df, the table shows the values of *t* corresponding to various levels of probability. Obtained *t* is significant at a given level if it is *larger than* the value shown in the table (ignoring the sign).

df	Level of Significance for Two-Tailed Test (α)					
	.20	.10	.05	.02	.01	.001
1	3.078	6.314	12.706	31.821	63.657	636.619
2	1.886	2.920	4.303	6.965	9.925	31.598
3	1.638	2.353	3.182	4.541	5.841	12.941
4	1.533	2.132	2.776	3.747	4.604	8.610
5	1.476	2.015	2.571	3.365	4.032	6.859
6	1.440	1.943	2.447	3.143	3.707	5.959
7	1.415	1.895	2.365	2.998	3.499	5.405
8	1.397	1.860	2.306	2.896	3.355	5.041
9	1.383	1.833	2.262	2.821	3.250	4.781
10	1.372	1.812	2.228	2.764	3.169	4.587
11	1.363	1.796	2.201	2.718	3.106	4.437
12	1.356	1.782	2.179	2.681	3.055	4.318
13	1.350	1.771	2.160	2.650	3.012	4.221
14	1.345	1.761	2.145	2.624	2.977	4.140
15	1.341	1.753	2.131	2.602	2.947	4.073
16	1.337	1.746	2.120	2.583	2.921	4.015
17	1.333	1.740	2.110	2.567	2.898	3.965
18	1.330	1.734	2.101	2.552	2.878	3.922
19	1.328	1.729	2.093	2.539	2.861	3.883
20	1.325	1.725	2.086	2.528	2.845	3.850
21	1.323	1.721	2.080	2.518	2.831	3.819
22	1.321	1.717	2.074	2.508	2.819	3.792
23	1.319	1.714	2.069	2.500	2.807	3.767
24	1.318	1.711	2.064	2.492	2.797	3.745
25	1.316	1.708	2.060	2.485	2.787	3.725
26	1.315	1.706	2.056	2.479	2.779	3.707
27	1.314	1.703	2.052	2.473	2.771	3.690
28	1.313	1.701	2.048	2.467	2.763	3.674
29	1.311	1.699	2.045	2.462	2.756	3.659
30	1.310	1.697	2.042	2.457	2.750	3.646
40	1.303	1.684	2.021	2.423	2.704	3.551
60	1.296	1.671	2.000	2.390	2.660	3.460
120	1.289	1.658	1.980	2.358	2.617	3.373
∞	1.282	1.645	1.960	2.326	2.576	3.291

Note: The bottom row (df = ∞) also equals critical values for *z*.

(continued)

TABLE C   Continued

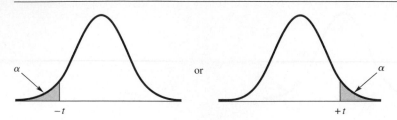

For any given df, the table shows the values of *t* corresponding to various levels of probability. Obtained *t* is significant at a given level if it is *larger than* the value shown in the table (ignoring the sign).

df	\.10	\.05	\.025	\.01	\.005	\.0005
			Level of Significance for One-Tailed Test (α)			
1	3.078	6.314	12.706	31.821	63.657	636.619
2	1.886	2.920	4.303	6.965	9.925	31.598
3	1.638	2.353	3.182	4.541	5.841	12.941
4	1.533	2.132	2.776	3.747	4.604	8.610
5	1.476	2.015	2.571	3.365	4.032	6.859
6	1.440	1.943	2.447	3.143	3.707	5.959
7	1.415	1.895	2.365	2.998	3.499	5.405
8	1.397	1.860	2.306	2.896	3.355	5.041
9	1.383	1.833	2.262	2.821	3.250	4.781
10	1.372	1.812	2.228	2.764	3.169	4.587
11	1.363	1.796	2.201	2.718	3.106	4.437
12	1.356	1.782	2.179	2.681	3.055	4.318
13	1.350	1.771	2.160	2.650	3.012	4.221
14	1.345	1.761	2.145	2.624	2.977	4.140
15	1.341	1.753	2.131	2.602	2.947	4.073
16	1.337	1.746	2.120	2.583	2.921	4.015
17	1.333	1.740	2.110	2.567	2.898	3.965
18	1.330	1.734	2.101	2.552	2.878	3.922
19	1.328	1.729	2.093	2.539	2.861	3.883
20	1.325	1.725	2.086	2.528	2.845	3.850
21	1.323	1.721	2.080	2.518	2.831	3.819
22	1.321	1.717	2.074	2.508	2.819	3.792
23	1.319	1.714	2.069	2.500	2.807	3.767
24	1.318	1.711	2.064	2.492	2.797	3.745
25	1.316	1.708	2.060	2.485	2.787	3.725
26	1.315	1.706	2.056	2.479	2.779	3.707
27	1.314	1.703	2.052	2.473	2.771	3.690
28	1.313	1.701	2.048	2.467	2.763	3.674
29	1.311	1.699	2.045	2.462	2.756	3.659
30	1.310	1.697	2.042	2.457	2.750	3.646
40	1.303	1.684	2.021	2.423	2.704	3.551
60	1.296	1.671	2.000	2.390	2.660	3.460
120	1.289	1.658	1.980	2.358	2.617	3.373
∞	1.282	1.645	1.960	2.326	2.576	3.291

Note: The bottom row (df = ∞) also equals critical values for *z*.

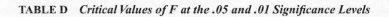

TABLE D *Critical Values of F at the .05 and .01 Significance Levels*

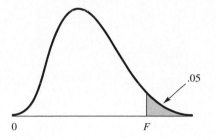

.05

0   F

df for the Denominator	df for the Numerator, $\alpha = .05$							
	1	2	3	4	5	6	8	12
1	161.4	199.5	215.7	224.6	230.2	234.0	238.9	243.9
2	18.51	19.00	19.16	19.25	19.30	19.33	19.37	19.41
3	10.13	9.55	9.28	9.12	9.01	8.94	8.84	8.74
4	7.71	6.94	6.59	6.39	6.26	6.16	6.04	5.91
5	6.61	5.79	5.41	5.19	5.05	4.95	4.82	4.68
6	5.99	5.14	4.76	4.53	4.39	4.28	4.15	4.00
7	5.59	4.74	4.35	4.12	3.97	3.87	3.73	3.57
8	5.32	4.46	4.07	3.84	3.69	3.58	3.44	3.28
9	5.12	4.26	3.86	3.63	3.48	3.37	3.23	3.07
10	4.96	4.10	3.71	3.48	3.33	3.22	3.07	2.91
11	4.84	3.98	3.59	3.36	3.20	3.09	2.95	2.79
12	4.75	3.88	3.49	3.26	3.11	3.00	2.85	2.69
13	4.67	3.80	3.41	3.18	3.02	2.92	2.77	2.60
14	4.60	3.74	3.34	3.11	2.96	2.85	2.70	2.53
15	4.54	3.68	3.29	3.06	2.90	2.79	2.64	2.48
16	4.49	3.63	3.24	3.01	2.85	2.74	2.59	2.42
17	4.45	3.59	3.20	2.96	2.81	2.70	2.55	2.38
18	4.41	3.55	3.16	2.93	2.77	2.66	2.51	2.34
19	4.38	3.52	3.13	2.90	2.74	2.63	2.48	2.31
20	4.35	3.49	3.10	2.87	2.71	2.60	2.45	2.28
21	4.32	3.47	3.07	2.84	2.68	2.57	2.42	2.25
22	4.30	3.44	3.05	2.82	2.66	2.55	2.40	2.23
23	4.28	3.42	3.03	2.80	2.64	2.53	2.38	2.20
24	4.26	3.40	3.01	2.78	2.62	2.51	2.36	2.18
25	4.24	3.38	2.99	2.76	2.60	2.49	2.34	2.16
26	4.22	3.37	2.98	2.74	2.59	2.47	2.32	2.15
27	4.21	3.35	2.96	2.73	2.57	2.46	2.30	2.13
28	4.20	3.34	2.95	2.71	2.56	2.44	2.29	2.12
29	4.18	3.33	2.93	2.70	2.54	2.43	2.28	2.10
30	4.17	3.32	2.92	2.69	2.53	2.42	2.27	2.09
40	4.08	3.23	2.84	2.61	2.45	2.34	2.18	2.00
60	4.00	3.15	2.76	2.52	2.37	2.25	2.10	1.92
120	3.92	3.07	2.68	2.45	2.29	2.17	2.02	1.83
∞	3.84	2.99	2.60	2.37	2.21	2.09	1.94	1.75

(continued)

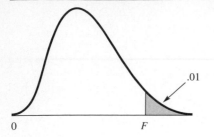

df for the Denominator	df for the Numerator, $\alpha = .01$							
	1	**2**	**3**	**4**	**5**	**6**	**8**	**12**
1	4052	4999	5403	5625	5764	5859	5981	6106
2	98.49	99.01	99.17	99.25	99.30	99.33	99.36	99.42
3	34.12	30.81	29.46	28.71	28.24	27.91	27.49	27.05
4	21.20	18.00	16.69	15.98	15.52	15.21	14.80	14.37
5	16.26	13.27	12.06	11.39	10.97	10.67	10.27	9.89
6	13.74	10.92	9.78	9.15	8.75	8.47	8.10	7.72
7	12.25	9.55	8.45	7.85	7.46	7.19	6.84	6.47
8	11.26	8.65	7.59	7.01	6.63	6.37	6.03	5.67
9	10.56	8.02	6.99	6.42	6.06	5.80	5.47	5.11
10	10.04	7.56	6.55	5.99	5.64	5.39	5.06	4.71
11	9.65	7.20	6.22	5.67	5.32	5.07	4.74	4.40
12	9.33	6.93	5.95	5.41	5.06	4.82	4.50	4.16
13	9.07	6.70	5.74	5.20	4.86	4.62	4.30	3.96
14	8.86	6.51	5.56	5.03	4.69	4.46	4.14	3.80
15	8.68	6.36	5.42	4.89	4.56	4.32	4.00	3.67
16	8.53	6.23	5.29	4.77	4.44	4.20	3.89	3.55
17	8.40	6.11	5.18	4.67	4.34	4.10	3.79	3.45
18	8.28	6.01	5.09	4.58	4.25	4.01	3.71	3.37
19	8.18	5.93	5.01	4.50	4.17	3.94	3.63	3.30
20	8.10	5.85	4.94	4.43	4.10	3.87	3.56	3.23
21	8.02	5.78	4.87	4.37	4.04	3.81	3.51	3.17
22	7.94	5.72	4.82	4.31	3.99	3.76	3.45	3.12
23	7.88	5.66	4.76	4.26	3.94	3.71	3.41	3.07
24	7.82	5.61	4.72	4.22	3.90	3.67	3.36	3.03
25	7.77	5.57	4.68	4.18	3.86	3.63	3.32	2.99
26	7.72	5.53	4.64	4.14	3.82	3.59	3.29	2.96
27	7.68	5.49	4.60	4.11	3.78	3.56	3.26	2.93
28	7.64	5.45	4.57	4.07	3.75	3.53	3.23	2.90
29	7.60	5.42	4.54	4.04	3.73	3.50	3.20	2.87
30	7.56	5.39	4.51	4.02	3.70	3.47	3.17	2.84
40	7.31	5.18	4.31	3.83	3.51	3.29	2.99	2.66
60	7.08	4.98	4.13	3.65	3.34	3.12	2.82	2.50
120	6.85	4.79	3.95	3.48	3.17	2.96	2.66	2.34
∞	6.64	4.60	3.78	3.32	3.02	2.80	2.51	2.18

Source: R. A. Fisher and F. Yates, *Statistical Tables for Biological, Agricultural, and Medical Research,* 4th ed., Longman Group Ltd., London (previously published by Oliver & Boyd, Edinburgh), Table V, by permission of the authors and the publisher.

TABLE E *Critical Values of Chi-Square at the .05 and .01 Levels of Significance (α)*

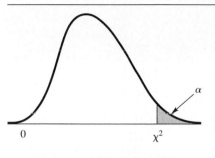

df	α		df	α	
	.05	**.01**		**.05**	**.01**
1	3.841	6.635	16	26.296	32.000
2	5.991	9.210	17	27.587	33.409
3	7.815	11.345	18	28.869	34.805
4	9.488	13.277	19	30.144	36.191
5	11.070	15.086	20	31.410	37.566
6	12.592	16.812	21	32.671	38.932
7	14.067	18.475	22	33.924	40.289
8	15.507	20.090	23	35.172	41.638
9	16.919	21.666	24	36.415	42.980
10	18.307	23.209	25	37.652	44.314
11	19.675	24.725	26	38.885	45.642
12	21.026	26.217	27	40.113	46.963
13	22.362	27.688	28	41.337	48.278
14	23.685	29.141	29	42.557	49.588
15	24.996	30.578	30	43.773	50.892

Source: R. A. Fisher and F. Yates, *Statistical Tables for Biological, Agricultural, and Medical Research,* 4th ed., Longman Group Ltd., London (previously published by Oliver & Boyd, Edinburgh), Table IV, by permission of the authors and the publisher.

TABLE F   *Critical Values of r at the .05 and .01 Levels of Significance (α)*

df	α .05	.01	df	α .05	.01
1	.99692	.999877	16	.4683	.5897
2	.95000	.990000	17	.4555	.5751
3	.8783	.95873	18	.4438	.5614
4	.8114	.91720	19	.4329	.5487
5	.7545	.8745	20	.4227	.5368
6	.7067	.8343	25	.3809	.4869
7	.6664	.7977	30	.3494	.4487
8	.6319	.7646	35	.3246	.4182
9	.6021	.7348	40	.3044	.3932
10	.5760	.7079	45	.2875	.3721
11	.5529	.6835	50	.2732	.3541
12	.5324	.6614	60	.2500	.3248
13	.5139	.6411	70	.2319	.3017
14	.4973	.6226	80	.2172	.2830
15	.4821	.6055	90	.2050	.2673

Source: R. A. Fisher and F. Yates, *Statistical Tables for Biological, Agricultural, and Medical Research,* 4th ed., Longman Group Ltd., London (previously published by Oliver & Boyd, Edinburgh), Table VI, by permission of the authors and the publisher.

TABLE G   *Critical Values of r_s at the .05 and .01 Levels of Significance (α)*

N	α .05	.01	N	α .05	.01
5	1.000	—	16	.506	.665
6	.886	1.000	18	.475	.625
7	.786	.929	20	.450	.591
8	.738	.881	22	.428	.562
9	.683	.833	24	.409	.537
10	.648	.794	26	.392	.515
12	.591	.777	28	.377	.496
14	.544	.714	30	.364	.478

Source: E. G. Olds, *The Annals of Mathematical Statistics.* "Distribution of the Sum of Squares of Rank Differences for Small Numbers of Individuals," 1938, vol. 9, and "The 5 Percent Significance Levels for Sums of Squares of Rank Differences and a Correction," 1949, vol. 20, by permission of the Institute of Mathematical Statistics.

TABLE II *Percentage Points of the Studentized Range (q) for the .05 and .01 Levels of Significance (α)*

df for MS$_{within}$	α	2	3	4	5	6	7	8	9	10	11
						k = Number of Means					
5	.05	3.64	4.60	5.22	5.67	6.03	6.33	6.58	6.80	6.99	7.17
	.01	5.70	6.98	7.80	8.42	8.91	9.32	9.67	9.97	10.24	10.48
6	.05	3.46	4.34	4.90	5.30	5.63	5.90	6.12	6.32	6.49	6.65
	.01	5.24	6.33	7.03	7.56	7.97	8.32	8.61	8.87	9.10	9.30
7	.05	3.34	4.16	4.68	5.06	5.36	5.61	5.82	6.00	6.16	6.30
	.01	4.95	5.92	6.54	7.01	7.37	7.68	7.94	8.17	8.37	8.55
8	.05	3.26	4.04	4.53	4.89	5.17	5.40	5.60	5.77	5.92	6.05
	.01	4.75	5.64	6.20	6.62	6.96	7.24	7.47	7.68	7.86	8.03
9	.05	3.20	3.95	4.41	4.76	5.02	5.24	5.43	5.59	5.74	5.87
	.01	4.60	5.43	5.96	6.35	6.66	6.91	7.13	7.33	7.49	7.65
10	.05	3.15	3.88	4.33	4.65	4.91	5.12	5.30	5.46	5.60	5.72
	.01	4.48	5.27	5.77	6.14	6.43	6.67	6.87	7.05	7.21	7.36
11	.05	3.11	3.82	4.26	4.57	4.82	5.03	5.20	5.35	5.49	5.61
	.01	4.39	5.15	5.62	5.97	6.25	6.48	6.67	6.84	6.99	7.13
12	.05	3.08	3.77	4.20	4.51	4.75	4.95	5.12	5.27	5.39	5.51
	.01	4.32	5.05	5.50	5.84	6.10	6.32	6.51	6.67	6.81	6.94
13	.05	3.06	3.73	4.15	4.45	4.69	4.88	5.05	5.19	5.32	5.43
	.01	4.26	4.96	5.40	5.73	5.98	6.19	6.37	6.53	6.67	6.79
14	.05	3.03	3.70	4.11	4.41	4.64	4.83	4.99	5.13	5.25	5.36
	.01	4.21	4.89	5.32	5.63	5.88	6.08	6.26	6.41	6.54	6.66
15	.05	3.01	3.67	4.08	4.37	4.59	4.78	4.94	5.08	5.20	5.31
	.01	4.17	4.84	5.25	5.56	5.80	5.99	6.16	6.31	6.44	6.55
16	.05	3.00	3.65	4.05	4.33	4.56	4.74	4.90	5.03	5.15	5.26
	.01	4.13	4.79	5.19	5.49	5.72	5.92	6.08	6.22	6.35	6.46
17	.05	2.98	3.63	4.02	4.30	4.52	4.70	4.86	4.99	5.11	5.21
	.01	4.10	4.74	5.14	5.43	5.66	5.85	6.01	6.15	6.27	6.38
18	.05	2.97	3.61	4.00	4.28	4.49	4.67	4.82	4.96	5.07	5.17
	.01	4.07	4.70	5.09	5.38	5.60	5.79	5.94	6.08	6.20	6.31
19	.05	2.96	3.59	3.98	4.25	4.47	4.65	4.79	4.92	5.04	5.14
	.01	4.05	4.67	5.05	5.33	5.55	5.73	5.89	6.02	6.14	6.25
20	.05	2.95	3.58	3.96	4.23	4.45	4.62	4.77	4.90	5.01	5.11
	.01	4.02	4.64	5.02	5.29	5.51	5.69	5.84	5.97	6.09	6.19
24	.05	2.92	3.53	3.90	4.17	4.37	4.54	4.68	4.81	4.92	5.01
	.01	3.96	4.55	4.91	5.17	5.37	5.54	5.69	5.81	5.92	6.02
30	.05	2.89	3.49	3.85	4.10	4.30	4.46	4.60	4.72	4.82	4.92
	.01	3.89	4.45	4.80	5.05	5.24	5.40	5.54	5.65	5.76	5.85
40	.05	2.86	3.44	3.79	4.04	4.23	4.39	4.52	4.63	4.73	4.82
	.01	3.82	4.37	4.70	4.93	5.11	5.26	5.39	5.50	5.60	5.69
60	.05	2.83	3.40	3.74	3.98	4.16	4.31	4.44	4.55	4.65	4.73
	.01	3.76	4.28	4.59	4.82	4.99	5.13	5.25	5.36	5.45	5.53
120	.05	2.80	3.36	3.68	3.92	4.10	4.24	4.36	4.47	4.56	4.64
	.01	3.70	4.20	4.50	4.71	4.87	5.01	5.12	5.21	5.30	5.37
∞	.05	2.77	3.31	3.63	3.86	4.03	4.17	4.29	4.39	4.47	4.55
	.01	3.64	4.12	4.40	4.60	4.76	4.88	4.99	5.08	5.16	5.23

Source: E. S. Pearson and H. O. Hartley, *Biometrika Tables for Statisticians*, Vol. 1, 3rd ed., Cambridge University Press, New York, 1966, by permission of the Biometrika Trustees.

Appendix D

List of Formulas

Name	Formula	Page
Proportion	$P = \dfrac{f}{N}$	37
Percentage	$\% = (100)\dfrac{f}{N}$	38
Ratio	$\text{Ratio} = \dfrac{f_1}{f_2}$	38
Rate of change	$\text{Rate of change} = (100)\left(\dfrac{\text{time}2f - \text{time}1f}{\text{time}1f}\right)$	40
Midpoint	$m = \dfrac{\text{lowest score value} + \text{highest score value}}{2}$	44
Cumulative percentage	$c\% = (100)\dfrac{cf}{N}$	46
Total percentage	$\text{total}\% = (100)\dfrac{f}{N_{\text{total}}}$	57
Row percentage	$\text{row}\% = (100)\dfrac{f}{N_{\text{row}}}$	57
Column percentage	$\text{column}\% = (100)\dfrac{f}{N_{\text{column}}}$	58

Name	Formula	Page
Position of the median value	Position of median $= \dfrac{N + 1}{2}$	85
Sample mean	$\overline{X} = \dfrac{\Sigma X}{N}$	85
Deviation from the sample mean	Deviation $= X - \overline{X}$	86
Mean of a simple frequency distribution	$\overline{X} - \dfrac{\Sigma fX}{N}$	92
Range	$R = H - L$	108
Inter-quartile range	$IQR = Q3 - Q1$	109
Variance	$s^2 = \dfrac{\Sigma(X - \overline{X})^2}{N}$	111
Standard deviation	$s = \sqrt{\dfrac{\Sigma(X - \overline{X})^2}{N}}$	112
Raw-score formula for variance	$s^2 = \dfrac{\Sigma X^2}{N} - \overline{X}^2$	114
Raw-score formula for standard deviation	$s = \sqrt{\dfrac{\Sigma X^2}{N} - \overline{X}^2}$	114
Variance of a simple frequency distribution	$s^2 = \dfrac{\Sigma fX^2}{N} - \overline{X}^2$	116
Standard deviation of a simple frequency distribution	$s = \sqrt{\dfrac{\Sigma fX^2}{N} - \overline{X}^2}$	116
Probability of an outcome or event	$P = \dfrac{\text{number of times an outcome can occur}}{\text{total number of times any outcome can occur}}$	138
z score	$z = \dfrac{X - \mu}{\sigma}$	157

(continued)

Name	Formula	Page
Raw-score equivalent of a z score	$X = \mu + z\sigma$	163
z ratio of a sampling distribution	$z = \dfrac{\overline{X} - \mu}{\sigma_{\overline{X}}}$	185
Standard error of the mean	$\sigma_{\overline{X}} = \dfrac{\sigma}{\sqrt{N}}$	186
95% confidence interval	$95\% \text{ CI} = \overline{X} \pm 1.96\sigma_{\overline{X}}$	189
99% confidence interval	$99\% \text{ CI} = \overline{X} \pm 2.58\sigma_{\overline{X}}$	190
Standard error of the mean	$s_{\overline{X}} = \dfrac{s}{\sqrt{N - 1}}$	195
t ratio	$t = \dfrac{\overline{X} - \mu}{s_{\overline{X}}}$	195
Confidence interval	$\text{CI} = \overline{X} \pm t s_{\overline{X}}$	197
Standard error of the proportion	$s_p = \sqrt{\dfrac{P(1 - P)}{N}}$	200
95% confidence interval for a proportion	$95\% \text{ CI} = P \pm (1.96)s_P$	200
z test of difference between means of independent samples	$z = \dfrac{\overline{X}_1 - \overline{X}_2}{\sigma_{\overline{X}_1 - \overline{X}_2}}$	222
Standard error of the difference between means (pooled)	$s_{\overline{X}_1 - \overline{X}_2} = \sqrt{\left(\dfrac{N_1 s_1^2 + N_2 s_2^2}{N_1 + N_2 - 2}\right)\left(\dfrac{N_1 + N_2}{N_1 N_2}\right)}$	229
t test of difference between means of independent samples	$t = \dfrac{\overline{X}_1 - \overline{X}_2}{s_{\overline{X}_1 - \overline{X}_2}}$	230
Standard error of the difference between means (nonpooled)	$s_{\overline{X}_1 - \overline{X}_2} = \sqrt{\dfrac{s_1^2}{N_1 - 1} + \dfrac{s_2^2}{N_2 - 1}}$	231

Name	Formula	Page
Standard deviation of differences between related samples	$s_D = \sqrt{\dfrac{\Sigma D^2}{N} - (\bar{X}_1 - \bar{X}_2)^2}$	237
Standard error of the difference between means of related samples	$s_{\bar{D}} = \dfrac{s_D}{\sqrt{N-1}}$	238
t-test of difference between means of related samples	$t = \dfrac{\bar{X}_1 - \bar{X}_2}{s_{\bar{D}}}$	239
z-test of difference between sample proportions	$z = \dfrac{P_1 - P_2}{s_{P_1 - P_2}}$	239
Pooled sample proportion	$P* = \dfrac{N_1 P_1 + N_2 P_2}{N_1 + N_2}$	239
Standard error of the difference between sample proportions	$s_{P_1 - P_2} = \sqrt{P*(1 - P*)\left(\dfrac{N_1 + N_2}{N_1 N_2}\right)}$	239
Partitioning sum of squares	$SS_{total} = SS_{between} + SS_{within}$	269
Total sum of squares from deviations	$SS_{total} = \Sigma(X - \bar{X}_{total})^2$	270
Within-groups sum of squares from deviations	$SS_{within} = \Sigma(X - \bar{X}_{group})^2$	270
Between-groups sum of squares from deviations	$SS_{between} = \Sigma N_{group}(X_{group} - X_{total})^2$	272
Total sum of squares from raw scores	$SS_{total} = \Sigma X_{total}^2 - N_{total}\bar{X}_{total}^2$	273
Within-groups sum of squares from raw scores	$SS_{within} = \Sigma X_{total}^2 - \Sigma N_{group}\bar{X}_{group}^2$	273
Between-groups sum of squares from raw scores	$SS_{between} = \Sigma N_{group}\bar{X}_{group}^2 - N_{total}\bar{X}_{total}^2$	273
Between-groups mean square	$MS_{between} = \dfrac{SS_{between}}{df_{between}}$	275

(continued)

Name	Formula	Page
Within-groups mean square	$MS_{within} = \dfrac{SS_{within}}{df_{within}}$	275
F ratio	$F = \dfrac{MS_{between}}{MS_{within}}$	276
Tukey's honestly significant difference	$HSD = q\sqrt{\dfrac{MS_{within}}{N_{group}}}$	279
Main effects sum of squares from deviations	$SS_A = \Sigma N_a(\overline{X}_a - \overline{X}_{total})^2; SS_B = \Sigma N_b(\overline{X}_b - \overline{X}_{total})^2$	290
Interaction sum of squares from deviations	$SS_{AB} = \Sigma N_{group}(\overline{X}_{group} - \overline{X}_a - \overline{X}_b + \overline{X}_{total})^2$	290
Main effects sum of squares from raw scores	$SS_A = \Sigma N_a\overline{X}_a^2 - N_{total}\overline{X}_{total}^2; SS_B = \Sigma N_b\overline{X}_b^2 - N_{total}\overline{X}_{total}^2$	291
Interaction sum of squares from raw scores	$SS_{AB} = \Sigma N_{group}\overline{X}_{group}^2 - \Sigma N_a\overline{X}_a^2 - \Sigma N_b\overline{X}_b^2 + N_{total}\overline{X}_{total}^2$	291
F ratio for testing the significance of main and interaction effects	$F_A = \dfrac{MS_A}{MS_{within}}; F_B = \dfrac{MS_B}{MS_{within}}; F_{AB} = \dfrac{MS_{AB}}{MS_{within}}$	291
Chi-square	$\chi^2 = \Sigma \dfrac{(f_o - f_e)^2}{f_e}$	309
Chi-square with Yates' correction	$\chi^2 = \Sigma \dfrac{(|f_o - f_e| - .5)^2}{f_e}$	326
Pearson's correlation	$r = \dfrac{SP}{\sqrt{SS_X SS_Y}}$	352
Pearson's *r* from deviations	$r = \dfrac{\Sigma(X - \overline{X})(Y - \overline{Y})}{\sqrt{\Sigma(X - \overline{X})^2 \Sigma(Y - \overline{Y})^2}}$	352
Pearson's *r* from raw scores	$r = \dfrac{\Sigma XY - N\overline{X}\overline{Y}}{\sqrt{(\Sigma X^2 - N\overline{X}^2)(\Sigma Y^2 - N\overline{Y}^2)}}$	353
t ratio for testing the significance of *r*	$t = \dfrac{r\sqrt{N - 2}}{\sqrt{1 - r^2}}$	354

Name	Formula	Page
Partial correlation coefficient	$r_{XY.Z} = \dfrac{r_{XY} - r_{XZ}r_{YZ}}{\sqrt{1 - r_{XZ}^2}\sqrt{1 - r_{YZ}^2}}$	363
Regression model	$Y = a + bX + e$	376
Regression coefficient	$b = \dfrac{\text{SP}}{\text{SS}_X}$	378
Regression coefficient from deviations	$b = \dfrac{\Sigma(X - \overline{X})(Y - \overline{Y})}{\Sigma(X - \overline{X})^2}$	378
Regression coefficient from raw scores	$b = \dfrac{\Sigma XY - N\overline{X}\,\overline{Y}}{\Sigma X^2 - N\overline{X}^2}$	378
Y-intercept	$a = \overline{Y} - b\overline{X}$	378
Regression line	$\hat{Y} = a + bX$	380
Error term	$e = Y - \hat{Y}$	384
Error sum of squares	$\text{SS}_{\text{error}} = \Sigma(Y - \dot{Y})^2$	385
Total sum of squares	$\text{SS}_{\text{total}} = \Sigma(Y - \overline{Y})^2$	385
Regression sum of squares	$\text{SS}_{\text{reg}} = \text{SS}_{\text{total}} - \text{SS}_{\text{error}}$	385
Coefficient of determination	$r^2 = \dfrac{\text{SS}_{\text{total}} - \text{SS}_{\text{error}}}{\text{SS}_{\text{total}}}$	387
Coefficient of nondetermination	$1 - r^2 = \dfrac{\text{SS}_{\text{error}}}{\text{SS}_{\text{total}}}$	387
Mean square regression	$\text{MS}_{\text{reg}} = \dfrac{\text{SS}_{\text{reg}}}{\text{df}_{\text{reg}}}$	389

(continued)

Name	Formula	Page
Mean square error	$MS_{error} = \dfrac{SS_{error}}{df_{error}}$	389
F ratio for testing the significance of the regression	$F = \dfrac{MS_{reg}}{MS_{error}}$	389
Regression coefficients for two predictors	$b_1 = \dfrac{s_Y}{s_X}\left(\dfrac{r_{YX} - r_{YZ}r_{XZ}}{1 - r_{XZ}^2}\right); b_2 = \dfrac{s_Y}{s_Z}\left(\dfrac{r_{YZ} - r_{XY}r_{XZ}}{1 - r_{XZ}^2}\right)$	395
Y-intercept for two predictors	$b_0 = \bar{Y} - b_1\bar{X} - b_2\bar{Z}$	395
Multiple coefficient of determination	$R^2 = \dfrac{r_{YX}^2 + r_{YZ}^2 - 2r_{YX}r_{YZ}r_{XZ}}{1 - r_{XZ}^2}$	396
Odds	$Odds = \dfrac{P}{1 - P}$	403
Log-odds or *logit*	$L = \log\left(\dfrac{P}{1 - P}\right)$	404
Spearman's rank-order correlation	$r_s = 1 - \dfrac{6\Sigma D^2}{N(N^2 - 1)}$	418
z ratio to test the significance of Spearman's correlation	$z = r_s\sqrt{N - 1}$	422
Goodman's and Kruskal's gamma	$G = \dfrac{N_a - N_i}{N_a + N_i}$	425
z ratio for testing the significance of gamma	$z = G\sqrt{\dfrac{N_a + N_i}{N(1 - G^2)}}$	426
Phi coefficient	$\phi = \sqrt{\dfrac{\chi^2}{N}}$	431
Contingency coefficient	$C = \sqrt{\dfrac{\chi^2}{N + \chi^2}}$	433
Cramér's V	$V = \sqrt{\dfrac{\chi^2}{N(k - 1)}}$	434

Glossary

accidental sampling A nonrandom sampling method whereby the researcher includes the most convenient cases in his or her sample.

addition rule The probability of obtaining any one of several different outcomes equals the sum of their separate probabilities.

alpha The probability of committing a Type I error.

analysis of variance A statistical test that makes a single overall decision as to whether a significant difference is present among three or more sample means.

area under the normal curve That area that lies between the curve and the base line containing 100% or all of the cases in any given normal distribution.

bar graph A graphic method in which rectangular bars indicate the frequencies or percentages for categories.

between-groups sum of squares The sum of the squared deviations of every sample mean from the total mean.

bimodal distribution A frequency distribution containing two or more modes.

box plot A graphic method for simultaneously displaying several characteristics of a distribution.

central tendency What is average or typical of a set of data; a value generally located toward the middle or center of a distribution.

chi-square A nonparametric test of significance whereby expected frequencies are compared against observed frequencies.

class interval A category in a group distribution containing more than one score value.

class limit The point midway between adjacent class intervals that serves to close the gap between them.

coefficient of determination Equal to the Pearson's correlation squared, the proportion of variance in the dependent variable that is explained by the independent variable.

coefficient of nondetermination Equal to 1 minus the Pearson's correlation squared, the proportion of variance in the dependent variable that is not explained by the independent variable.

column percent In a cross-tabulation, the result of dividing a cell frequency by the number of cases in the column. Column percents sum to 100% for each column of a cross-tabulation.

confidence interval The range of mean values (proportions) within which the true population mean (proportion) is likely to fall.

contingency coefficient Based on chi-square, a measure of the degree of association for nominal data arranged in a table larger than 2×2.

converse rule The probability of an event not occurring equals 1 minus the probability that it does.

correlation The strength and direction of the relationship between two variables.

correlation coefficient Generally ranging between 1.00 and 1.00, a number in which both the strength and direction of correlation are expressed.

Cramér's V An alternative to the contingency coefficient that measures the degree of association for nominal data arranged in a table larger than 2×2.

critical region (rejection region) The area in the tail(s) of a sampling distribution that dictates that the null hypothesis be rejected.

cross-tabulation A frequency and percent table of two or more variables taken together.

cumulative frequency The total number of cases having any given score or a score that is lower.

cumulative frequency polygon A graphic method in which cumulative frequencies or cumulative percentages are depicted.

cumulative percentage The percent of cases having any score or a score that is lower.

curvilinear correlation A relationship between X and Y that begins as either positive or negative and then reverses direction.

deciles Percentile ranks that divide the 100-unit scale by 10s.

degrees of freedom In small sample comparisons, a statistical compensation for the failure of the sampling distribution of differences to assume the shape of the normal curve.

deviation The distance and direction of any raw score from the mean.

error term (disturbance term) The residual portion of a score that cannot be predicted by the independent variable. Also the distance of a point from the regression line.

expected frequencies The cell frequencies expected under the terms of the null hypothesis for chi-square.

F ratio The result of an analysis of variance, a statistical technique that indicates the size of the between-groups mean square relative to the size of the within-groups mean square.

five percent level (.05) of significance A level of probability at which the null hypothesis is rejected if an obtained sample difference occurs by chance only 5 times or less out of 100.

frequency distribution A table containing the categories, score values, or class intervals and their frequency of occurrence.

frequency polygon A graphic method in which frequencies are indicated by a series of points placed over the score values or midpoints of each class interval and connected with a straight line that is dropped to the base line at either end.

Goodman's and Kruskal's gamma An alternative to the rank-order correlation coefficient for measuring the degree of association between ordinal-level variables.

grouped frequency distribution A table that indicates the frequency of occurrence of cases located within a series of class intervals.

histogram A graphic method in which rectangular bars indicate the frequencies or percentages for the range of score values.

hypothesis An idea about the nature of social reality that is testable through systematic research.

independent outcomes Two outcomes or events are independent if the probability of one occurring is unchanged by whether the other occurs.

interaction effect The effect of two factors combined that differs from the sum of their separate (main) effects.

inter-quartile range The range (or difference) between the first and third quartiles; the range containing the middle 50% of a distribution.

interval level of measurement The process of assigning a score to cases so that the magnitude of differences between them is known and meaningful.

judgment sampling (purposive sampling) A nonrandom sampling method whereby logic, common sense, or sound judgment is used to select a sample that is presumed representative of a larger population.

kurtosis The peakedness of a distribution.

leptokurtic Characteristic of a distribution that is quite peaked or tall.

level of confidence How certain we are that a confidence interval covers the true population mean (proportion).

level of significance A level of probability at which the null hypothesis can be rejected and the research hypothesis can be accepted.

line chart A graph of the differences between groups or trends across time on some variable(s).

logistic regression A statistical procedure for predicting a dichotomous dependent variable on the basis of one or more independent variables.

logit The logarithm of the odds of an event, also known as log-odds.

mam effect In analysis of variance, the effect of a factor independent of other factors.

margin of error The extent of imprecision expected when estimating the population mean or proportion, obtained by multiplying the standard error by the table value of z or t.

marginal distribution In a cross-tabulation, the set of frequencies and percents found in the margin that represents the distribution of one of the variables in the table.

mean The sum of a set of scores divided by the total number of scores in the set. A measure of central tendency.

mean deviation The sum of the absolute deviations from the mean divided by the number of scores in a distribution. A measure of variability that indicates the average of deviations from the mean.

mean square A measure of variation used in an F test obtained by dividing the between-group sum of squares or within-group sum of squares (in analysis of variance) or the regression sum of squares or error sum of squares (in regression analysis) by the appropriate degrees of freedom.

measurement The use of a series of numbers in the data analysis stage of research.

median The middle-most point in a frequency distribution. A measure of central tendency.

median test A nonparametric test of significance for determining the probability that two random samples have been drawn from populations with the same median.

mesokurtic Characteristic of a distribution that is neither very peaked nor very flat.

midpoint The middle-most score value in a class interval.

mode The most frequent, typical, or common value in a distribution.

multiple coefficient of determination The proportion of variance in the dependent variable that is explained by the set of independent variables in combination.

multiple regression A statistical procedure for predicting a dependent variable on the basis of several independent variables.

multiplication rule The probability of obtaining a combination of independent outcomes equals the product of their separate probabilities.

multistage sampling A random sampling method whereby sample members are selected on a random basis from a number of well-delineated areas known as clusters (or primary sampling units).

mutually exclusive outcomes Two outcomes or events are mutually exclusive if the occurrence of one rules out the possibility that the other will occur.

negative correlation The direction of relationship wherein individuals who score high on the X variable score low on the Y variable; individuals who score low on the X variable score high on the Y variable.

negatively skewed distribution A distribution in which more respondents receive high than low scores, resulting in a longer tail on the left than on the right.

95% confidence interval The range of mean values (proportions) within which there are 95 chances out of 100 that the true population mean (proportion) will fall.

99% confidence interval The range of mean values (proportions) within which there are 99 chances out of 100 that the true population mean (proportion) will fall.

nominal level of measurement The process of placing cases into categories and counting their frequency of occurrence.

nonparametric test A statistical procedure that makes no assumptions about the way the characteristic being studied is distributed in the population and requires only ordinal or nominal data.

nonrandom sampling A sampling method whereby each population member does not have an equal chance of being drawn into the sample.

normal curve A smooth, symmetrical distribution that is bell-shaped and unimodal.

null hypothesis The hypothesis of equal population means. Any observed difference between samples is seen as a chance occurrence resulting from sampling error.

observed frequencies In a chi-square analysis, the results that are actually observed when conducting a study.

odds The probability that an event occurs divided by the probability that the event does not occur.

odds ratio The odds for one group divided by the odds for some comparison group.

one percent (.01) level of significance A level of probability at which the null hypothesis is rejected if an obtained sample difference occurs by chance only 1 time or less out of 100.

one-tailed test A test in which the null hypothesis is rejected for large differences in only one direction.

ordinal level of measurement The process of ordering or ranking cases in terms of the degree to which they have any given characteristic.

parametric test A statistical procedure that requires that the characteristic studied be normally distributed in the population and that the researcher have interval data.

partial correlation coefficient The correlation between two variables when one or more other variables are controlled.

Pearson's correlation coefficient A correlation coefficient for interval data.

percentage A method of standardizing for size that indicates the frequency of occurrence of a category per 100 cases.

percentage distribution The relative frequency of occurrence of a set of scores or class intervals.

percentile rank A single number that indicates the percent of cases in a distribution falling at or below any given score.

phi coefficient Based on chi-square, a measure of the degree of association for nominal data arranged in a table.

pie chart   A circular graph whose pieces add up to 100%.

platykurtic   Characteristic of a distribution that is rather flat.

population (universe)   Any set of individuals who share at least one characteristic.

positive correlation   The direction of a relationship wherein individuals who score high on the X variable also score high on the Y variable; individuals who score low on the X variable also score low on the Y variable.

positively skewed distribution   A distribution in which more respondents receive low than high scores, resulting in a longer tail on the right than on the left.

power of a test   The ability of a statistical test to reject the null hypothesis when it is actually false and should be rejected.

primary sampling unit (cluster)   In multistage sampling, a well-delineated area considered to include characteristics found in the entire population.

probability   The relative frequency of occurrence of an event or outcome. The number of times any given event could occur out of 100.

proportion   A method for standardizing for size that compares the number of cases in any given category with the total number of cases in the distribution.

quartiles   Percentile ranks that divide the 100-unit scale by 25s.

quota sampling   A nonrandom sampling method whereby diverse characteristics of a population are sampled in the proportions they occupy in the population.

random sampling   A sampling method whereby every population member has an equal chance of being drawn into the sample.

range   The difference between the highest and lowest scores in a distribution. A measure of variability.

rate   A kind of ratio that indicates a comparison between the number of actual cases and the number of potential cases.

ratio   A method of standardizing for size that compares the number of cases falling into one category with the number of cases falling into another category.

regression analysis   A technique employed in predicting values of one variable *(Y)* from knowledge of values of another variable *(X)*.

regression line   A straight line drawn through the scatter plot that represents the best possible fit for making predictions of Y from X.

research hypothesis   The hypothesis that regards any observed difference between samples as reflecting a true population difference and not just sampling error.

row percent   In a cross-tabulation, the result of dividing a cell frequency by the number of cases in the row. Row percents sum to 100% for each row of a cross-tabulation.

sample   A smaller number of individuals taken from some population (for the purpose of generalizing to the entire population from which it was taken).

sampling distribution of differences between means   A frequency distribution of a large number of differences between random sample means that have been drawn from a given population.

sampling distribution of means   A frequency distribution of a large number of random sample means that have been drawn from the same population.

sampling error   The inevitable difference between a random sample and its population based on chance alone.

scatter plot   A graph that shows the way scores on any two variables X and Y are scattered throughout the range of possible score values.

simple random sampling   A random sampling method whereby a table of random numbers is employed to select a sample that is representative of a larger population.

skewness   Departure from symmetry.

slope   In regression, the change in the regression line for a unit increase in X. The slope is interpreted as the change in the Y variable associated with a unit change in the X variable.

Spearman's rank-order correlation coefficient   A correlation coefficient for data that have been ranked or ordered with respect to the presence of a given characteristic.

spurious relationship   A noncausal relationship between two variables that exists only because of the common influence of a third variable. The spurious relationship disappears if the third variable is held constant.

standard deviation   The square root of the mean of the squared deviations from the mean of a distribution. A measure of variability that reflects the typical deviation from the mean.

standard error of the difference between means   An estimate of the standard deviation of the sampling distribution of differences based on the standard deviations of two random samples.

standard error of the mean   An estimate of the standard deviation of the sampling distribution of means based on the standard deviation of a single random sample.

standard error of the proportion   An estimate of the standard deviation of the sampling distribution of proportions based on the proportion obtained in a single random sample.

statistically significant difference   A sample difference that reflects a real population difference and not just a sampling error.

straight-line correlation   Either a positive or negative correlation, so that the points in a scatter diagram tend to form a straight line through the center of the graph.

stratified sampling   A random sampling method whereby the population is first divided into homogeneous subgroups from which simple random samples are then drawn.

strength of correlation   Degree of association between two variables.

sum of squares   The sum of squared deviations from a mean.

systematic sampling   A random sampling method whereby every *n*th member of a population is included in the sample.

t **ratio**   A statistical technique that indicates the direction and degree that a sample mean difference falls from zero on a scale of standard error units.

total percent   In a cross-tabulation, the result of dividing a cell frequency by the total number of cases in the sample. Total percents sum to 100% for the entire cross-tabulation.

total sum of squares   The sum of the squared deviations of every raw score from the total mean of the study.

Tukey's HSD (honestly significant difference)   A procedure for the multiple comparison of means after a significant *F* ratio has been obtained.

two-tailed test   A test that is used when the null hypothesis is rejected for large differences in both directions.

Type I error   The error of rejecting the null hypothesis when it is true.

Type II error   The error of accepting the null hypothesis when it is false.

unimodal distribution   A frequency distribution containing a single mode.

unit of observation   The element that is being studied or observed. Individuals are most often the unit of observation, but sometimes collections or aggregates, such as families, census tracts, or states, are the unit of observation.

variability   The manner in which the scores are scattered around the center of the distribution. Also known as dispersion or spread.

variable   Any characteristic that varies from one individual to another. Hypotheses usually contain an independent variable (cause) and a dependent variable (effect).

variance   The mean of the squared deviations from the mean of a distribution. A measure of variability in a distribution.

weighted mean   The "mean of means" that adjusts for differences in group size.

within-groups sum of squares   The sum of the squared deviations of every raw score from its sample group mean.

Y-intercept   In regression, the point where the regression line crosses the *Y* axis. The *Y*-intercept is the predicted value of *Y* for an *X* value of zero.

Yates's correction   In the chi-square analysis, a factor for small expected frequencies that reduces the overestimate of the chi-square value and yields a more conservative result (only for tables).

z **score (standard score)**   A value that indicates the direction and degree that any given raw score deviates from the mean of a distribution on a scale of standard deviation units.

z **score for sample mean differences**   A value that indicates the direction and degree that any given sample mean difference falls from zero (the mean of the sampling distribution of differences) on a scale of standard deviation units.

Answers to Problems

Chapter 1

1. d
2. e
3. b
4. c
5. b
6. (a) Content analysis, IV = gender, DV = aggressiveness in description of Super Bowl; (b) Experiment, IV = type of sport, DV = aggressiveness of play; (c) Participant observation, IV = whether team wins or loses, DV = extent of arguing and fighting; (d) Survey, IV = aggressiveness, DV = preferred sporting events
7. (a) Nominal; (b) Interval; (c) Interval; (d) Ordinal; (e) Interval; (f) Nominal; (g) Ordinal; (h) Nominal; (i) Interval (assuming equal intervals between points on scale)
8. (a) Interval; (b) Interval; (c) Ordinal; (d) Nominal; (e) Nominal; (f) Ordinal; (g) Interval; (h) Ordinal
9. (a) Secondary analysis; (b) Survey; (c) Participant observation; (d) Experiment
10. b
11. a
12. (a) Ordinal; (b) Interval; (c) Interval (assuming equal intervals between points on scale); (d) Nominal
13. d
14. e
15. (a) Nominal; (b) Interval; (c) Ordinal; (d) Interval; (e) Interval (assuming equal intervals between points on scale); (f) Nominal; (g) Ordinal
16. c

17. (a) Survey; (b) Participant observation; (c) Content analysis

Chapter 2

1. a
2. e
3. d
4. a
5. d
6. b
7. c
8. c
9. c
10. (a) 50.8%; (b) 26.6%; (c) $P = .51$; (d) $P = .27$
11. (a) 53.1%; (b) 73.9%; (c) $P = .53$; (d) $P = .74$
12. (a) 14.9%; (b) 6.6%; (c) $P = .15$; (d) $P = .07$; (e) Left-handedness is more prevalent among men.
13. 156.25
14. $15/20 = 3/4$
15. There are 85.71 live births per every 1,000 women of childbearing age.
16. 66.67%
17. 15.63%
18. (a) 2; (b) 5.5–7.5, 3.5–5.5, 1.5–3.5, –.5–1.5; (c) m: 6.5, 4.5, 2.5, 0.5; (d) %: 31.1, 37.8, 15.6, 15.6; (e) cf: 45, 31, 14, 7; (f) c%: 100, 69.0, 31.2, 15.6
19. (a) 5; (b) 34.5–39.5, 29.5–34.5, 24.5–29.5, 19.5–24.5, 14.5–19.5; (c) m: 37, 32, 27, 22, 17; (d) %: 13.5, 9.5, 29.7, 31.1, 16.2; (e) cf: 74, 64, 57, 35, 12; (f) c%: 100, 86.5, 77.0, 47.3, 16.2

20. (a) 59.38; (b) 14.58

21. (a) 84.82; (b) 29.64

22. (a) IV = social class, DV = housing status; (b) 77.5%, 22.5%, 100.0%, 42.7%, 57.3%, 100.0%, 22.0%, 78.0%, 100.0%, 50.0%, 50.0%, 100.0%; (c) 50.0%; (d) 50.0%; (e) 22.5%; (f) 42.7%; (g) Lower class; (h) Upper class; (i) The higher the social class, the greater the tendency to own rather than rent.

23. (a) IV = Voter age, DV = Vote; (b) 33.8%, 31.3%, 63.0%, 30.6%, 38.8%, 28.8%, 43.8%, 18.5%, 27.8%, 31.2%, 37.5%, 25.0%, 18.5%, 41.7%, 30.0%, 100.0%, 100.0%, 100.0%, 100.0%, 100.0%; (c) 38.8%; (d) 28.8%; (e) 45–59; (f) 30–44; (g) 60+

24. (a) Because neither opinion is clearly the result of the other; (b) 35.0%, 16.1%, 51.1%, 38.9%, 10.0%, 48.9%, 73.9%, 26.1%, 100.0%; (c) 73.9%; (d) 51.1%; (e) 35.0%; (f) 10.0%; (g) 55.0%; (h) People who favor the death penalty are more likely to oppose mercy killing, whereas people who oppose the death penalty are more likely to favor mercy killing.

25. (a) 61.7%; (b) 54.9%; (c) P = .62; (d) P = .55

26. (a) No, there is no IV or DV because gender does not cause sexual orientation; (b) 81.3%, 89.9%, 85.8%, 13.1%, 7.6%, 10.2%, 5.6%, 2.5%, 4.0%, 100.0%, 100.0%, 100.0%; (c) 85.8%; (d) 4.0%; (e) 4.0%; (f) 38.7%; (g) Both males and females tend to be heterosexual.

27. (a) 6.8%, 10.5%, 3.1%, .9%, 21.3%, 32.1%, 23.3%, 19.3%, 4.0%, 78.7%, 38.9%, 33.8%, 22.4%, 4.8%, 100.0%; (b) 21.3%; (c) 78.7%; (d) 19.3%; (e) 6.8%; (f) Married.

28. (a) Because opinion on one subject does not cause opinion on the other; (b) 26.5%, 15.6%, 42.2%, 11.8%, 46.0%, 57.8%, 38.4%, 61.6%, 100.0%; (c) 42.2%; (d) 38.4%; (e) 26.5%; (f) 46.0%; (g) 27.5%; (h) People who favor bilingual education tend to also favor affirmative action, and people who oppose bilingual education also tend to oppose affirmative action.

29. Draw a pie chart.

30. Draw a bar graph.

31. Draw a bar graph and a frequency polygon.

32. Draw a histogram and a line chart.

33. (a) size = 50; midpoints: 775, 720, 670, 620, 570, 520, 470, 420, 370; upper and lower limits: 745–805, 695–745, 645–695, 595–645, 545–595, 495–545, 445–495, 395–445, 345–395; cf: 38, 37, 35, 32, 27, 17, 9, 5, 2; %: 2.6, 5.3, 7.9, 13.2, 26.3, 21.1, 10.5, 7.9, 5.3; c%: 100, 97.5, 92.2, 84.3, 71.1, 44.8, 23.7, 13.2, 5.3; (b) Draw a histogram and a frequency polygon; (c) Draw a cumulative frequency polygon.

34. Use a blank map to show unemployment rates.

Chapter 3

1. a
2. a
3. c
4. b
5. c
6. b
7. a
8. c
9. (a) R; (b) R; (c) Because the variable is not measured at the interval level.
10. IQ: mean = 105.5; Gender: mode = Female; Ethnicity: mode = White; Age: mean = 31; Frequency of Appearance: Median = between "sometimes" and "rarely."
11. Age: mean = 78.7; Health Status: Median = Fair; Type of Abuse: Mode = Financial Abuse; Duration of Abuse: Mean = 5.4 months
12. (a) 10 lb; (b) 10 lb
13. (a) 2 times; (b) 2.55 times
14. (a) 3 years; (b) 4 years; (c) 11 years; (d) Median, because of the skewness of the distribution.
15. (a) $12; (b) $14; (c) $14.57
16. (a) $12; (b) $15; (c) $15.75
17. (a) 0 times; (b) 2.5 times; (c) 2.5 times

18. (a) Mo = 7; Mdn = 5; Mean = 4.7

(b)

Households	f
9	2
8	1
7	3
6	2
5	2
4	2
3	2
2	2
1	2
0	1

Mo = 7; Mdn = 5; Mean = 4.7

19. deviations from mean: –7; –6; –8; –8; 29
These deviations indicate that the distribution is skewed.

20. X = 18; deviation = +3.43; received $3.43 more per hour than average
X = 16; deviation = +1.43; received $1.43 more per hour than average
X = 20; deviation = +5.43; received $5.43 more per hour than average
X = 12; deviation = –2.57; received $2.57 less per hour than average
X = 14; deviation = –.57; received $.57 less per hour than average
X = 12; deviation = –2.57; received $2.57 less per hour than average
X = 10; deviation = –4.57; received $4.57 less per hour than average
The deviations indicate a fairly symmetric distribution.

21. (a) Mo = 2; Mdn = 2; Mean = 2.2

(b)

Children	f
6	1
5	1
4	2
3	3
2	6
1	4
0	3

Mo = 2; Mdn = 2; Mean = 2.2

22. $\overline{X}_w$ = 18.21
23. (a) 4; (b) 4; (c) 4.23
24. Median = upper middle
25. (a) 4; (b) 4; (c) 3.7
26. (a) 7; (b) 4.5; (c) 4.25; assumes equal intervals between points on scale
27. (a) 6; (b) 6; (c) 6.26
28. (a) 8; 7.5; 6.8
(b) 5; 5.5; 5.6
(c) knowledge
29. Mean = 15.82; Mdn = 14.55; No mode
30. Mo = completed high school; Mdn = completed high school; cannot calculate mean because the data is ordinal.

Chapter 4

1. d
2. b
3. d
4. c
5. a
6. c
7. (a) Student A; (b) Student B
8. (a) A = 5; B = 4; (b) A = 2.5; B = 1.00; (c) A = 1.89; B = 1.10. Class A has greater variability of attitude scores.
9. (a) 7; (b) 5; (c) s^2 = 6.00; s = 2.45
10. (a) 6; (b) 3.5; (c) s^2 = 3.24; s = 1.80
11. (a) 7; (b) 3; (c) s^2 = 4.25; s = 2.06
12. (a) 9; (b) 4; (c) s^2 = 7.35; s = 2.71
13. (a) 13; (b) 10; (c) s^2 = 22.16; s = 4.71
14. (a) 16; (b) 7.5; (c) s^2 = 20.44; s = 4.52
15. (a) 750; (b) 412.5; (c) s^2 = 56931.25; s = 238.60
16. (a) 3.6; (b) 1.2; (c) s^2 = .78; s = .88
17. range = 7; mean deviation = 2.44; s = 2.75; s^2 = 7.56
18. s = 2.08; s^2 = 4.34
19. s = 1.19
20. s^2 = 2.37; s = 1.54
21. s^2 = 3.89; s = 1.97
22. s^2 = 3.44; s = 1.86

Chapter 5

1. c
2. b
3. c
4. b
5. a
6. b
7. a
8. b
9. (a) .5; (b) .6; (c) .7; (d) .8
10. (a) .4; (b) .4; (c) .6; (d) .4
11. (a) .84; (b) .59
12. (a) .75; (b) .06
13. .03
14. (a) .50; (b) .20; (c) .10; (d) .40; (e) .40; (f) .10
15. (a) .11; (b) .11; (c) .89; (d) .01; (e) .04; (f) .0005
16. (a) .50; (b) .30; (c) .70; (d) .027
17. (a) .019; (b) .038; (c) .077; (d) .50; (e) .231; (f) .308; (g) .308
18. (a) .533; (b) .467; (c) .336; (d) .145; (e) .049; (f) .039
19. (a) .4; (b) .36
20. (a) .98; (b) .96; (c) .94; (d) .92
21. (a) 34.13%; (b) 68.26%; (c) 47.72%; (d) 95.44%
22. (a) 34.13%; (b) 68.26%; (c) 47.72%; (d) 95.44%; (e) 15.87%; (f) 2.28%
23. (a) 0; (b) −1; (c) 1.5; (d) .7; (e) 2.5; (f) −1.2
24. (a) 50%; (b) 15.87%; (c) 43.32%; (d) 24.20%; (e) 98.76%; (f) 88.49%
25. (a) 40.82%; (b) .4082; (c) 28.81%; (d) .2881; (e) 25.14%; (f) .2514; (g) .063; (h) 95.25% (95th percentile)
26. (a) 11.51%; (b) .1151; (c) 28.81%; (d) .2881; (e) 34.46%; (f) .3446; (g) .4238; (h) .25; (i) .075
27. (a) .1525; (b) .0228; (c) .00000027
28. (a) 52.22%; (b) .5222; (c) 48.38%; (d) .4838; (e) .21%; (f) .0021; (g) .000004; (1st percentile)
29. (a) .1056; (b) .000013; (c) Can conclude that their claim is false, since the probability was less than .05 that we would randomly select five cars that perform poorly by chance.
30. (a) .3849; (b) .1151

Chapter 6

1. d
2. c
3. c
4. d
5. a
6. 1.8
7. (a) 9.47 − 16.53; (b) 8.36 − 17.64
8. (a) SE = 1.73; (b) 9.32 − 16.68; (c) 7.92 − 18.08
9. SE = 0.27
10. (a) 2.38 − 3.49; (b) 2.19 − 3.68
11. SE = 0.35
12. (a) 5.09 − 6.50; (b) 4.84 − 6.74
13. SE = .33
14. SE = .34
15. (a) 3.43 − 4.79; (b) 3.20 − 5.02
16. (a) .24; (b) 5.51 − 6.49; (c) 5.35 − 6.65
17. 16.17 − 17.43
18. 35.09 − 40.92
19. 2.65 − 3.35
20. 2.71 − 3.29
21. 2.57 − 3.03
22. (a) 1.23; (b) 39.47 − 44.53; (c) 38.57 − 45.43
23. (a) .03; (b) .29 − .43; (c) .27 − .45
24. (a) .02; (b) .63 − .73; (c) .62 − .74
25. (a) .01; (b) .42 − .48; (c) .41 − .49
26. (a) .04; (b) .53 − .70; (c) .50 − .73
27. No, the pollster is not justified, since .50 falls within the 95% confidence interval
28. (a) .04; (b) .27 − .44; (c) .24 − .46
29. (a) .07; (b) .56 − .82

Chapter 7

1. b
2. d
3. a
4. a
5. b
6. a
7. c

8. b

9. c

10. a

11. d

12. .0668

13. $t = 2.5$, df $= 78$, reject the null hypothesis at .05

14. $t = -9.5$, df $= 91$, reject the null hypothesis at .05

15. $t = -0.93$, df $= 16$, retain the null hypothesis at .05

16. $t = 2.43$, df $= 8$, reject the null hypothesis at .05

17. $t = -2.98$, df $= 8$, reject the null hypothesis at .05

18. $t = 1.90$, df $= 13$, retain the null hypothesis at .05

19. $t = -1.52$, df $= 18$, retain the null hypothesis at .05

20. $t = 3.11$, df $= 20$, reject the null hypothesis at .05

21. $t = 0.41$, df $= 15$, retain the null hypothesis at .05

22. $t = 4.64$, df $= 21$, reject the null hypothesis at .05

23. $t = 4.76$, df $= 18$. High-anomie areas have significantly higher suicide rates than low-anomie areas. Reject the null hypothesis at .05

24. $t = -2.85$, df $= 18$. Students' friendship choices in a cooperative approach are significantly different than students' friendship choices in a competitive approach. Reject the null hypothesis at .05

25. $t = 2.05$, df $= 18$. The new course was not significantly different from the customary course. Retain the null hypothesis at .05

26. $t = -0.30$, df $= 19$, retain the null hypothesis at .05

27. $t = 5.22$, df $= 25$, reject the null hypothesis at .05

28. $t = 5.12$, df $= 11$, reject the null hypothesis at .05

29. $t = 9.55$, df $= 5$, reject the null hypothesis at .05

30. $t = 3.12$, df $= 7$. The film resulted in a significant reduction of racist attitudes. Reject the null hypothesis at .05

31. $t = 4.02$, df $= 9$. Aggression does significantly differ as a result of participation in the conflict-resolution program. Reject the null hypothesis at .05

32. $t = 3.43$, df $= 9$. The antidrug lecture makes a significant difference in students' attitudes. Reject the null hypothesis at .05

33. $t = -6.00$, df $= 4$, reject the null hypothesis at .05

34. $t = -3.75$, df $= 9$, reject the null hypothesis at .05

35. $z = 3.88$, reject the null hypothesis at .05

36. $z = -0.60$, retain the null hypothesis at .05

37. $z = -0.55$, retain the null hypothesis at .05

Chapter 8

1. a

2. a

3. b

4. c

5. d

6. $F = 2.72$, df $= 3, 12$, retain the null hypothesis at .05

7. $F = 46.33$, df $= 2, 9$, reject the null hypothesis at .05

8. HSD $= 1.71$, all mean differences are significant

9. $F = 6.99$, df $= 2, 12$, reject the null hypothesis at .05

10. HSD $= 3.59$, significant difference between HMO (1) and None (3)

11. $F = 13.14$, df $= 2, 21$, reject the null hypothesis at .05

12. HSD $= 1.85$, significant difference between punks (2) and jocks (3), and between geeks (1) and jocks (3).

13. $F = 2.82$, df $= 2, 27$, retain the null hypothesis at .05

14. $F = 11.41$, df $= 2, 12$, reject the null hypothesis at .05

15. HSD = 2.54, significant differences drugs (2) and control (3) groups, and between the alcohol (1) and control (3) groups.

16. $F = 19.73$, df = 2, 12, reject the null hypothesis at .05

17. $F = 122.85$, df = 2, 12, reject the null hypothesis at .05

18. HSD = 13.43, significant difference between treatment C (3) and treatments A (1) and B (2).

19. $F = 0.75$, df = 3, 16, retain the null hypothesis at .05

20. $F = 33.87$, df = 2, 15, reject the null hypothesis at .05

21. HSD = 1.36, significant difference between all groups

22. $F = 16.76$, df = 2, 18, reject the null hypothesis at .05

23. HSD = 1.84, significant difference between workers (1) and managers (2), and between workers (1) and owners (3)

24. $F = 1.38$, df = 2, 18, retain the null hypothesis at .05

25. $F = 4.60$, df = 2, 12, reject the null hypothesis at .05

26. HSD = 4.14, significant difference between high school graduates (2) and college graduates (3)

27. $F = 3.17$, df = 3, 16, retain the null hypothesis at .05

28. $F = 24.98$, df = 2, 12, reject the null hypothesis at .05

29. HSD = 2.02, significant difference between all groups

30. Gender of teacher effect: $F = 44.13$, significant
Gender of student effect: $F = 3.72$, not significant
Interaction effect: $F = 10.33$, significant

31. (a) Group means: 13.8, 8.8, 10.4, 12.6, 9.2, 10.0
(b) Gender effect: $F = .75$, not significant
Degree effect: $F = 14.62$, significant
Interaction effect: $F = .50$, not significant

Chapter 9

1. c
2. a
3. c
4. b
5. a
6. d
7. $\chi^2 = 8.81$, df = 3, reject the null hypothesis at .05
8. $\chi^2 = 3.90$, df = 2, retain the null hypothesis at .05
9. $\chi^2 = 4.94$, df = 2, retain the null hypothesis at .05
10. $\chi^2 = 62.34$, df $-$ 1, reject the null hypothesis at .05
11. $\chi^2 = 12.46$, df = 1, reject the null hypothesis at .05
12. $\chi^2 = 12.49$, df = 1, reject the null hypothesis at .05
13. $\chi^2 = .052$, df = 1, retain the null hypothesis at .05
14. $\chi^2 = 16.55$, df = 1, reject the null hypothesis at .05
15. $\chi^2 = 2.18$, df = 1, retain the null hypothesis at .05
16. $\chi^2 = 0.09$, df $-$ 1, retain the null hypothesis at .05
17. $\chi^2 = 0.78$, df = 2, retain the null hypothesis at .05
18. $\chi^2 = 3.29$, df = 3, retain the null hypothesis at .05
19. $\chi^2 = 13.19$, df = 1, reject the null hypothesis at .05
20. $\chi^2 = 20.92$, df = 3, reject the null hypothesis at .05
21. $\chi^2 = 1.69$, df = 2, retain the null hypothesis at .05
22. $\chi^2 = 17.74$, df = 3, reject the null hypothesis at .05
23. $\chi^2 = 6.27$, df = 2, reject the null hypothesis at .05
24. $\chi^2 = 10.96$, df = 4, reject the null hypothesis at .05

25. Table 9.5: $\chi^2 = 18.31$, df $= 3$, reject the null hypothesis at .05; Table 9.6: $\chi^2 = 11.91$, df $= 2$, reject the null hypothesis at .05. The χ^2 value decreases when the categories are collapsed.

26. Mdn $= 5$, $\chi^2 = 2.07$ (with Yates), df $= 1$, retain the null hypothesis at .05

27. Mdn $= 6$, $\chi^2 = 19.57$ (with Yates), df $= 1$, reject the null hypothesis at .05

28. Mdn $= 3$, $\chi^2 = 4.29$ (with Yates), df $= 1$, reject the null hypothesis at .05

Chapter 10

1. b

2. a

3. c

4. c

5. b

6. a

7. $r = 0.86$, df $= 4$, significant at .05

8. $r = 0.84$, df $= 8$, significant at .05

9. $r = 0.62$, df $= 8$, not significant at .05

10. $r = -0.69$, df $= 10$, significant at .05

11. $r = 0.68$, df $= 5$, not significant at .05

12. $r = -0.08$, df $= 4$, not significant at .05

13. $r = 0.97$, df $= 6$, significant at .05

14. $r = 0.98$, df $= 5$, significant at .05

15. $r = 0.98$, df $= 6$, significant at .05

16. $r = 0.68$, df $= 6$, not significant at .05

17. $r = 0.69$, df $= 8$, reject the null hypothesis at .05

18. $r = 0.61$, df $= 6$, not significant at .05 (cannot generalize)

19. $r = -0.73$, df $= 8$, significant at .05

20. $r_{xy} = -0.69$, $r_{xz} = 0.85$, $r_{yz} = -0.61$, $r_{xyz} = -0.42$

21. $r_{xy} = 0.68$, $r_{xz} = 0.47$, $r_{yz} = 0.87$, $r_{xyz} = 0.63$

22. (a) r_{xy} and r_{yz} are significant at the .05 level; (b) $r_{xyz} = 0.58$; the simple correlation r_{yz} just looks at the correlation between x and y without considering other variables; the partial correlation r_{xyz} looks at the correlation between x and y while controlling for the influence of the third variable, z.

Chapter 11

1. b

2. d

3. c

4. b

5. b

6. (a) Draw a scatter plot; (b) $b = -0.41$, $a = 7.58$; (c) Draw the regression line: $\hat{Y} = 7.58 - 0.41X$; (d) $\hat{Y} = 3.07$; (e) $r^2 = 0.75$, $1 - r^2 = 0.25$; (f)

Source	SS	df	MS	F
Regression	17.78	1	17.78	24.43 sig.
Error	5.82	8	0.73	
Total	23.60	9		

7. (a) Draw a scatter plot; (b) $b = 0.98$, $a = 3.91$; (c) Draw the regression line: $\hat{Y} = 3.91 + 0.98X$; (d) $\hat{Y} = 3.91$, $\hat{Y} = 5.87$; (e) $r^2 = 0.35$, $1 - r^2 = 0.65$ (f)

Source	SS	df	MS	F
Regression	11.01	1	11.01	3.16 not sig.
Error	20.87	6	3.48	
Total	31.88	7		

8. (a) Draw a scatter plot; (b) $b = 1.03$, $a = 1.91$; (c) Draw the regression line: $\hat{Y} = 1.91 + 1.03X$; (d) $\hat{Y} = 8.09$; (e) $r^2 = 0.41$, $1 - r^2 = 0.59$ (f)

Source	SS	df	MS	F
Regression	34.03	1	34.03	5.45 sig.
Error	49.97	8	6.25	
Total	84.00	9		

9. (a) Draw a scatter plot; (b) $b = 0.46$, $a = 12.66$; (c) Draw the regression line: $\hat{Y} = 12.66 + 0.46X$; (d) $\hat{Y} = \$14.50$; (e) $r^2 = 0.40$, $1 - r^2 = 0.60$
(f)

Source	SS	df	MS	F
Regression	6.73	1	6.73	5.30 not sig.
Error	10.17	8	1.27	
Total	16.90	9		

(g) The typical starting wage is $12.66 and the typical increase in wage each additional year is $.46

10. (a) Draw a scatter plot; (b) $b = 0.48$, $a = 0.84$; (c) Draw the regression line: $\hat{Y} = 0.84 + 0.48X$; (d) $\hat{Y} = 2.28$; (e) $r^2 = 0.59$, $1 - r^2 = 0.41$
(f)

Source	SS	df	MS	F
Regression	13.52	1	13.52	17.37 sig.
Error	9.34	12	0.78	
Total	22.86	13		

11. (a) Draw a scatter plot; (b) $b = 0.59$, $a = 1.30$; (c) Draw the regression line: $\hat{Y} = 1.30 + 0.59X$; (d) $\hat{Y} = 5.73$; (e) $r^2 = 0.93$, $1 - r^2 = 0.07$
(f)

Source	SS	df	MS	F
Regression	55.81	1	55.81	93.20 sig.
Error	4.19	7	0.60	
Total	60.00	8		

12. (a) Draw a scatter plot; (b) $b = 0.84$, $a = 0.67$; (c) Draw the regression line: $\hat{Y} = 0.67 + 0.84X$; (d) $\hat{Y} = 3.19$; (e) $r^2 = 0.84$, $1 - r^2 = 0.16$
(f)

Source	SS	df	MS	F
Regression	42.02	1	42.02	31.58 sig.
Error	7.98	6	1.33	
Total	50.00	7		

13. (a) Draw a scatter plot; (b) $b = -0.67$, $a = 8.62$; (c) Draw the regression line: $\hat{Y} = 8.62 - 0.67X$; (d) $\hat{Y} = 6.61$; (e) $r^2 = 0.83$, $1 - r^2 = 0.17$
(f)

Source	SS	df	MS	F
Regression	26.22	1	26.22	25.18 sig.
Error	5.21	5	1.04	
Total	31.43	6		

14. (a) Draw a scatter plot; (b) $b = -1.50$, $a = 22.08$; (c) Draw the regression line: $\hat{Y} = 22.08 - 1.50X$; (d) $\hat{Y} = 9.33$; (e) $r^2 = 0.78$, $1 - r^2 = 0.22$
(f)

Source	SS	df	MS	F
Regression	94.50	1	94.50	24.06 sig
Error	27.50	7	3.93	
Total	122.00	8		

15. (a) Draw a scatter plot; (b) $b = -1.32$, $a = 39.58$; (c) Draw the regression line: $\hat{Y} = 39.58 - 1.32X$; (d) $\hat{Y} = 29.02$; (e) $r^2 = 0.06$, $1 - r^2 = 0.94$
(f)

Source	SS	df	MS	F
Regression	72.65	1	72.65	0.32 not sig.
Error	1130.21	5	226.04	
Total	1202.86	6		

16. (a) Draw a scatter plot; (b) $b = 0.79$, $a = 0.29$; (c) Draw the regression line: $\hat{Y} = 0.29 + 0.79X$; (d) $\hat{Y} = 2.66$; (e) $r^2 = 0.60$, $1 - r^2 = 0.40$
(f)

Source	SS	df	MS	F
Regression	2.37	1	2.37	14.84 sig.
Error	1.59	10	0.16	
Total	3.96	11		

17. $R^2 = 0.86$

18. $r^2_{xy} = 0.36$, $r^2_{zy} = 0.09$, $R^2 = 0.39$

19. (a) TOPFRESH alone explains 37.6%; SMALLCLS alone explains 29.6% (b) TOPFRESH and SMALLCLS together explain 48.1% (c) TOPFRESH and SMALLCLS overlap in their explained variance

20. (a)

Source	SS	df	MS	F
Regression	127,510	2	63755.00	40.72
Error	82,990	53	1565.85	
Total	210,500			

(b) N = 56
(c) 60.6%

21. (a) The baseline pain level when both variables are zero is $b_0 = 1.321$
(b) For each increase of one year of age, the expected pain level increases by $b_1 = .065$, holding constant gender. The expected level of pain for males is $b_2 = .857$ higher than for females, holding constant age.
(c) Neither age nor gender has a significant effect on pain level.

22. (a) Odds:

Democrat	0.70
Republican	0.29
Independent	0.40

(b) Odds ratios:

Dem v Rep	2.41
Dem v Ind	1.76
Rep v Ind	0.73

23. (a) Because the dependent variable is a dichotomy (2 categories)
(b) 1.056 0.98 1.201

Chapter 12

1. c
2. c
3. c
4. $r_s = 0.58$, $N = 5$, not significant at .05
5. $r_s = 0.84$, $N = 10$, significant at .05
6. $r_s = 0.83$, $N = 8$, significant at .05
7. $r_s = -.07$, $N = 10$, significant at .05
8. $r_s = 0.80$, $N = 9$, significant at .05
9. $r_s = 0.65$, $N = 12$, significant at .05
10. $r_s = 0.89$, $N = 10$, significant at .05
11. $r_s = 0.33$, $N = 12$; not necessary to test the significance
12. $G = 0.35$, $z = 1.92$, not significant at .05
13. $G = -0.30$, $z = -2.13$, significant at .05
14. $G = -0.39$, $z = -1.93$, not significant at .05
15. $\chi^2 = 17.14$, $\phi = 0.54$
16. $\chi^2 = 16.78$, $C = 0.52$
17. $\chi^2 = 8.42$, $C = 0.35$
18. $\chi^2 = 23.53$, $V = 0.38$, significant at .05
19. $\chi^2 = 2.08$, $V = 0.17$, not significant at .05
20. $\chi^2 = 0.84$, $V = 0.10$, not significant at .05
21. $\chi^2 = 75.05$, $V = 0.62$, significant at .05
22. $\chi^2 = 14.12$, $V = 0.45$, significant at .05

Index